Garden and Landscape History

ERASMUS DARWIN'S GARDENS

Garden and Landscape History

ISSN 1758–518X

General Editor
Tom Williamson

This exciting series offers a forum for the study of all aspects of the subject. It takes a deliberately inclusive approach, aiming to cover both the 'designed' landscape and the working, 'vernacular' countryside; topics embrace, but are not limited to, the history of gardens and related subjects, biographies of major designers, in-depth studies of key sites, and regional surveys.

Proposals or enquiries may be sent directly to the editor or the publisher at the addresses given below; all submissions will receive prompt and informed consideration.

Professor Tom Williamson, School of History, University of East Anglia, Norwich, Norfolk NR4 7TJ, UK.

Boydell & Brewer, PO Box 9, Woodbridge, Suffolk, England, UK IP12 3DF, UK.

Previous publications are listed at the back of this volume.

ERASMUS DARWIN'S GARDENS

MEDICINE, AGRICULTURE AND THE SCIENCES IN THE EIGHTEENTH CENTURY

PAUL A. ELLIOTT

THE BOYDELL PRESS

First published 2021
The Boydell Press, Woodbridge

ISBN 978–1–78327–610–3

The Boydell Press is an imprint of Boydell & Brewer Ltd
PO Box 9, Woodbridge, Suffolk IP12 3DF, UK
and of Boydell & Brewer Inc.
668 Mt Hope Avenue, Rochester, NY 14620–2731, USA
website: www.boydellandbrewer.com

A CIP catalogue record for this book is available
from the British Library

The publisher has no responsibility for the continued existence or accuracy of URLs
for external or third-party internet websites referred to in this book, and does
not guarantee that any content on such websites is, or will remain, accurate or
appropriate

CONTENTS

FIGURES

The author and publisher are grateful to all the institutions and individuals listed for permission to reproduce the materials in which they hold copyright. Every effort has been made to trace the copyright holders; apologies are offered for any omission, and the publisher will be pleased to add any necessary acknowledgement in subsequent editions.

ACKNOWLEDGEMENTS

I would like to pay tribute to Desmond King-Hele FRS before anyone else, because his support over many years has been an important source of inspiration. I was fortunate to have his advice with my doctoral research on the Derby Philosophers during the 1990s and beyond, and I will always appreciate the effort he took to send encouraging letters and copies of his publications. It was heartening to have someone of his eminence as scientist, writer and historian of science (who already had a distinguished career as a space scientist at Farnborough) taking an interest in my own research, and I was pleased to help in a small way with the updated edition of Darwin's correspondence. Anyone who investigates Erasmus Darwin and the Lunar Society circle knows how much we are all indebted to him for his enthusiasm and for the painstaking and industrious research he conducted over many decades, as his biographies – *Doctor of Revolution* (1977) and *Erasmus Darwin: a Life of Unequalled Achievement* (1999) – edition of the *Collected Letters of Erasmus Darwin* (2007) and other works make abundantly clear. As unfortunately I did not manage to complete this book in his lifetime, I would like to dedicate it to his memory.

I won't spell out the main academic inspirations in detail here as this will be fully evident in each chapter. However, as regards Darwin himself I would single out, in addition to Desmond, the work of Roy Porter and Maureen McNeil, particularly for showing me the value of taking his medical practice and theories seriously in period context (and using the larger third edition of *Zoonomia*, 1801), and I have tried to apply and extend this insight further here by exploring how Darwin applied these to the treatment – and understanding of – animal and plant diseases. Another key academic inspiration generally has come from the work of historical geographers and historians of science such as Stephen Daniels, Charles Watkins, Charles Withers, David Livingstone, Jan Golinski and Vladimir Jankovic, who have taken a cultural-geographical approach to the history of the sciences, gardens, parks, agriculture and woodland. Research in the history of gardens and agriculture has also, not surprisingly, offered many insights, and I would particularly highlight the research of Tom Williamson and John Dixon Hunt in this respect. John Beckett's writing and encouragement have been another great help, particularly in relation to east midlands history, the histories of the aristocracy and agriculture – as have our discussions concerning the history of science and green spaces.

The book has also benefited from discussions with colleagues in geography and history at various universities, especially Charles Watkins, Stephen Daniels, Susan Seymour and Mike Heffernan of the School of Geography, Nottingham University. Discussions with Emily Sloan about her research and the completion and publication of her book *The Landscape Studies of Hayman Rooke (1723–1806): Antiquarianism, Archaeology and Natural History in the Eighteenth Century* (2019), in the same series as this volume, helped encourage me to complete this work as well as providing useful points of comparison between two rather different Georgian natural history enthusiasts. The current volume grew out of an essay on Darwin's gardens that I wrote for *Enlightenment, Modernity and Science* (2010) and a short chapter on 'Erasmus Darwin's trees' that I contributed to Laura Auricchio, Elizabeth Heckendorn Cook and Giulia Pacini eds, *Invaluable Trees: Cultures of Nature, 1660–1830* (2012), and I am grateful to the editors and peer reviewers of these for critical comments. Some of the material on Darwin's farm and animals was delivered at the 'Lunarticks, Linnaeus and Lichfield' conference at The Old Stables, Cathedral Close in October 2016, and I am grateful to the staff of Erasmus Darwin House Museum, my fellow presenter Jeremy Barlow and other participants for fruitful critical observations and discussion. Although not directly related to the Darwin book, I have also learnt much generally about the cultural history of landscapes and environments from involvement in the Arts and Humanities Research Council-funded Social World of Nottingham's Historic Green Spaces community history project (2013–14, 2016), and would like to thank the other members of the academic team, John Beckett, Judith Mills and Jonathan Coope, and the many contributors and partners for providing such a rewarding experience.

This book was largely written during my time at the University of Derby and I am very grateful to friends and colleagues for their support. This has included those in history – Ian Barnes, Robert Hudson, Ruth Larsen, Ian Whitehead, Tom Neuhaus, Cath Feely, Oliver Godsmark and Kathleen McIlvenna – and others in Humanities, Arts, Education and elsewhere. My work has benefited too from encouragement and criticism from members of our eighteenth-century research group, Erin Lafford, Freya Gowerly, Joe Harley, Ruth Larsen and Paul Whickman, and also help from our Research Impact Officer, Victoria Barker, and Research Support Officer Christine Selden. Another constant source of inspiration has come from undergraduate and postgraduate history students at Derby, especially those who have taken my courses on landscape and environmental history, the history of medicine and our co-taught Enlightenment option, and likewise my doctoral students, especially Luke Schoppler, Mark Knight and Matt Winfield. Likewise, I'm pleased to acknowledge sustained support and encouragement from successive Deans at Derby – Huw Davies, Malcolm Todd and Keith McLay – and also from Neil Campbell, my predecessor as Head of Research for Humanities and Journalism. The University of Derby has provided financial support too for research visits to archives, for which I am grateful. I'd also like to show my appreciation for the stimulation provided by Dennis Hayes, Ruth Mieschbuehler, Vanessa Pupavac and other regular contributors to the East Midland Salon, who offered comment and criticism on talks I have given in the convivial atmosphere of Derby's characterful Brunswick Tavern on Darwin and the Derby Philosophers, the

Enlightenment and landscape and environmental history, in true Georgian clubbable fashion. Jonathan Powers, whose series of 'mini monographs' provides a scholarly and approachable introduction to writers, philosophers and scientists with Derbyshire connections, has also been helpful.

The experience of giving public talks on Darwinian topics has been another continuing fount of ideas and stimulation. I would like to thank the dedicated and enthusiastic staff, volunteers and audiences at the brilliant Erasmus Darwin House Museum in Lichfield, Derby Museum and Art Gallery, Buxton Crescent and Thermal Spa Heritage Trust and Derby Local Studies Library for making my various presentations so rewarding. My involvement in co-designing and co-curating an exhibition on the 'Moonstruck Philosophers' with Mark Young, Mandy Henchliffe and Victoria Barker at Derby Local Studies Library as part of the Being Human Festival funded by the British Academy in November 2019 – and leading associated events such as public walks around the city – was likewise a fulfilling and enjoyable experience that made me think differently about the relationship between places, practices and ideas associated with Darwin and the Derby Philosophers.

I am very grateful to Tom Williamson, Caroline Palmer, Elizabeth McDonald, the anonymous peer reviewers and publishers Boydell and Brewer for commissioning the book and for their patience despite the completion taking much longer than initially expected. Thanks to Nick Bingham, Senior Production Editor and Sarah Harrison as Copyeditor, who guided it through the production stage, contributed ideas, helped resolve ambiguities and saved me from some errors (any that remain of course, are my sole responsibility!). Colleagues and friends associated with the Centre for Urban History at Leicester University and the Urban History Group conference, particularly Roey Sweet, Richard Rodger, Bob Morris and Barry Doyle, have also provided support and help over the years. The book has also greatly benefited from assistance provided by the staff of many library and record offices, especially those in Nottinghamshire, Derbyshire, Staffordshire (in Stafford and the now unfortunately closed Lichfield branch) and Birmingham, and I would particularly like to thank Sarah Chubb and colleagues at Derbyshire County Record Office, Lucy Bamford and the staff of Derby Museums, Ros Westwood and colleagues of Derbyshire museums and Mark, Mandy and colleagues at Derby Local Studies Library past and present for the dedicated and enthusiastic support they have provided for my research over many years. My involvement with the Derwent Valley Mills World Heritage Site Research and Research Framework committees has provided insights concerning Enlightenment and industrialisation more broadly, and I am grateful to David Knight of Trent and Peak Archaeology and the members of these for their encouragement. Similarly, my experience serving on the board of the innovative *History West Midlands Magazine* supplied other perspectives concerning Darwin, the Lunar circle and their regional milieu, and I am grateful to Malcolm Dick and those associated with the Centre for West Midlands History at the University of Birmingham in different ways for these experiences.

From a personal perspective, my friends Dan, Katy, Arthur and Lawrence have offered enthusiastic support. The microbiologist Martha Clokie of the University of Leicester and Julian Clokie provided helpful advice when I had great difficulty finding much academic work on the history of phytopathology, which remains a neglected subject. I am also very grateful to my mother Kathleen and other members of my family: Magdalen, Christian, Nikki, Harry, Jack, Bernadette, Nick, Jude, Marcus, Joe, Beverley, Peter, Matthew, Hollie, Ashley, Molly, Christian, Eleanor and, especially, Stuart for their love and encouragement.

A NOTE ON PLANT NAMES AND
IDENTIFICATION

I have retained the original botanical or vernacular tree and plant names from sources such as Erasmus Darwin's works and the published Linnaean translations of the Lichfield botanical society and sometimes the fifth edition of William Withering, *Systematic Arrangement of British Plants* (1812), unless it seems to me that this would be too confusing. I have also used the herbals of John Hill and Robert Thornton, John Lindley's *Flora Medica* (1838) and other texts to help identify medicinal plants.

Figure 1.   Stipple engraving of Erasmus Darwin by J. Joll, after Joseph Wright of Derby.

INTRODUCTION

The physician Dr Erasmus Darwin (1731–1802) (Figure 1) is still most well known as Charles Darwin's grandfather, a major evolutionary thinker and a natural philosopher whose ideas partly prefigured those of his grandson. A genuine larger-than-life character of robust physical size and generous but sometimes sarcastic humour, whose medical practice took him travelling around the midland counties for decades, Erasmus Darwin obtained fame in his own lifetime as the author of *The Botanic Garden* (1791), an epic poem with lengthy philosophical notes published in two parts, *The Loves of the Plants* (1789), which amused Georgian society with its poetic portrayal of vegetable amours, and the longer *Economy of Vegetation* (1791). Darwin also published *Zoonomia* (1794/96), a major study of human physiology and medicine, *The Temple of Nature* (1803), a grand epic poetical celebration of the wonders of life with philosophical notes, and *Phytologia* (1800), on the philosophy of agriculture and gardening. He came from Nottinghamshire but spent most of his life in Lichfield, Staffordshire and Derby.

Darwin grew up on the low-lying family estate in Elston in east Nottinghamshire, close to the river Trent and its tributaries (Figure 2). The family were gentry who also owned land in other parts of the county and in Lincolnshire, some of which Darwin inherited on his marriage to his first wife Mary Howard (1740–1770) in 1757. His father Robert (1682–1754), a lawyer, and his mother Elizabeth (Hill) (1702–97) had six other children: Elizabeth (1725–1800), Anne (1727–1813), Susannah (1729–1789), William Alvey (1726–1783), John (1730–1805), rector of Elston, and Robert Waring (1724–1816), a lawyer and botanist who inherited Elston Hall (Figure 3). Darwin attended Chesterfield school, Derbyshire, when it had a strong reputation under William Burrow (1683–1758), which conjured up memories of 'a thousand pleasing circumstances' and where his classmates included the antiquarian Rev. Samuel Pegge (1733–1800) and Lord George Cavendish (1728–94), second son of William Cavendish, 3rd duke of Devonshire.[1]

After receiving a medical education at St John's College, Cambridge, and Edinburgh University during the 1750s and unsuccessfully trying to practise at Nottingham in 1756, Darwin quickly built up a reputation and income as a physician at Lichfield and Derby, travelling around the midland counties tending to patients and thriving in the highly competitive, burgeoning medical marketplace of Georgian England (Figure 4).[2]

Figure 2. The midland counties, showing roads, canals and towns before the coming of the railways, from Bell, *A New and Comprehensive Gazetteer of England and Wales*, 4 vols (1834), vol. 1.

According to one of its Georgian historians, John Jackson, Lichfield was a 'place of very little mercantile business' situated in a 'pleasant and healthful valley' in central England and surrounded by moderate hills with 'fine springs'. It was 'chiefly inhabited by gentry' and 'ancient and numerous' families. Overseen by the graceful spires of its cathedral, known as the 'Ladies of the Vale' and visible for miles around, its buildings had 'assumed the air and taste of modern times' and embodied the 'improving spirit of the age' (Figure 5). As a market town, diocesan capital and coaching centre, Lichfield's prosperity depended upon local trade, agriculture, leisure, travel and church society, with coteries formed of professionals such as diocesan clergy, lay officials and lawyers, although the bishop did not reside in the palace during the eighteenth century, but at Eccleshall castle. Although it functioned in some respects as a second Staffordshire county town, the city remained fairly small, with a population of 3,088

ure 3.   Pen and ink sketch of Elston Hall, from K. Pearson, *The Life, Labours and Letters of Francis Galton*, . 1 (1914).

in 1695 rising to 4,842 in 1801.[3] With its gentry and 'ancient' families, bookshops, charitable institutions, clubs and associations, Lichfield also had pretensions to polite society and learned culture. In a much quoted jesting remark to James Boswell (which tacitly acknowledged that the city was being left behind by its burgeoning, industrious neighbour), Samuel Johnson (1709–1784) claimed that Lichfield was a 'city of Philosophers' who worked 'with our heads, and make the boobies of Birmingham work for us with their hands'.[4] It was here that Erasmus and Mary Darwin had five children: Charles (1758–1778), Erasmus junior (1759–1799), Robert (1766–1848) (who became Waring after the death of his uncle), William Alvey (1767–1767) and Elizabeth (1763–1764), who both died in their first year. Darwin's near neighbour, his close friend and biographer the poet Anna Seward (1742–1809) (Figure 6), who will feature much in these pages, believed that his 'talents and social virtues' had from his family home in The Close 'shed their lustre' over the city, and it was 'to this *rus in urbe*' that a 'knot of philosophic friends' often 'resorted', some of whom were to become members of the three philosophical associations Darwin was largely responsible for founding: the Lunar Society of Birmingham, the Derby Philosophical Society and Lichfield Botanical Society. He nurtured these groups as a highly enthusiastic socialite, conversational wit,

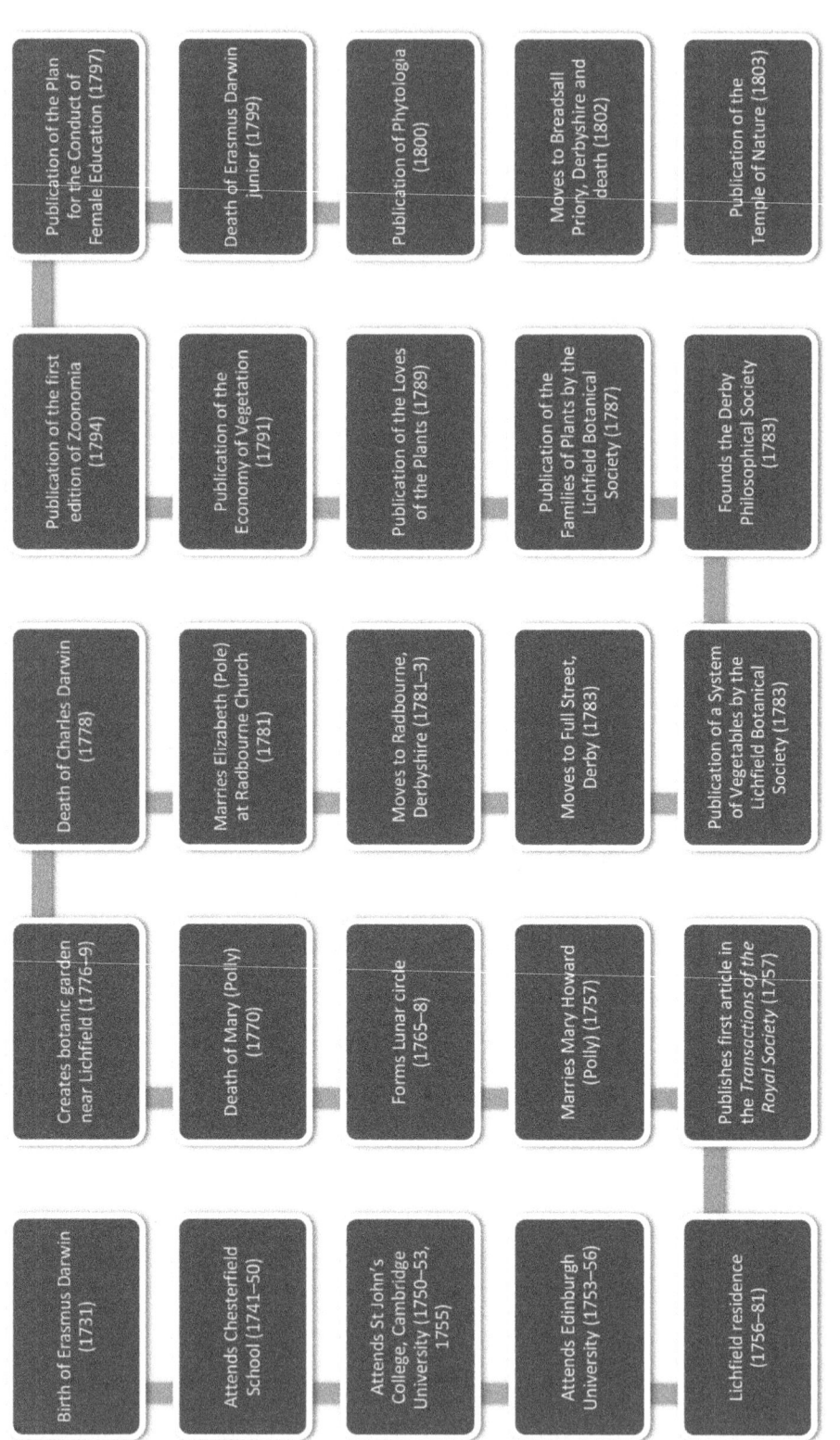

Figure 4. Timeline of Erasmus Darwin.

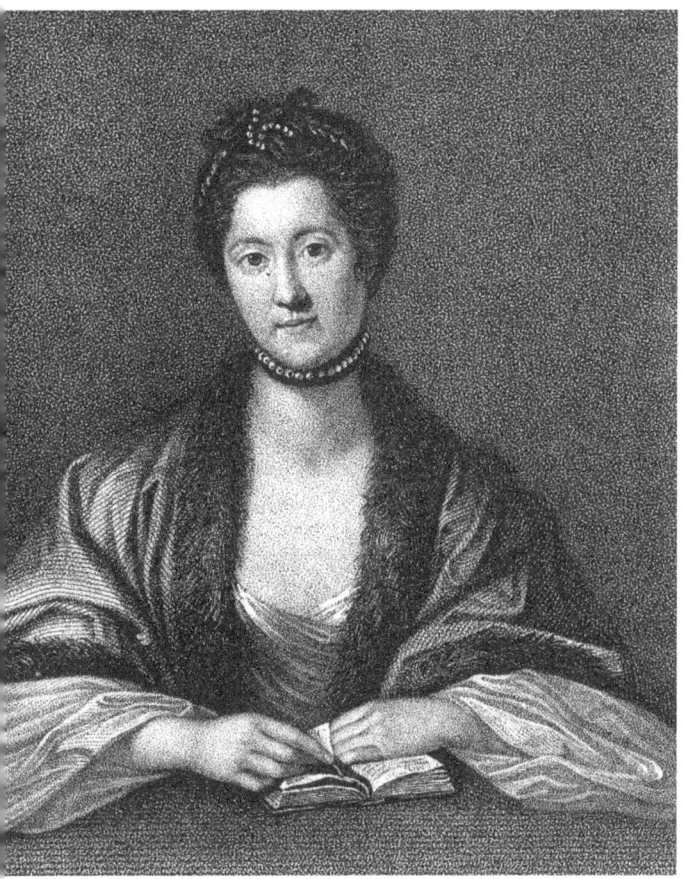

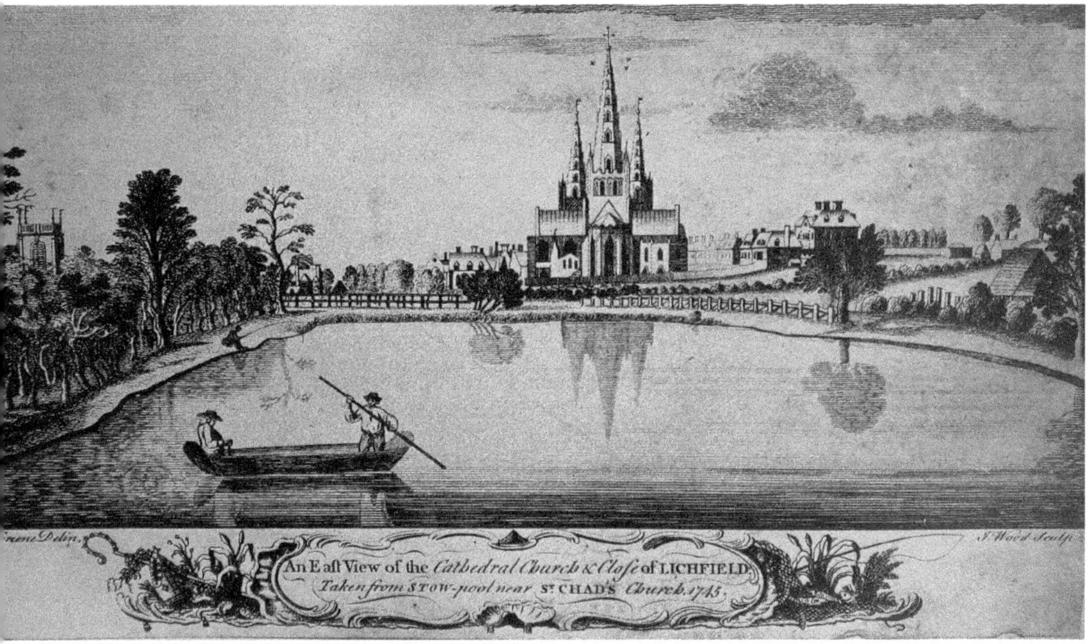
An Eaſt View of the Cathedral Church & Cloſe of LICHFIELD
Taken from STOW-pool near St. CHAD'S Church.1745.
J. Wood Sculp.

Figure 5. (above) Engraved view across Stowe Pool towards Lichfield cathedral (1745), with the bishop's palace, the residence of Anna Seward, to the right; from A. S. Turberville ed., *Johnson's England* (1933), vol. 1.

Figure 6. (left) Anna Seward, engraving by A. Cardon after painting by T. Kettle (1762).

omnivorous reader and inveterate writer. As we will see, the scientific organisations that Darwin helped to found and lead and his circles of philosophical friends provide an excellent demonstration of the centrality of sociability and clubbability to the intellectual excitement that characterised the British Enlightenment.[5]

Darwin experienced two major tragedies during the 1770s. The first of these was the death of his wife Mary in 1770 after a long and painful illness which her husband had only been able to alleviate with opium but never cure. The second was the death of their son Charles as a student at medical school in Edinburgh in 1778 after a cut acquired during the dissection of a dead child's brain became infected. We know from his correspondence how devastated Darwin was by both these events. Following his return from Edinburgh he composed or part-composed an elegy in memory of Charles and in 1780 published his son's prize-winning essay on pus and mucus in a volume with a biographical memoir.[6] However, after the death of Mary and an affair with Mary Parker resulting in the birth of two daughters, Susan and Mary Parker (junior), Darwin became close to Elizabeth Pole (1747–1832), the wife of Colonel Edward Sacheverell Pole (1718–1780) of Radburn Hall, Derbyshire. When the colonel died Darwin married Elizabeth and came to live at the hall in 1781, after which he acquired a townhouse on Full Street in the centre of Derby just off the market square the next year. The Darwins took up residence in summer 1783 after undertaking some improvements and stayed there for nearly twenty years before moving to the Priory near Derby in 1802 (at Breadsall).[7]

By the later eighteenth century Derby was a flourishing medium-size county town and agricultural and manufacturing centre with brewing, textile and china industries. Its population grew from 8,563 in 1791 to 10,828 in 1801. The transformation brought by the eighteenth-century urban renaissance was strongly evident in the town's neo-classical buildings, such as the reconstruction of All Saints Church (designed by James Gibbs), the fashionable residences of Friargate and elegant inns, as well as improved roads and pavements, although a growing concentration of poorer folk lived in more overcrowded conditions beside Markeaton Brook.[8] The town also benefited from the improvements occurring in road, river and canal communications during the 'long' eighteenth century, which connected it to national networks and major centres such as Birmingham and ports like Gainsborough, Kingston-upon Hull and Liverpool.[9] There were also mills for slitting and rolling iron, a copper smelting and rolling works, china-making and mineral and spa industries, while coal- and lead-mining brought wealth, business and mechanical knowledge.[10] John and Thomas Lombe's original silk mills remained a tourist attraction, but local textile industries underwent major innovation and expansion from 1771, especially at Derby and along the Derwent Valley at Darley Abbey, Cromford, Milford and Belper. Moving from Richard Arkwright's spinning frame and carding engine, the Arkwrights, Strutts and other textile manufacturers demonstrated how a range of mechanical processes could be accomplished in one location with improvements in water power technologies, use of steam power to improve water supplies and directly power the machinery, innovations in stove-vent heating systems and experiments with fire-resistant building methods for floors and roofs, including the use of hollow pots and iron plates to cover wooden beams, tiled floors placed upon brick jack-arches, cast-iron beams and iron columns.[11] There were also major developments in the formation of industrial communities in Derbyshire and the Derwent Valley through provision of domestic houses, public buildings, gardens, allotments, plantations and farms.[12] Local lead-mining was at its most productive between 1600 and 1800, and there were associated spa and petrifaction industries that produced worked objects such as ornaments, vases and urns from materials such as gypsum and blue john, which were originally by-products of mining. Demand increased, encouraged by visiting tourists, collectors and natural philosophers in search of Peak wonders, picturesque prospects, mysterious subterranean caverns and the healing waters of springs at Matlock, Buxton and other county locations.[13]

While living in Derby, Darwin continued his scientific and mechanical experiments alongside medical practice; in this he was supported by the members of the Philosophical Society, producing his scientific and poetical works with the aid of the society's library. Inspired by experimental observations of trees and shrubs, *The Botanic Garden* included long additional notes on vegetable perspiration, placentation (provision for the nourishment of young buds, bulbs or seeds), circulation, respiration, glandulation (glands for nutriment of seeds, bulbs and buds) and impregnation that were to grow into much longer equivalent chapters in *Phytologia*, his study of the philosophy of agriculture and gardening. Darwin's botanical work was importantly shaped by his experiences translating the *Systema Vegetabilium* and *Genera Plantarum* of Swedish naturalist Carl Linnaeus, published under the auspices of Lichfield botanical society in 1783. Encouraged by philosophical friends, he also published *Zoonomia*, a major study of human physiology and medicine, and *The Temple of Nature*, a poetical celebration of the wonders of life. *The Botanic Garden* and *The Temple of Nature* adopted aspects of fashionable pastoral, epic and mock epic poetry exemplified by Alexander Pope's *Rape of the Lock* (1714), which turned everyday trivial occurrences into ostensibly momentous triumphs and tribulations. Darwin used his epic poems to convey scientific or 'philosophical' ideas, distinguishing between poetic language and prose that was more appropriate for presenting and analysing scientific ideas and likening the scenes that he painted in them to those of a landscape artist. In the interlude of *The Loves of the Plants* he self-deprecatingly and humorously used an imagined conversation between the bookseller and poet to distinguish between the 'pure description' of poetry and the 'sense' of the notes. While the poet was a 'flower painter' or 'landskip' artist, the principal distinction between poetry and prose was that the former 'admits of but few words expressive of very abstract ideas, whereas prose abounds with them'; poets, with their personifications and allegories, write 'principally for the eye', while prose writers employ 'more abstracted terms'. Hence in 'graver' philosophical works for instruction rather than amusement it became 'tedious' if too many descriptive pictorial words were employed.[14] While some poetry could be didactic, 'science' was 'best delivered in prose' because 'its mode of reasoning is from stricter analogies than metaphors or similies'.

Since 2000 Darwin's residence in Cathedral Close, Lichfield, has been opened as the Erasmus Darwin House museum and study centre, celebrating his life and achievements, while a wealth of new manuscript material that has been uncovered and preserved at Cambridge is transforming our understanding of his work. While *The Botanic Garden* has received much academic scrutiny in, for instance, literary and gender history, there has never been a systematic study of the landscapes, environments, animals and plants that inspired Darwin's work or a full attempt to reconstruct and analyse his gardens and plantations using evidence from manuscript papers and other historical sources. It is well known that *The Botanic Garden* was partly inspired by a botanic garden that Darwin created near Lichfield and informed his efforts to translate the works of Linnaeus into English, aided by the members of the botanical society at Lichfield. This book, therefore, explores the relationships between the landscapes that Darwin encountered at various stages of his life and his medical, scientific and literary activities, trying to reconstruct, for instance, what the Lichfield botanic garden and Derby garden and orchard were like as places, examining how encounters with plants and trees challenged Darwin's understanding of Linnaeus's 'artificial' system of plant classification and demonstrating how these shaped his works on natural history, including *The Botanic Garden*, *Phytologia* and his Linnaean translations. Inspired by his medical practice, he saw many parallels between animals, plants and humans, which informed his work on vegetable anatomy and physiology. From observations of subterranean caverns to the use of plant 'bandages' and electrical machines to hasten seed germination to what were seen as philosophically and ethically controversial studies of vegetable 'brains', nerves and sensations, Darwin's landscape and garden experiences had a profound impact upon his career. They provided him with insights into medicine, taxonomy, chemistry, geology, soil creation, evolution, the 'economy' of the natural world and much more. His medical practice and ideas as well his industrial interests encouraged him to perceive and experience landscapes in dynamic terms and he took a close interest, for instance, in the application of natural philosophy to agriculture, especially in *Phytologia*.

In these publications, and encouraged by his medical practice, Darwin presented what could be regarded as a biological, geological and cosmological developmental theory in which, as we shall see, constant analogies were made between the animal and vegetable worlds. Although he is often described as an evolutionist, this was not a term employed by Darwin and it risks 'whiggish' anachronism, conflating him as it does with nineteenth-century naturalists, including his grandson, and seeing his work as part of an inevitable progression towards 'modern' Darwinian evolutionary scientific consensus when in fact many of his ideas looked as much backwards as forwards. Exploring the question of the relationship between the evolutionary ideas of Erasmus Darwin, his grandson Charles and Jean Baptiste Lamarck, Howard Gruber suggested that it would be more accurate to describe the elder Darwin's 'general world view' as one of 'pan-transformism', because of his emphasis upon the changes of individual growth rather than species transmutations.[15] Erasmus Darwin does refer, however, to 'progressive improvement' and the gradual acquisition of 'new powers' by animal life to 'preserve their existence', considering that 'innumerable successive reproductions for some thousands, or perhaps millions of ages, may at length have produced many of the vegetable and animal inhabitants which now people the earth'.[16]

The places in which Darwin lived, worked and socialised helped to shape the content and character of his philosophical work. His approach to professional practice was in many ways traditional and retained elements of neo-humoural medicine, with its emphasis upon individuals having particular 'temperaments' or dominant characteristics and illnesses being caused by imbalances in the 'humours' arising from external environmental changes, which was believed to have originated in the work of ancient Greek philosopher and physician Hippocrates of Kos, active during the fifth and fourth centuries BC. Darwin's medical practice was likewise fairly traditional by eighteenth-century standards, as evidenced by the range of treatments it utilised, such as bleeding, purging and vomiting, although, as we will see, he could be sparing in employing such methods, rather allowing recovery to take place on its own. Drawing upon the education he received at Edinburgh, Darwin's medical theory combined elements drawn from the physiology of Swiss anatomist Albrecht Haller (1708–77), especially the notion of sensory and nervous stimulus and response, combined with the synthesis of associationism, moral psychology and physiology detailed by the English physician and philosopher David Hartley (1705–1757). The general developmental theory that Darwin explored in his major works drew upon his medical practice and natural philosophy, emphasising the degree to which life engaged in dialectical interplay with its environment, having a difficult – even adversarial – but ultimately useful relationship with its situation. He asserted that life was governed by the 'spirit of animation', an ethereal force of energy acting through the nerves on the muscles, which had some apparent similarities with electricity. Promulgated especially by the natural philosopher Isaac Newton (1642–1727), the concept of ethers was much used during the eighteenth century as a means of seeking to explain how phenomena in the universe and natural world acted at a distance upon each other. Ethers were considered to be fluids that were composed of very small, fine or subtle particles invisible to the eye, which could be known by their effects, such as light, heat, sound, electricity, magnetism and gravity. The operation of gravity, for example, seemed to explain both the positions and motions of planets, comets and other heavenly bodies and why objects fell to earth from above.[17]

According to Darwin, with every contraction of each 'fibre' (that is, muscle or sense organ) there was an 'expenditure of the sensorial power' and, where this power had been increasing and the 'muscles or organs of sense' were therefore operating with 'greater energy', the 'propensity to activity' was therefore reduced in proportion because of the 'exhaustion or diminution' in the amount that had occurred. What he called the 'sensorium' or totality of brain, organs, muscles and living spirit had four 'faculties', which produced all bodily motions through 'irritation' caused by 'external bodies', through 'sensation' caused by feelings of pain or pleasure, through 'volition' occasioned by 'desire or aversion' and through 'association', which was caused by other 'fibrous motion' and the powers of irritation, sensation, volition and association. Darwin believed that understanding the spirit of animation to be ethereal enabled him to combine psychology, or the philosophy of the mind, with physiology, the understanding of how the body worked, and cross the barrier between mind and matter, thoughts and actions, sensory information and how it was understood by living beings. However, because he tried to do this, as we will see in this book, his ideas were attacked for mechanical reductionism, infidelity and the interpolation of unnecessary principles. Neverthless, Darwin's concept of the spirit of animation served a crucial psychological, physiological and heuristic (theoretically illuminating) purpose underpinning his developmentalism and emphasis upon life as responsive, tenacious and above mere laws of motion, animal chemistry or hydraulics. Although 'sensorial motions' perhaps comes closest, no single word or short phrase adequately encapsulates Darwin's efforts to interlock psychology with physiology, and so, even though there was no such term in the eighteenth century, the term 'psychophysiology' will be generally used as a convenient shorthand to represent his endeavour.[18]

Largely forgotten during the nineteenth and early twentieth centuries once his poetry had gone out of fashion, Darwin regained fame posthumously as a precursor or anticipator of his grandson's evolutionary theories. For Desmond King-Hele, who did more than anyone else to reinvigorate Darwin's reputation, he was an enlightenment polymath, successful physician, mechanical genius and robust character who anticipated numerous modern inventions and inspired those around him.[19] While much Darwin scholarship has continued to be fascinated by his evolutionary ideas, attempts have been made in recent decades to provide different and less 'whiggish' perspectives.[20] Robert Schofield, Maureen McNeil, Jenny Uglow, Peter Jones and this author have situated Darwin within the industrialising intellectual communities of the English midlands, focusing upon his leading role in the Lunar Society of Birmingham and enthusiastic friendships with industrialists such as Matthew Boulton (1728–1809), James Watt (1736–1819), Josiah Wedgwood (1730–1795) and William Strutt (1756–1830). McNeil has argued that Darwin was also, to some extent, an apologist for industrialisation, who 'celebrated the powers of scientists, industrialists, and machines over nature', while, in a classic essay, Roy Porter refocused attention upon his medical career, emphasising how much he provided an original secularised 'physician's vision' of Hartley's psychophysiology (summarised above).[21]

At the same time literary scholars, gender historians, historians of science and others have taken a keen interest in Darwin's poetry and ideas, including their impact on the romantic poets and the gender aspects of his interpretation of the Linnaean botanical system.[22] Janet Browne and Londa Schiebinger have argued that, by privileging the stamens as 'male' sexual characters above the 'female' pistils within the flower as definers of classes, Darwin's poetic personified presentation of the Linnaean system in *Loves of the Plants* reflected and reinforced late Georgian gender perceptions and differences.[23] Martin Priestman authoritatively situates Darwin's poetry in relation to Enlightenment concerns such as Rosicrucianism and his literary and scientific networks, exploring his relationship with fellow-poets Richard Payne Knight (1751–1824) and Thomas Jones in particular and the reasons why his work attracted such hostility in the revolutionary period.[24] Patricia Fara, likewise, in her entertaining and refreshingly candid account of her encounters with Darwin, reminds us how influential his ideas were and – like Priestman, especially – how concerned political opponents became at some of the ideas, even those in the apparently playful *Loves of the Plants* as well as the more serious *Economy of Vegetation* and *The Temple of*

Nature, as the satirical poem 'Loves of the Triangles', ostensibly written by 'Mr. Higgins' but in fact by George Canning, John Hookham Frere and George Ellis, and published in the *Anti-Jacobin* (1797–8), clearly demonstrates.[25] The essays in Christopher Smith and Robert Arnott's *Genius of Erasmus Darwin* (2005) explore some of these themes too while also providing more information about his professional, medical, scientific and midlands context.[26]

Despite the amount of academic attention that Darwin has received, the geographies of his career and the impact of landscapes and environment upon his medical and intellectual development have received much less analysis. For example, while McNeil's stimulating academic analysis of Darwin's intellectual development is adept at resituating him within some key Enlightenment discourses, such as industrialisation, progress and agricultural improvement, she pays much less attention to the geographical dimensions of his work or the landscapes and environments that shaped his ideas and practices, leading him to come across as a placeless intellectual entity – ironically so, as Darwin was particularly attuned to the impact of landscapes and environment upon the development of the individual and the interface between human beings and their surroundings, and all his works are replete with references to landscapes and places as evidence for his arguments. While his mind traversed the globe through reading and communications with philosophical friends such as the Lunar brethren, however, his experience of place was largely confined to an English midland triangle between Nottinghamshire to the east, the Derbyshire Peak in the north-west and Birmingham in the south (see Figure 1).[27]

Encouraged by the medical potential of botany as well as the aesthetic enjoyment it provided, and particularly after 1778 by his second wife Elizabeth, a keen gardener, Darwin owned and developed a series of gardens, advised his friends on green matters and enthusiastically utilised and promoted plant-based medicines. The first two chapters of this book explore the relationships between Darwin's gardens at Lichfield and Derby and his employment of herbal medicines and work on natural history, primarily represented by *The Botanic Garden* and *Phytologia*. Botanic or physic gardens were originally planted as a resource for medical practitioners and others to supply the plant materials needed for healing, but during the seventeenth and eighteenth centuries they increasingly became places where plants were collected (and often labelled) for education, study and experimentation too. It is well known that Darwin's *Botanic Garden* poems were partly inspired by one such place that Darwin created near Lichfield. However, as we shall see, Darwin's encounters with this and his other gardens and orchards in Lichfield and Derby and his use of vegetable-based substances in medical practice challenged his understanding of the botanical system of Carl Linnaeus and encouraged him to try and adapt or improve it to take better account of the complexities of plants growing in the ground. In effect, he undertook three Linnaean translations: an initial attempt to represent the Linnaean system in an English picturesque botanic garden; the translation of the works of Linnaeus into English prose; and, finally, the translation of the Linnaean system into epic popular poetry. Darwin's gardens were more than just places where plants grew, however, and provided him with insights into geology (such as the principles behind artesian wells bubbling up by natural pressure), rock formations and soil creation and composition; hence it was the creation of the Lichfield botanic garden that inspired the second two translations. Through the various studies they encouraged and the combination of physic garden with landscape beauty, Darwin's gardens aspired to be, as Anna Seward maintained, places that combined 'the Linnean science with the charm of landscape'.[28]

Encouraged by medical practice, including his employment of plants in treatments and the botanical interests of friends in the Lunar Society such as William Withering (1741–1799), Darwin immortalised his Lichfield botanic garden in his epic scientific poems *The Loves of the Plants* and *The Economy of Vegetation*. He also, as noted, founded a small Lichfield botanical society with local friends Brooke Boothby and Andrew Jackson, which worked to produce the Linnaean translation and make it more accessible for gardeners, nurserymen and planters who did not read Latin. Through his poetical successes, Darwin's Lichfield creation became one of the most influential botanical gardens in British history. Fed by various springs, the garden featured a 'mossy fountain' and a cold bath developed by a previous local physician, Dr John Floyer. Most importantly, Darwin strove, as Seward emphasised, to take full advantage of the natural topography of the site to help him combine science with beauty, planting trees, shrubs and flowers to adorn the vale. Weighing Seward's laudatory descriptions against the poems themselves and other manuscript and published evidence, the first chapter of this book attempts to recreate Darwin's original botanic garden as it really was and assesses its significance as a place in his own career and intellectual development and in that of his friends, especially Seward.

The 'long' eighteenth century, from around the restoration of the monarchy after the civil wars and Cromwellian Commonwealth onwards, has been seen in Peter Borsay's memorable phrase as a period of 'urban renaissance' in which the built environment and economic, social and cultural life of towns were transformed in an age of comparative stability.[29] Darwin was a strong supporter of urban improvement and believed that it fostered social, economic, political and intellectual progress. He became a keen advocate of the enclosure of common lands (with shared rights) on the periphery of Derby, which drew him into significant political controversies and forced him to articulate his vision of the relationship between progress and urban improvement. Darwin nurtured his garden at Full Street and kept an orchard across the river and, fortunately, a list of trees and plants in both still exists in a notebook, which includes descriptions of other garden features and indications of where everything was positioned. The notebook shows that Darwin delighted in plants such as delphiniums, phlox, saxifrage, antirrhinums, peonies and cistus, and lists changes where some plants had died or been changed. The sections are not defined according to botanical categories but by situation – for instance: 'beyond the shed', 'beyond the fish pond' and 'turn up towards the summer house' – meaning that the plan can be reproduced. Using the copious notebook information, correspondence and other sources, such as maps and Linnaean botanical works, the first chapter tries to reconstruct Darwin's Derby garden as it would have been experienced by him and his family and assesses its significance in his work.

The eighteenth century was the age of improvement, particularly in terms of agriculture, which saw and celebrated the application of new techniques and technology to farming and the breeding of many new varieties of trees, plants and animals. This was supported by philosophical and agricultural societies, including the Royal Society and the Society of Arts, which offered prizes for agricultural and horticultural improvement, and, especially from the second half of the eighteenth century, the county agricultural societies. From the 1790s county surveys were published by various investigators under the auspices of the government's Board of Agriculture. Darwin's love of plants and gardens is manifest in the interest that he took in agriculture and his most detailed work on this subject, *Phytologia*, drew upon a detailed knowledge of midlands horticulture as well as voracious philosophical reading and plant observations. Darwin took much inspiration from the burgeoning British literature on gardening, horticulture and farming, which included works such as Philip Miller's *Gardener's Dictionary* and Jethro Tull's *Horse Hoeing Husbandry*. He was also able to converse with numerous landowners over decades, including gentry and aristocracy, many of whom were patients and friends, and took a keen interest in the impact of disease upon plants and animals, drawing parallels with his medical work and using his garden to make observations. For instance, he grew different cereal crops to observe the impact of different factors upon growth and was able to observe the effects of flooding on plant growth after the Derwent burst its banks and inundated his garden. The third and fourth chapters explore Darwin's contribution to agriculture and horticulture, including his analyses of agricultural experiment, land drainage, breeding, soil improvement and cultivation, arguing that agricultural improvement and practical horticulture were central to his progressive enlightenment worldview.

Darwin's medical practice and understanding of the interface between psychology and physiology also played an important role in shaping his agricultural philosophy and frequently encouraged him to draw parallels between humans, animals and plants, and the fourth, fifth, sixth and seventh chapters examine how this shaped the character of his recommendations for improvement. His renown as a medical practitioner and emphasis upon environmental factors as causes of disease encouraged him to apply his ideas and practices to crops and livestock, which was why he believed, as he emphasised to Joseph Banks (1743–1820), the president of the Royal Society, that *Phytologia*, his 'philosophy of agriculture and gardening', was a 'supplement' to *Zoonomia* that applied his theories of the 'animal economy' to the vegetable world.[30] Coming from a gentry family, Darwin moved in close proximity to other landowners, including the 'new' landed wealth of prosperous traders, manufacturers and industrialists as well as the more established aristocracy, gentry and farming communities. Through such connections and his own observations he gained much information concerning farming practices, which was supplemented by his voracious reading on the subject and experience keeping livestock on his own orchard and allotment in Derby, which he designated his 'farm'. Darwin sought to apply medical ideas and practices to his own animals and plants as well as those of friends, patients and family, closely observing how all living creatures in his gardens interacted with each other and coped with the struggle for existence. One manifestation, as chapters 6 and 7 maintain, was a keen appreciation of the agency, vitality and fecundity of animals, which he was able to utilise when devising treatments for diseases that commonly plagued them and which led him to appreciate the dangers and opportunities presented by inter-species parallels and transmissions.

Despite the central importance of Darwin's many different encounters with trees, there has never been a study of his arboriculture, which is explored in the eighth and ninth chapters. Many of his tree investigations were inspired by the economic utility of agricultural and horticultural improvement, but Darwin was able to unite, as Anna Seward claimed, practical botany with picturesque landscapes by creating and experiencing both planted and poetical botanic gardens. His romantic appreciation of landscape beauty was partly founded upon the picturesque qualities of trees and plants as well as his psychological and physiological theories, and it was local woodland (and amorous joy) that inspired Darwin to express himself poetically for the first time since his student days as he personified the spirits of a grove of trees in his garden under threat from the axe. Darwin's love of trees also stemmed from a general interest in their qualities, beauty and botany that was widespread within his intellectual circle, represented by the work of close friends such as Seward, Brooke Boothby (1744–1824), Thomas Gisborne (1758–1846), the artist Joseph Wright of Derby (1734–97) and Francis Noel Clarke Mundy (1739–1815) of Markeaton, Derbyshire, a landowner, magistrate and writer. Mundy's poem on 'Needwood Forest' in Staffordshire, for instance, inspired Darwin to compose an 'Address to Swilcar Oak', which he inserted into the chapter on the production of leaves and wood in *Phytologia*. Furthermore, the trees in Darwin's orchards and gardens from Lichfield to Derbyshire and his knowledge and experience of local woodland strongly informed his analyses of vegetable physiology and anatomy. The size and longevity of trees provided the scale necessary for Darwin to observe vegetable physiology in detail, encouraging parallels with animal physiology. Observations of local estate and forest trees informed his suggestions for the improvement of timber production, which he believed were vital to strengthen the defence of the country against foreign invasion. Visits to local estate plantations such as those at Shugborough Hall and important woodlands such as Needwood, Staffordshire and Sherwood Forests demonstrated where trees flourished best and underscored their cultural status as 'monarchs of the forest' in the economy of nature. Observations of estate woodlands on the edge of the Staffordshire moors suggested that large coniferous plantations might be nurtured on exposed mountainous and boggy moors such as those of the Pennines for the benefit of the economy, industry and navy. In *Phytologia* Darwin provided detailed recommendations concerning the most efficacious means of growing, nurturing, straightening, curing, transplanting and felling trees.

While revelling in the beauty and peace afforded by plants, following in the footsteps of enterprising British plant experimenters such as Nehemiah Grew and Rev. Stephen Hales, Darwin also used them as a means to observe vegetable physiology, anatomy, respiration and nutrition and to conduct experiments. The final two chapters explore how Darwin used his arboricultural observations to try to determine the optimum conditions for growth, which he hoped would lead to improvements in agriculture and horticulture. Having accepted the French chemist Antoine Lavoisier's concept of oxygen in preference to his friend Joseph Priestley's phlogiston theory, which posited the release of a universal combustable material by burning substances, Darwin explored plant physiology, respiration and nutrition. He also tried to determine the types of soil and manure, moisture and other factors that best enabled trees to thrive. Darwin drew constant analogies between humans, animals, trees and plants, believing that their bodies and senses functioned in similar ways, operating through what he called the 'spirit of animation', a very subtle 'fluid' akin to electricity that mediated between brains, nerves, senses and muscular fibres. For this reason he was fascinated with the idea that applying electricity might facilitate plant growth and conducted experiments on plants with electrical machines, suggesting that the erection of numerous metallic points in gardens or fields might cause 'quicker vegetation' by providing plants 'more abundantly with the electric ether' and increasing rainfall. However, the constant analogies that Darwin drew between plants and animals, encouraged by his medical experience, attracted hostility and sometimes ridicule in the fraught political climate of the revolutionary 1790s, which we will explore further. Despite praise for his work from Sir Joseph Banks and Sir John Sinclair, Darwin and his 'Jacobin' or revolutionary plants were satirised in publications and prints as the tranquillity of the garden, woodland and country vale were shattered by the noises of industrial and political upheavals.

This book shows how Darwin, as passionate about nature as his contemporary Rev. Gilbert White of Selbourne (1720–1793) or his grandson Charles, was intrigued by everything from swarming insects and warring bees in his gardens through domestic pets, pigs and livestock on his Derby 'farm' to fungi growing from horse dung in local tan-yards. His landscape and garden experiences, from investigations of plant bodies and the use of vegetable 'bandages' in his orchard and electrical machines to hasten seed germination to provocative speculations concerning plant 'brains', nerves and sensations, transformed his understanding of nature. They offered insights into medicine and the environmental causes of diseases, taxonomy, chemistry, evolution, potential new medicines and foodstuffs and mutual rivalry and interdependency in the economy of nature. Like the erotic vegetables of *The Loves of the Plants* (1789), which appalled and delighted his readers, the multifarious living beings of Darwin's works were real, dynamic, interacting and evolving creatures rather than merely poetic abstractions, who reacted to their external environments and internal stimuli. The chapters of this book provide a grand tour of Darwin's gardens, horticulture and agriculture, demonstrating how his plants and animals wrestled with adversity, experiencing pleasures and pains, loving, competing, adapting and combatting disease (or succumbing to their mortality), taking nourishment from earth and air which was transformed within their organs and tissues into vital fluids that propelled their lives and loves.

NOTES

1   E. Darwin, letter to W. Burrow, 11 December 1750 in D. King-Hele ed., *The Collected Letters of Erasmus Darwin* (Cambridge, 2007), 17–19; D. King-Hele, *Erasmus Darwin: a Life of Unequalled Achievement* (London, 1999), 1–10; P. Riden, *A History of Chesterfield Grammar School* (Cardiff, 2017).

2   A. Seward, *Memoirs of the Life of Dr. Darwin* (London, 1804), 1–154; C. Darwin, *The Life of Erasmus Darwin*, edited by D. King-Hele (Cambridge, 2003), 7–91; King-Hele, *Erasmus Darwin*, 25–49; D. Gibbs, 'Physicians and physic in seventeenth and eighteenth-century Lichfield', and G. C. Cook, 'Dr. Erasmus Darwin MD FRS (1731–1802): England's greatest physician?' in C. U. M. Smith and R. Arnott eds, *The Genius of Erasmus Darwin* (Aldershot, 2005), 35–46, 47–62.

3   J. Jackson, *History of the City and Cathedral of Lichfield* (London, 1805), 18–19; T. Harwood, *History and Antiquities of the Church and City of Lichfield* (London, 1806); J. Nightingale, *The Beauties of England and Wales*, vol. 13, part 2, Somersetshire and Staffordshire (London, 1813), 786–819; L. Schwarz, 'On the margins of industrialisation: Lichfield, 1700–1840', in J. Stobart and N. Raven eds, *Towns, Regions and Industries: Urban and Industrial Change in the Midlands, c1700–1840* (Manchester, 2005), 177–8, 185–6, 189; L. Schwarz, 'Residential leisure towns in England towards the end of the eighteenth century', *Urban History*, 27 (2000), 51–61.

4   M. A. Hopkins, *Dr Johnson's Lichfield* (London, 1956); C. Upton, *A History of Lichfield* (Chichester, 2001), 19.

5   Seward, *Memoirs of Dr. Darwin*, xiii, 16; P. Clark, *British Clubs and Societies, 1580–1800: The Origins of an Associational World* (Oxford, 2000); J. E. McClellan III, *Science Reorganised: Scientific Societies in the Eighteenth Century* (New York, 1985).

6   A. Duncan (and E. Darwin attrib.), *An Elegy on the Much-lamented Death of a Most Ingenious Young Gentleman* (London, 1778); C. Darwin, *Experiments Establishing a Criterion between Mucaginous and Purulent Matter* (Lichfield, 1780); King-Hele, *Erasmus Darwin*, 89–93, 142–5.

7   King-Hele, *Erasmus Darwin*, 170–3, 192–4, 325–6, 329–30, 340–2.

8   J. Pilkington, *A View of the Present State of Derbyshire*, 2 vols (Derby, 1789), vol. 2, 134–96; W. Hutton, *The History of Derby from the Remote Ages of Antiquity to the Year MDCCXCI*, 2nd edn (London, 1817); R. Simpson, *A Collection of Fragments Illustrative of the History and Antiquities of Derby*, 2 vols (Derby, 1826); M. Craven, *Illustrated History of Derby*, 2nd edn (Derby, 2006),78–85, 96–134; P. A. Elliott, *The Derby Philosophers: Science and Culture in British Urban Society, 1700–1850* (Manchester, 2009), 18–23.

9   F. Nixon, *The Industrial Archaeology of Derbyshire* (Newton Abbot, 1969), 135–51; B. Cooper, *Transformation of a Valley: The Derbyshire Derwent* (London, 1983), 162–99; A. E. and M. Dodd, *Peakland Roads and Trackways* (Ashbourne, 2000), 134–81; T. Brighton, *The Discovery of the Peak District* (Chichester, 2004), 9–18, 59–70; D. Hey, *Derbyshire: A History* (Lancaster, 2008), 302–3; M. Pawelski, 'Turnpikes and local industry: a study of the relationship between the lead industry and the turnpike system in eighteenth-century Derbyshire', in C. Wrigley ed., *The Industrial Revolution: Cromford, the Derwent Valley and the Wider World* (Cromford, 2015), 55–71.

10   Pilkington, *Derbyshire*, vol. 2, 169–77; J. Britton and E. W. Brayley, *The Beauties of England and Wales*, vol. 3, Cumberland, Isle of Man and Derbyshire (London, 1802), 364–75, 377–8; Hutton, *History of Derby*, 154–81; R. S. Fitton and A. P. Wadsworth, *The Strutts and Arkwrights, 1758–1830: A Study of the Factory System* (Manchester, 1958); Nixon, *Industrial Archaeology of Derbyshire*, 173–97; A. Calladine, 'Lombe's Mill: an exercise in reconstruction', *Industrial Archaeology Review*, 16 (1993), 82–99; S. Chapman, 'The Derby silk mill revisited', in C. Wrigley ed., *The Changing Lives of Working People during the Industrial Revolution: The Derwent Valley and Beyond* (Cromford, 2018), 27–54.

11 Britton and Brayley, *Derbyshire*, 364; A. Menuge, 'The cotton mills of the Derbyshire Derwent and its tributaries', *Industrial Archaeology Review*, 16 (1993), 38–61 at 40–56; H. R. Johnson and A. W. Skempton, 'William Strutt's cotton mills, 1793–1812', *Transactions of the Newcomen Society*, 30 (1955–7), 179–205; Fitton and Wadsworth, *Strutts and Arkwrights*, 196–205; K. A. Falconer, 'Fireproof mills – the widening perspectives', *Industrial Archaeology Review*, 16 (1993), 11–26; D. Romaine, 'Water power and mechanics: Cromford', in C. Wrigley ed., *The Changing Lives of Working People During the Industrial Revolution: The Derwent Valley and Beyond* (Cromford, 2018), 111–26.

12 Fitton and Wadsworth, *Strutts and Arkwrights*, 224–60; D. Peters, *Darley Abbey: From Monastery to Industrial Community* (Buxton, 1974), 39–45, 68–80; Cooper, *Transformation of a Valley*; R. S. Fitton, *The Arkwrights: Spinners of Fortune* (Manchester, 1989), 187–90; Derwent Valley Mills Partnership, *The Derwent Valley Mills and their Communities* (Matlock, 2001), 16–70; Hey, *Derbyshire*, 342–55; C. Hartwell, N. Pevsner and E. Williamson, *The Buildings of England: Derbyshire* (New Haven, 2016), 62–4, 158–64, 288–92, 358–60.

13 Britton and Brayley, *Derbyshire*, 372–5; J. Mawe, *The Mineralogy of Derbyshire* (London, 1802); W. Watson, *A Delineation of the Strata of Derbyshire* (Sheffield, 1811), reprinted with an introduction by T. D. Ford (Hartington, 1973); T. D. Ford, 'White Watson, 1760–1835 and his geological tablets', *Mercian Geologist*, 13 (1995), 157–64; H. S. Torrens, 'John Mawe (1766–1829) and a note on his travels in Brazil', *Bulletin of the Peak District Mining Museum*, 11 (1992), 267–71; H. S. Torrens, 'Erasmus Darwin's contributions to the geological sciences', in C. U. M. Smith and R. Arnott eds, *The Genius of Erasmus Darwin* (Aldershot, 2005), 259–72; Elliott, *Derby Philosophers*, 142–5, 148–51, 192.

14 The Lichfield Botanical Society, *A System of Vegetables … translated from the thirteenth edition of the Systema Vegetabilium of the late Professor Linneus*, 2 vols (Lichfield, 1783); E. Darwin, *The Loves of the Plants*, 4th edn (London, 1799), 'interlude'; *The Economy of Vegetation*, 4th edn (London, 1799); *Zoonomia; or the Laws of Organic Life*, 3rd edn, 4 vols (London, 1801); *Phytologia; or, the Philosophy of Agriculture and Gardening* (London, 1800); E. Darwin, *The Temple of Nature; or the Origin of Society* (London, 1803).

15 H. E. Gruber, *Darwin on Man: A Psychological Study of Scientific Creativity* (London, 1974), 148–9.

16 Darwin, *Temple of Nature*, 29.

17 H. S. Thayer ed., *Newton's Philosophy of Nature: Selections from his Writings* (New York, 1974), 112–16, 141–5; R. S. Westfall, *The Life of Isaac Newton* (Cambridge, 1994), 103–5, 144–8; B. J. Teeter Dobbs and M. C. Jacob, *Newton and the Culture of Newtonianism* (Atlantic Highlands, 1995), 12–15, 53–5; J. L. Heilbron, *Electricity in the 17th and 18th Centuries: A Study in Early Modern Physics*, 2nd edn (New York, 1999), 53–70, 239–41.

18 Darwin, *Zoonomia*, vol. 1, 5–14, 92–3; G. S. Rousseau, 'Psychology', in G. S. Rousseau and R. Porter eds, *The Ferment of Knowledge: Studies in the Historiography of Eighteenth-century Science* (Cambridge, 1980), 143–210; R. Porter, 'Erasmus Darwin: doctor of evolution?' in J. R. Moore ed., *History, Humanity, and Evolution: Essays for John C. Greene* (Cambridge, 1989), 39–69; G. S. Rousseau ed., *The Languages of Psyche: Mind and Body in Enlightenment Thought* (Berkeley, 1990), 31–4; C. U. M. Smith, 'All from fibres: Erasmus Darwin's evolutionary psychobiology', in C. U. M. Smith and R. Arnott eds, *The Genius of Erasmus Darwin* (Aldershot, 2005), 133–43; P. Elliott, '"More subtle than the electric aura": Georgian medical electricity, the spirit of animation and the development of Erasmus Darwin's psychophysiology', *Medical History*, 52 (2008), 195–220.

19 Seward, *Memoirs of Dr. Darwin*; Darwin, *Life of Erasmus Darwin*; King-Hele, *Erasmus Darwin*; A. Seward, *Anna Seward's Life of Erasmus Darwin*, edited by P. K. Wilson, E. A. Dolan and M. Dick (Studley, 2010).

20 S. J. Gould, *Dinosaur in a Haystack: Reflections on Natural History* (New York, 1995), 427–57; M. Ruse, *Mystery of Mysteries: Is Evolution a Social Construction?* (Cambridge MA, 1999), 37–53.

[21]   R. E. Schofield, *The Lunar Society of Birmingham: A Social History of Provincial Science and Industry in Eighteenth-century England* (Oxford, 1963); M. McNeil, *Under the Banner of Science: Erasmus Darwin and his Age* (Manchester, 1987), 180; Porter, 'Erasmus Darwin: doctor of evolution?', 39–69; J. Uglow, *The Lunar Men: the Friends who made the Future* (London, 2002); P. M. Jones, *Industrial Enlightenment: Science, Technology and Culture in Birmingham and the West Midlands, 1760–1820* (Manchester, 2008); Elliott, *Derby Philosophers*; essays by G. Rousseau, M. Green, T. H. Levere and G. Budge in G. Budge ed., 'Science and the Midlands Enlightenment', *Journal for Eighteenth-Century Studies*, 30 (2007), issue 2, pp. 157–308.

[22]   D. Worrall, 'William Blake and Erasmus Darwin's *Botanic Garden*', *Bulletin of the New York Public Library*, 79 (1975), 397–417; D. C. Leonard, 'Erasmus Darwin and William Blake', *Eighteenth-Century Life*, 4 (1978), 79–81; D. King-Hele, *Erasmus Darwin and the Romantic Poets* (London, 1986); N. Trott, 'Wordsworth's Loves of the Plants', in N. Trott and S. Perry eds, *1800: The New Lyrical Ballads* (Basingstoke, 2001), 141–68.

[23]   J. Browne, 'Botany for gentlemen: Erasmus Darwin and the *Loves of the Plants*', *Isis*, 80 (1989), 593–620; L. Schiebinger, 'The private life of plants: sexual politics in Carl Linnaeus and Erasmus Darwin', in M. Benjamin ed., *Science and Sensibility: Gender and Scientific Enquiry, 1780–1945* (Oxford, 1991); T. Fulford, 'Coleridge, Darwin, Linnaeus: the sexual politics of botany', *The Wordsworth Circle*, 28 (1997), 124–30; D. Coffey, 'Protecting the botanic garden: Seward, Darwin and Coalbrookdale', *Women's Studies*, 31 (2002), 141–64.

[24]   M. Priestman, *The Poetry of Erasmus Darwin: Enlightened Spaces, Romantic Times* (Farnham, 2013).

[25]   P. Fara, *Erasmus Darwin: Sex, Science and Serendipity* (Oxford, 2012).

[26]   C. U. M. Smith and R. Arnott eds, *The Genius of Erasmus Darwin* (Aldershot, 2005).

[27]   McNeil, *Under the Banner of Science*.

[28]   Seward, *Memoirs of Dr. Darwin*, 127.

[29]   J. H. Plumb, *The Growth of Political Stability in England: 1675–1725* (London, 1967); P. Borsay ed., *The Eighteenth-century Town* (London, 1990); P. Borsay, *The English Urban Renaissance: Culture and Society in the Provincial Town, 1660–1770* (Oxford, 1991); R. Sweet, *The English Town, 1680–1840* (London, 1999); P. Clarke ed., *The Cambridge Urban History*, vol. 2, *1549–1840* (Cambridge, 2000).

[30]   Darwin, *Phytologia*, dedication, vii–viii; Schofield, *Lunar Society*, 398; McNeil, *Under the Banner of Science*, 168–9; Erasmus Darwin's interest in plant and animal diseases does not feature in P. Ayres, *The Aliveness of Plants: The Darwins at the Dawn of Plant Science* (London, 2008).

CHAPTER 1

LICHFIELD AND DERBY GARDENS

Erasmus Darwin loved seeing, touching, smelling, using, studying and celebrating plants and wanted others to share in his enjoyment, from members of his family to his patients, who utilised their virtues or powers as drugs to combat their illnesses and improve their lives. In *Phytologia*, his treatise upon agriculture and gardening, he exclaimed that 'the beautiful colours of the petals of flowers with their polished surfaces' were 'scarcely rivalled' by those of shells, feathers or 'precious stones'. Many of these 'transient beauties' that gave such 'brilliancy to our gardens' delighted 'the sense of smell with their odours' and were employed extensively 'as articles either of diet, medicine, or the arts'.[1] Knowledge of plants was always important for Darwin, given that vegetable products were the basis of so much of the *materia medica* (substances prescribed as treatments), and botany was taught in the medical schools in order that practitioners could distinguish useful from useless or even dangerous plants.

During the 1770s and 1780s, however, botany and gardening assumed greater significance in Darwin's life, encouraged by a community of plant enthusiasts and what Sylvia Bowerbank has described as literary 'defenders' of 'environmental stewardship' who were centred upon Lichfield and invested it with 'deep feeling and value' as a multilayered 'storied place'.[2] Among them were the poet Anna Seward, Rev. John Saville (1736–1803), Darwin's future wife Elizabeth Pole and the members of the botanical society that Darwin formed, as noted above, with Brooke Boothby and William Jackson (1743/5–1798), who were engaged upon a project to translate some of the works of Carl Linnaeus (1707–1778) from Latin into English. Encouraged by his medical and scientific botanical concerns, his growing love of gardening and his experience of the countryside around Lichfield and the local coterie of plant lovers, Darwin developed a botanic garden in a small well-watered valley near the city, which inspired the two parts of his epic poem *The Botanic Garden* (1791). The formation of the Lichfield botanical garden was a crucial moment in both Darwin and Seward's poetical careers. Through the success of the *Botanic Garden*, which, as the poet Samuel Taylor Coleridge (1772–1834) remarked, was 'for some years greatly extolled, not only by the reading public in general' but by poets of his generation, Darwin's Lichfield creation became one of the most influential British botanical gardens.[3] After moving to Derby, he made full use of his back garden and an 'orchard' on the opposite bank of the River Derwent to observe how plants grew and undertake experiments, while his botanic garden remained in place much as he left it for nearly twenty years, managed by Jackson.[4]

Much recent work in literary criticism and the history of science has focused on the sex and gender dimensions of Darwin's *Botanic Garden* and less on the relationship between poem and place. Analysis of his gardening practices and gardens at Lichfield and Derby, however, demonstrates that the materiality of place and the practical lived experience of designing and planting gardens and then nurturing, observing, harvesting and minutely comparing plants in their changing environments helped to inspire both his poetical works and earthen green places. These experiences created stimulating tensions between abstract Enlightenment systems and messy, dirty realities on the ground, encouraging Darwin to challenge his understanding of Linnaean botany and devise more 'natural' systems of taxonomy and undertake a series of experiments and observations on vegetable anatomy. Adherence to the Linnaean system remained, however, a central guiding inspiration as he sought to realise it as a picturesque botanic garden, accessible English-language tool and epic, poetical vestibule to the temple of scientific knowledge. Darwin's gardens were much more than repositories for living plants. Just as medical science and his own observations shaped his professional practice and understanding of human bodies and behaviours, so his gardens were collective social and cultural creations that informed his insights into the economy (interdependency) of the natural world, medical botany, vegetable and animal diseases, horticulture, geology, mechanics, hydrostatics (the study of water), meteorology, the principles of landscape beauty and much more.[5] Using Darwin's published works and manuscript materials and a variety of other sources, this chapter seeks to uncover what the Lichfield and Derby gardens were like as living, dynamic, material and social places and explore the ways in which they inspired his medical and scientific work.

TWO BOTANIC GARDENS

Darwin's Lichfield house on Beacon Street had small gardens to the front and rear and was separated from the road at the front by 'a narrow, deep dingle' that had originally formed part of the moat protecting The Close but was now largely drained of water and 'overgrown with tangled arbours and knot grass'. Darwin placed a bridge across this ditch from the front of the house to the road and planted the area with trees and bushes. Later he sacrificed two parts of his front garden to support local road improvements, providing a further inducement to develop other gardens.[6] The Close was enclosed within a wall and the 'deep dry ditch, on all sides, except towards the city', where it was 'defended by a great lake, or marsh, formed by its brook', and contained the large bishop's palace, where Anna Seward and her father Thomas Seward (1708–90) resided; the dean's residence and prebendaries' houses were 'in a court on the hill', with the bishop away at Eccleshall.[7] According to Anna Seward, Darwin added a 'handsome new front' to his dwelling, with 'Venetian windows, and commodious apartments'. The new front faced towards Bacon Street, but did not go straight onto the road, and the 'tangled and hollow' moat bottom was 'cleared away into lawny smoothness, and made a terrace on the bank', making it 'level with the floor of his apartments', while the planting of the 'steep declivity' with lilacs and rose bushes provided screens from 'passers-by and the sunshine', enabling him to observe a fine, extensive prospect across the road through an opening of 'pleasant and umbrageous fields'.[8]

However, it was the opportunities presented by developing a botanic garden on a new site that provided the greatest inspiration for Darwin. The fullest descriptions of how the Lichfield botanic garden appeared are found in *The Botanic Garden* and Seward's *Memoirs of the Life of Dr. Darwin* (1804). Although she was understandably keen to magnify her role in his life, and the Darwin family were angry at some of the calumnies and falsehoods they believed she promulgated, as a close friend and, as Sylvia Bowerbank has convincingly argued, 'defender of local environments' and proponent of Lichfield's 'storied' beauty against the 'contemptuous assaults of London-based criticism' (including those from fellow-townsman, lexicographer and metropolitan literary colossus, Samuel Johnson), Seward was an enthusiastic supporter of the botanic garden as interlinked place and poem, helping invest it with meaning and mystique.[9]

In striving to combine the British naturalistic gardening style with systematic planting, Darwin's botanic garden was a major development in landscape gardening as well as a source of poetic inspiration. In this he was encouraged by a small community of passionate Lichfield-focused garden enthusiasts, especially Seward, John Saville, Brooke Boothby, Francis Noel Clarke Mundy and William Jackson, a proctor in the cathedral jurisdiction and fellow member of the botanical society. Just as this 'Lichfield coterie ... worked and re-worked each other's texts' while recognising that the 'declared author' was 'generally the originator and main author', as Teresa Barnard has argued, so their gardens and the writings inspired by them were likewise partly collaborative ventures, although, as Seward admitted, she was herself rather 'ignorant of horticulture' and botany and more interested in the beauties of plants and gardens, though growing more interested in the former later in life.[10] The practice of co-authorship is clearly articulated by Seward in the case of Mundy's 'Needwood Forest' (1776), for example; in a letter of 1777 to her close friend Mary Powys of Shrewsbury and later Clifton the contributions of Darwin and Seward are specified but the attribution of Mundy as the public author of the work is unquestioned.[11]

Although gardening and botany were not a primary concern, some members of the Lunar Society circle around Darwin also had green interests that likewise encouraged the development of the botanic garden, but the knowledge of plants Darwin gained in his medical practice and his efforts to translate the Linnaean botanical system into English were more important factors in shaping its creation. Just as the Linnaean system provided a model and inspiration for the medical system propounded in Darwin's comprehensive medical treatise *Zoonomia*, so he saw it as a useful tool for medical practitioners to help identify and utilise plants in their treatments. Vegetable identification was required for medical education and the prescription of many medicines, and Darwin was taught aspects of botany at Edinburgh University, using and developing this knowledge for treating patients in his midlands medical practice. The dramatic expansion in global trade, the spur of international rivalries with other

European powers and activities of the East India Company and other merchant concerns during the seventeenth and eighteenth centuries, greatly increased the availability and variety of imported plant knowledge and vegetable-based drugs such as senna and Jesuit's bark in the medical marketplace, but much remained home-grown.[12]

The botanic garden was developed from the original conception of a physic garden as living medical resource that was more useful than a dried herbarium of preserved plants or plant parts, which often failed to preserve vegetable virtues (the active and useful qualities or ingredients). Darwin utilised Linnaean taxonomy in the arrangement of his botanic garden, combined with picturesque planting and landscaping. Plant-based treatments were essential in his neo-humoural medical practice, which aimed to return the body to what he called the 'natural state', where all the 'irritative motions' were behaving normally, and though, as Roy Porter argued in his transformative essay on Darwin's medicine, it might appear somewhat arcane or esoteric, it is worth taking time to understand his nosology and treatments sympathetically and with sensitivity to period context.[13]

The interventions that Darwin utilised are detailed in a series of articles summarised in the *materia medica* of *Zoonomia* that had existed in drafts since the 1770s, when the botanic garden was being created. These treatments were varied and included recommendations for changes in diet and behaviour, medicines to be applied internally and externally (often based upon organic substances) and treatments such as venesection or bleeding. For 'Nutrientia', or things that preserved 'in the natural state the due exertions' of all irritative motions by producing 'growth' and restoring the vigour of the bodily system, Darwin recommended various animal and vegetable products, water, fresh air and oxygen. He believed that these were effective because the nourishment from animal flesh such as venison and beef, as well as fish, shell fish, dairy products, fruit and vegetables, stimulated the 'absorbent and secerning [secreting] vessels', thus inducing feelings of warmth and strength.[14] Treatments utilised by him to increase the 'exertions of all the irritative motions' towards a more 'natural state' or 'Incitantia' included alcohol or the 'spiritous part of fermented liquors' and various plant substances, many, including opium, tobacco and Indian berry (*Menispernum cocculus*), procured by apothecaries through international trade networks; other substances used exhilarated the 'passions of the mind', such as joy, love, play (even anger), and external applications of 'heat, electricity, ether, essential oils, friction and exercise' were also thought beneficial. Darwin thought that these worked by promoting 'secretions and absorptions', increasing the 'natural heat' and removing pains caused by 'defect of irritative motions' or 'nervous pains' and their associated 'convulsions'.[15] Treatments that increased the irritative motions by promoting secretion and stimulating glands 'into action' or 'Secerentia' worked in various ways to increase bodily heat and relieve suffering, which Darwin believed originated from 'a defect of motion' in secretion vessels. These treatments operated diversely as diaphoretics (inducing perspiration), sialogogues (increasing saliva production), expectorants (bringing up mucous from the lungs, bronchi and trachea), diuretics (increasing urine production) or cathartics (purgatives), or through stimulating mucus production in the bladder, rectum, cellular membranes, nostrils or tear ducts. Plant-based examples procured from around the globe included ginger, pepper, cardamom, essential oils of cinnamon, nutmeg, gum Arabic, tobacco, copaifera balsam from Central and South America (*Copaifera officinalis* or *C. coriacea*, Lindley) and olibanum resin from the *Boswellia thurifera* tree of India.[16]

The large number of treatments that increased irritative motions constituting absorption or 'Sorbentia' in Darwin's system worked by stimulating the 'absorbent vessels' through different parts of the body, including the skin, urethra, intestines and liver. They caused vomiting, purging or perspiration in stronger doses and included opium, Peruvian bark (*Cinchona officinalis*), orange peel, cinnamon, nutmeg and false hemp (*Datisca cannabina*), originally from the eastern Mediterranean and Central Asia.[17] Interventions that inverted the 'natural order' of 'successive irritative motions' or 'Inverentia' worked as emetics, strong cathartics or errhines (increasing nasal discharge), or by producing urine, cold sweats, heart palpitations or other effects that inverted the 'natural order' of 'vascular motions', and included some poisons or even the bringing on of fear or anxiety.[18] Those that restored the 'natural order' of 'inverted irritative motions' or 'Reverentia' in Darwin's medical scheme operated by reclaiming 'the inverted motions', ideally without increasing bodily heat above the normal level to combat those associated with the 'hysteric disease' or particular parts of the body including the stomach, intestines and urinary tracts.[19] Finally, treatments that diminished the 'exertion of the irritative motions' or 'Torpentia' in his scheme included foodstuffs and drinks that possessed lower stimulus value than normal dietary staples, diminishing heat, light, oxygen and other stimuli, while phlebotomy (bleeding), nausea and even anxiety could have the same effect. Other Torpentia included substances which 'chemically' destroyed or prevented acrimony, annihilated worms or extraneous bodies in the system such as bladder stones, lubricated the effected vessels or 'soften[ed] or extend[ed] the cuticle over tumour or phlegmons'.[20]

Some of the plant-based substances recommended by Darwin were procured in the field by apothecaries or imported from abroad, while others were grown in the physic or botanic gardens established by universities and practitioners during the seventeenth and eighteenth centuries as medical resources and teaching aids. Wealthy individuals such as Bishop Henry Compton (1632–1713) at Fulham Palace, Charles Hamilton (1704–1744) at Painshill and Henry Fox (1705–1774) at Holland Park acquired formidable tree and plant collections as others garnered artworks or antiquities. In his 'Elysium Britannicum' John Evelyn (1620–1706) included a design for a pyramid mount with a flat top and geometric beds designed to accommodate a 'philosophical' or botanical garden for the reception and study of vegetables from around the world, while the 'botanick grounds' at John Stuart (1713–1792), third earl of Bute's Luton Hoo estate in Bedfordshire, combined with royal patronage, helped to stimulate the development of Kew Gardens as a systematic national collection under the curatorship of Joseph Banks.[21]

Physic gardens were usually developed by medical practitioners or universities on practical, accessible and level sites within or close to towns, with walls or fences to manage and protect valuable and rare specimens, and did not require beautiful locations or features; any feelings of interest and beauty aroused were usually thought to be intrinsic to the plants themselves rather than to their situation. As increasing numbers of plants were imported from around the world and botanists' collecting aspirations became increasingly ambitious, incorporation of the various species together became more difficult for all but the wealthy and institutions. Within site limitations, most botanic collections were planned geometrically, like other contemporary formal gardens, and some, such as the Oxford physic garden (1622), were divided into four zones to represent the quarters of the globe, each section accommodating plants thought to be closely related in their healing virtues and in terms of the states of humoral medicine. The first British physic gardens, such as those at Oxford and Chelsea (1673), grouped plants or 'simples' by their medicinal qualities so that apothecaries and medical students could learn to identify them.[22] From the 1760s botanical gardens such as those at Cambridge and Edinburgh began to incorporate geometric beds with labelled plants laid out taxonomically according to the Linnaean system so that examples in flower could be examined, which accorded well with the aspirations of Enlightenment medicine, science and natural theology towards order, regularity and demonstration of the laws underpinning the divinely ordained economy of nature. The botanic garden developed by the physicians William Beeston (1671–1732), William Coyte (1708–1775) and William Beeston Coyte (c. 1741–1810) at Ipswich during the eighteenth century, for instance, was replanted to reflect Linnaean taxonomy. Just as Darwin's botanic garden was partly stimulated by his Linnaean plant studies, which helped inspire the *Botanic Garden*, so Beeston Coyte produced a full catalogue of his collection and then an 'alphabetical arrangement' of all the genera and plant species represented, which included 'references to original authorities' for each of these, with plates, an 'index of the natural order' and an appendix on the 'medicinal virtues of British plants'. This utilised advice from James Edward Smith (1759–1828), founder of London's Linnean Society, who had famously purchased the library, papers and herbarium of Linnaeus in 1784 before they were barred from leaving their homeland or acquired by other interested parties (who included the French and Empress Catherine III of Russia); this collection served as the foundation for the new Society's collection and activities, including its regular *Transactions*.[23]

Darwin gained knowledge of Linnaeus through books such as James Lee's *Introduction to the Science of Botany* (1760), which included translated extracts from the Swedish botanist's works. He took a closer interest in the challenges of systematic botany during the 1770s, obtaining a copy of the first volume of Linnaeus' *Systema Naturae* by March 1775 and twenty-seven volumes of Comte de Buffon's *Histoire Naturelle* for his library at around the same time.[24] Darwin was encouraged by William Withering's greater involvement with the Lunar circle during the 1770s. Like Darwin, Withering was educated at Edinburgh University, where he was taught botany by John Hope (1725–86), professor of Medical Botany and one of the first academics to lecture on Linnaeus at a British university. Withering visited Lichfield often and established a medical practice in Stafford, serving as physician to the Stafford Infirmary from 1766. He also collected botanical specimens, encouraging one of his patients (whom he subsequently married) to undertake plant drawings, and by 1772 had begun a treatise on British botany on a Linnaean plan. This 'Botanical arrangement' appeared in 1775 with encouragement and help from Darwin and other Lunar associates and featured 'descriptions of the genera and species' presented according to the 'celebrated' Linnaean system. Darwin offered advice to Withering concerning terminology and symbolism, although his suggestion to use a short and 'easily remember'd and distinguishable title' was rejected in favour of a twenty-four-line designation. Although initial reviews of Withering's tome were enthusiastic, describing it as 'the most elaborate and complete' national flora 'that any country can boast of', and it remained a standard text into the nineteenth century, the first edition consisted largely of Linnaean translations relating to indigenous British genera and species. Plates depicted plants and plant sections and botanical instruments and there were instructions concerning the preservation of specimens, while some changes to terminology were recommended, primarily to render sexual distinctions between classes and orders in polite language so as not to offend British sensibilities.[25] Withering and Darwin fell out seriously on various issues, including the production of their rival Linnaean translations, but to begin with they stimulated each other's botanical studies.[26]

One of Darwin's closest friends and collaborators in botanical, philosophical and literary pursuits during his time at Lichfield and Derby was Brooke Boothby. The nephew of Dr Johnson's friend Hill Boothby, Boothby lived in London as a gentleman of leisure in the circle of Alleyne Fitzherbert, Lord St Helens, before spending most of his time in Lichfield from around 1772, when he is listed as a city resident and was among those – including Darwin and Canon Thomas Seward (1708–1790) at the bishop's palace – who subscribed for street lamps. A poet and later political writer during the revolutionary era, who in 1789 inherited a title, Ashbourne Hall and a large income from his great-uncle Sir William, Boothby was an admirer of French philosopher Jean-Jacques Rousseau (1712–1778) who in 1776 presented him with the manuscript of his autobiographical First Dialogue which he later published at Lichfield in 1780. Boothby was portrayed by Joseph Wright in one of the most famous portraits of the Enlightenment era as a gentleman of feeling and learning, lying in woodland beside a stream reading a copy of the Rousseau dialogue, and Seward thought he combined 'true genius, brilliant wit and the last polish of highlife society' with 'benevolence' and 'sweat temper', though with a tendency towards 'excessive instability'. A founding member of the Lichfield botanical society and Derby Philosophical Society (1783), Boothby was an ardent plant collector who constructed a conservatory at his Lichfield house to display exotic plants. On moving to Ashbourne Hall, which the family had occupied since the seventeenth century, he used money obtained from his wife's family following their marriage in 1784 to refashion the villa in the latest neo-classical style and collect tasteful works of art by Wright of Derby (such as some Italian views and paintings of Matlock) and other artists; he also commissioned a portrait of himself from the same artist. These changes were paralleled by improvements in the park, which was landscaped and planted in a more picturesque manner, with exotic trees and shrubs. He constructed another conservatory with sash windows to house his large collection of rare flowers and plants procured from around the world.[27]

Jackson, the final member of the Lichfield botanical society, was not highly regarded by Seward who claimed he was 'sprung from the lowest possible origin, and wholly uneducated', and that he had by 'force of literary ambition and unwearied industry' become a proctor in the ecclesiastical law courts, obtaining 'a profitable share of their emoluments' with a 'tolerable proficiency in the Latin and French languages'. She believed that his life had been 'shortened by late acquired habits of ebriety'. For Seward, Jackson was a 'would-be philosopher, a turgid and solemn coxcomb, whose morals were not the best, and who was vain of lancing his pointless sneers at Revealed Religion'; he 'admired' Boothby and 'worshipped and *aped* Dr. Darwin', serving as a 'useful drudge' to both men, his 'illustrious coadjutors' in their Linnaean translation, who exacted of him 'fidelity to the *sense* of their author' and corrected his 'inelegant English, weeding it of its pompous coarseness'. According to the *Gentleman's Magazine*, however, Jackson was a 'man of literature' and a 'useful assistant to Dr. Darwin in his ingenious publication of the System of Vegetables', and the doctor continued to seek Jackson's advice on plant identification matters during the 1790s.[28]

Seward believed Darwin was disappointed that 'no recruits flocked to his botanical standard at Lichfield', which she attributed to the fact that young local genteel males were little interested in natural history – however 'useful, entertaining, and creditable' it might be – feeling 'little desire to deck the board of the session, the pulpit, or the ensigns of war, with the Linnean wreaths and the chemical crystalines'. For these reasons the 'original triumvirate received no augmentation' and she was amused that 'scientific travellers' passing through Lichfield enquired 'after the state of the botanical society' when it in fact had little but a nominal existence.[29] Despite Seward's observations and the small number of society members, however, they did succeed in publishing their translations, while Boothby and Jackson were closely associated with the development of the botanic garden and the composition of Darwin's *Botanic Garden*, an association which survived until 1798.[30]

The other major member of Lichfield literary, landscape and botanical circles was of course Seward herself and she vividly portrayed various local scenes in some of her poems, including the original work that she composed at Darwin's botanic garden. Encouraged by their residence at the Derbyshire Peak village of Eyam, her editor, the Scottish novelist Sir Walter Scott, believed that Rev. Thomas Seward and his daughter 'imbibed a strong and enthusiastic partiality for mountainous scenery and in general for the pleasures of landscape', which was 'a source of enjoyment' during the rest of the latter's life and strongly reflected in her poetry.[31] Seward acknowledged how significant Darwin's encouragement had been in directing her towards poetry as a vocation and how she had 'acquired a taste for scenic beauty and poetic imagery' when young through 'listening to ingenious observations upon their nature' from his lips.[32] In 1798 She proudly described the beauties of Lichfield's gardens and prospects to Lady Eleanor Butler (1739–1829) and Sarah Ponsonby (1755–1831), the 'Ladies of Llangollen', in prose reminiscent of a layered landscape painting in the style of Claude Lorrain (1600–1682) or influential topographical poems such as James Thomson's *The Seasons* (1730). Looking across her 'small lawn, gently sloping upwards' and the 'sweet valley' beyond, she saw a 'semicircle of gentle hills, luxuriantly foliaged', with a 'lake in its bosom, and a venerable old church, with its grey and moss-grown tower on the water's edge'. Around Stowe Pool and towards the city there were 'elegant' villas 'interspersed with gardens and trees', a valley 'bursting into bloom', with fruits trees in a 'large public garden' 'now in full blossom', and a 'grove of silver amidst the lively and tender green of the fields and hedge-rows', so that Lichfield appeared 'like an umbraged village' with 'nothing ... more quiet and rural than the landscape' (Figure 7).[33]

John Saville, a Vicar Choral of Lichfield Cathedral and another local garden enthusiast, came to know Seward originally as her music teacher.[34] She considered him one of the 'brightest ornaments' of the city, a man 'consecrated by native talents, by science of many species', and passed over him in her memoirs of Darwin only because of his dislike of 'public attention towards himself'. Like Darwin and Seward, Saville was 'devoted' to 'scenic', 'botanic studies' and 'poetic ardours', but declined to be a member of the Lichfield botanical society because he did not want to be involved in writing the Linnaean translations.[35] Saville's first garden was next to his family house in Vicar's Close, Lichfield, and was called 'Damon's Bower' by Honora Sneyd (1751–1780). After separating from his wife Saville moved into a smaller house next door and leased land north of Stowe Pool in the grounds of Parchment House, where he created a botanic garden that was probably laid out in the early 1780s and obtained an 'immense' number of 'rare plants and flowers', engrossing' himself in their 'difficult, troublesome and expensive' culture. Saville's 'taste for botanical amusements' was well known in Lichfield; by 1785 his botanic garden included 'above 70 specimens of rare and elegant plants' and eventually, according to Seward, encompassed a collection of some 2,000 plants.[36]

One source of inspiration for Darwin, Seward and their Lichfield circle was the poetry and gardening of William Shenstone (1714–63), which informed their celebrations of landscapes and plants and approaches to Georgian landscape gardening more generally.

ure 7.   Stowe Pool, Lichfield, from J. Jackson, *History of Lichfield* (1805).

Shenstone designed a renowned garden with waterfalls at the Leasowes, Shropshire, which he extolled in his poetry, emphasising how he had pleased 'the imagination by scenes of grandeur, beauty or variety'. Formed on a relatively small site with two wooded valleys and pools, Shenstone's garden was shaped to achieve the maximum effect, introducing very varied planting with irregularly interspersed groups of trees and shrubs and green slopes. He deliberately tried to generate historical associations by fashioning enticing walks, gothic ruins, follies, alcoves, cascades of water, grottoes and inscriptions, and all on a relatively modest income of only about £300 per annum.[37] One of the reasons Darwin liked the location of his final house at the Priory so much was because it reminded him of Shenstone's valley, as his description of it to Richard Lovell Edgeworth (1744–1817) demonstrates.[38]

In around 1777 Darwin purchased 'a little, wild, umbrageous valley a mile from Lichfield' from the estates of Thomas Weld (1750–1810) of Lulworth Castle, Dorset. The sloping site, of about eight acres, was specially chosen as containing among the only rocks close to the city and provided opportunities for landscape gardening and planting similar to those available to Shenstone at the Leasowes. Darwin, who had conducted investigations into the springs, streams and water supply of The Close in Lichfield for the Lichfield Conduit Trust on behalf of the Cathedral and Chapter, described his land as 'irriguous from various springs, and swampy from their plenitude'. This, along with a 'mossy fountain of the purest and coldest water imaginable', attracted him to the place and had also originally encouraged Sir John Floyer, a physician who practised in Lichfield from 1676, to construct a 'cold bath' and bathhouse 'in the bosom of the vale', which Darwin acquired shortly after purchasing the rest of the vale around 1780. According to Seward, this was the 'only mark of human industry which could be found in the tangled and sequestered scene', although, with the bathers and others such as herself who already favoured it for walks, it was hardly unknown to locals.[39]

Within this 'beautiful piece of ground' was a 'grotto' some six yards wide and ten yards long with 'projecting rocks' that Darwin had excavated into the hillside to expose the layers of stone more clearly. The geology of the site consisted of 'silicious sandstone', the 'upper stratum' of which was only about five feet thick and 'divided from the lower stratum', which was just a 'sheet of clay, not more than 3 or 4 inches' thick. On top of the clay, between the 'lips of these rocks, a perpetual dribbling of water' oozed 'quite round the grotto, like a shower from a weeping rock'.[40] According to Seward, the dripping rock in the centre of the glen dropped 'perpetually', thrice each minute, and never ceased despite winter frosts, summer droughts and heavy rains, while aquatic plants bordered 'its top and branch from its fissures'. Darwin presented the grotto in *The Economy of Vegetation* as 'adapted to love scenes' and as a 'proper residence for the modern goddess of Botany', and described the features he had introduced and enhanced in the garden, referring to a 'glade' whose 'arching cliffs depending alders shade' and to the 'lawns' that 'to Peace and Truth belong' with 'steep slopes' as the 'Genius' or gardener had 'led with modest skill'. Seward described how Darwin made various changes to the landscape 'scenery', widening the brook in places into 'small lakes that mirrored the valley' and teaching 'it to wind between shrubby margins'. In her original poem, begun in the botanic garden in 1779, Seward emphasised how her friend 'first cultivated and adorned' this 'tangled and swampy vale'. He formed the land into 'smooth'd' 'lawns' with 'young woodland' and directed paths down 'steep slopes' through the 'wavy green' and 'marshy vale' and over the 'willowy mound':

> The willing pathway, and the truant rill,
> Stretch'd o'er the marshy vale yon willowy mound,
> Where shines the lake amid the tufted ground,
> Raised the young woodland, smooth'd the wavy green,
> And gave to Beauty all the quiet scene.[41]

According to the Lichfield surveyor, artist, painter and historian Charles Edward Stringer (1777–1855), here under Darwin's 'skilful hands' (with presumably a little help from those of his nameless gardeners!) contrivance was deftly blended with naturalness and the botanic garden 'assumed a form of the greatest beauty', with falls of 'highly picturesque appearance', shrubberies, thickets, pools and a 'mazy path', altogether having 'the effect of an extensive wilderness'.[42] It seems likely that Darwin would have purchased the plants and sought advice from a local nurseryman, perhaps John Bramall (d. 1807), steward of the Lichfield Friendly Society of Florists and Gardeners, who ran a nursery on Cherry Orchard and advertised in local newspapers with a stock including vegetables, hedging shrubs, fruit trees, grass seed and flowers. Darwin's landscape gardening, botanical and planting ventures were probably also partly inspired by various improvements undertaken by the Corporation, Cathedral and Chapter and Conduit Lands trustees, which coincided with the development of his botanic garden and with which Bramall was associated. These were designed to provide opportunities for polite promenading and to underline Lichfield's fashionable status, which was more important because of its relative economic eclipse by Birmingham and other regional towns. In 1783 Bramall was engaged by the corporation to plant trees on Borrowcop Hill and the island in Stowe Pool after a line of trees to the summit of the former had been planted by the corporation in 1756.[43] In 1772 a gravel New Walk was laid out on the south side of Minster Pool between Dam Street and Bird Street, with a gate at either end; encouraged by Seward and inspired by the Serpentine laid out in Hyde Park at the instigation of Queen Caroline, the Conduit Trust altered the north bank of the Minster Pool to create a more serpentine effect in 1773. The Pool was cleared and narrowed, and by 1776 an island had been positioned on the west side. Subsequently, in 1789, an island in Stowe Pool was planted with fir trees by Bramall, who carried out further planting in 1792.[44]

No list of plants or map showing the design of Darwin's botanic garden appears to survive and we cannot be sure how many of the species featured in the *Loves of the Plants* and *Economy of Vegetation* were growing there; Seward merely described how trees 'of various growth' adorned the 'borders of the fountain, the brook, and the lakes', as did examples of 'various classes of plants, uniting the Linnean science with the charm of landscape'. Another late Georgian collection, Coyte's *Hortus Botanicus Gippovicensis* (1796), listed at least 120 species of hardy trees and shrubs in his botanical garden and it is likely, despite the somewhat boggy terrain in Lichfield and although the Ipswich venture was more a conventional physic garden, that some of the same plants were in Darwin's garden. Some of Coyte's plants were certainly utilised in eighteenth-century medicine, although they cannot all have been growing on the site at the same time and were probably acquired for aesthetic as well as medical and collecting purposes. They included five different kinds of fig tree; ash and weeping ash; common, Norway and American maple; yew; common juniper; Phoenician cedar; white and black poplar; ten kinds of willow; the American and Chinese Arbor-vitae or thuja; hornbeam; chestnut; beech; American and oriental plane; walnut; both evergreen and common oak; mulberry; and birch alder.[45]

According to Darwin's friend John Jackson, who printed the first edition of *The Loves of the Plants* (1789) in Lichfield, after Darwin's improvements and the publication of *The Botanic Garden* the 'Bath and Botanic Garden' recovered their 'pristine celebrity' and were 'much used'.[46] The botanical society at Lichfield continued to exist after Darwin moved away to Derbyshire and remained associated with his garden through the activities of Jackson, who 'maintained' it 'on the original plan' until his death in 1798. While there were only three members of the society, people continued to visit the garden. In 1798, demonstrating a continued attachment to the place underscored by the success of his poem, Darwin was still referring to 'our botanical society at Lichfield' in correspondence with Robert Thornton in which he suggested that the Society would like to subscribe to his *Temple of Flora* (1798).[47] In August 1799, with Jackson gone, Darwin explained to his son Robert that the Lichfield botanic garden was 'now let by me to Mr. Bond' although it remained in the name of Erasmus junior for legal reasons; the latter was thinking of retiring from business and had talked of 'building a cottage, and going to live there, retiring from business!' suggesting a shared family sense of it as a place of repose, but Darwin disapproved, hoping that Robert would not mention it and that it would 'all fall through'.[48] In November 1799 Darwin told Robert he intended to sell the lease and in February 1800, not long after Erasmus junior's death, that he intended to dispense with 'the cold bath at Lichfield' for £100 if he could get it, as he thought it 'no longer worth my keeping', perhaps because the tragic demise of his son helped encourage his own retirement and the family move to the Priory. However, the lease was less valuable than it had been as it only had 'two lives in it', his own and Robert's.[49] In June 1800 he reported to Robert that Bond now had the botanic garden.[50]

According to a correspondent to the *Gardener's Magazine* in 1838, after leaving the hands of the Darwin family and Jackson the botanic garden 'gradually fell away from its former beauty', until it became merely a 'wilderness' with ponds 'choked with weeds, the cascades broken down, the walks overgrown with rank grass, and the "trim parterres" converted into pasture for cattle'. Occasionally in spring a 'stray snowdrop or a clump of daffodils' that had 'survived the general wreck' remained, but nothing else to 'tell of its high and palmy days, when the high-priest of Flora stood surrounded by the blossoms of a thousand climes'.[51] In his grandfather's memoir Charles Darwin followed a local guidebook in describing the site as 'a wild spot' that was nevertheless 'very picturesque', with 'many of the old trees remaining' and a few

'Darwinian' flowers.[52] The most detailed nineteenth-century Ordnance Survey map of Staffordshire, surveyed in 1882 and published in 1884 (Figure 8), shows a site measuring approximately seven to eight acres, which equates to the size given in other sources. Various garden features are represented and some of the layout appears to have remained intact as it can be differentiated from the relatively treeless surrounding fields and what were probably self-seeded trees growing along nearby watercourses. It was sometimes complained that Darwin's gardens were full of plants perceived by others as weeds and it may be that the *Gardener's Magazine* correspondent's comments reflect their expectations of botanical gardens rather than providing a reliable assessment of how much survived. The 1884 Ordnance Survey map shows the bath house developed by Floyer and preserved by Darwin and a double stream running along the slope of the vale towards the two 'lakes', which are labelled 'Fish Ponds' and surrounded by marshier ground. One footpath is shown passing between the two ponds and another from the bath to the ponds, while the outline of the 'grotto' excavated into the side of the hill that served as the 'proper residence for the Goddess of Botany' in *The Loves of the Plants* seems to be still evident. There are deciduous and a few coniferous trees around the boundaries and scattered across the grass and along the banks of the stream, as Seward and Darwin described.[53]

Although the 1884 Ordnance Survey map showing the botanic garden site provides some clues, it would be interesting to know exactly how Darwin was able to successfully unite 'the Linnean science with the charm of landscape' as Seward claimed, and how much there were sections planted systematically according to Linnaean plant family and genus like in contemporary university botanical gardens and, if so, whether shrubs or trees were accommodated within this arrangement or not. Likewise, it would be

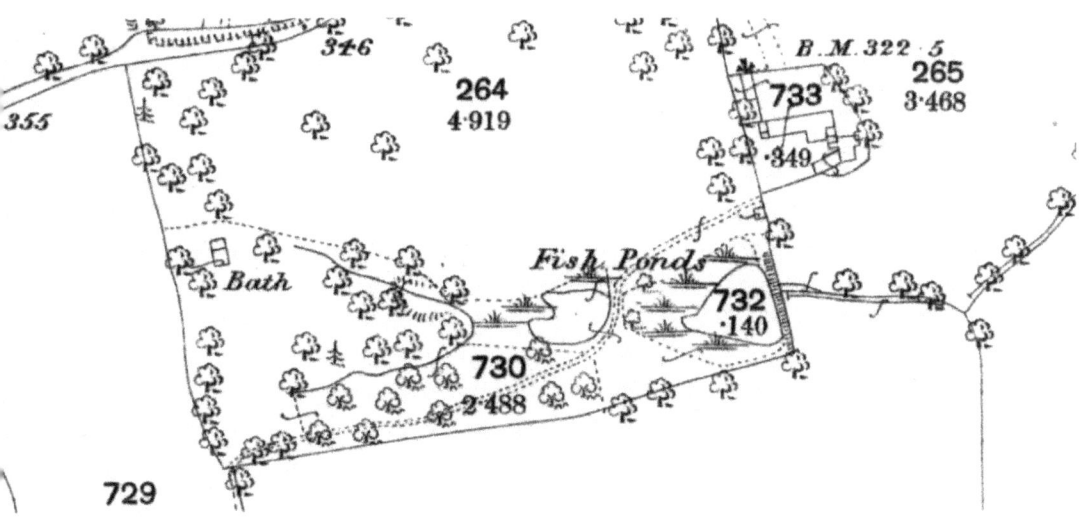

ure 8. Site of Erasmus Darwin's botanic garden, shown on the Ordnance Survey map, 25 inches to the le, 1884.

fascinating to understand if the plants as a whole were intended to be taxonomically representative of Linnaean categories with examples that would survive outside in the British climate. Seward's statement that the borders of the fountain, brook and lakes were adorned with 'various classes of plants' as well as trees 'of various growth' certainly suggests a partial systematic arrangement as does the subject matter of *The Botanic Garden* itself of course. But as the 1884 OS map indicates, this was combined with planting designed to maximise the landscape beauty of the sloping site and frame views from the grassy footpaths inspired, as we have emphasised, by gardens such as Shenstone's Leasowes with its winding walks, babbling dingle, green slopes, clumps of trees and contrived scenes celebrated in verses presented to the poet. The most probable configuration of the systematic collection in Darwin's botanic garden would be groupings around some or all of the twenty-four Linnaean 'classes' in the borders, which were sometimes likened spatially to 'streets and squares within a populous city' or counties within a state at national level (with genera and species as parishes and villages).[54]

As their collaborative poems show, Darwin and Seward encouraged each other in their literary enterprises, including the latter's *Elegy on Captain Cook* (1779). As Darwin formed the botanic garden Seward's curiosity grew, although he had 'restrained' her from visiting 'her always favourite scene' until it had acquired its 'new beauties from cultivation'. The intention was for him to accompany her on the first visit, but Darwin was called away on medical practice so Seward went on her own with 'her tablets and pencil'. Here, 'seated on a flower-bank, in the midst of that luxuriant retreat', she claimed to have composed some verses – 'while the sun was gilding the glen' and 'birds, of every plume, poured their song from the boughs' – in which Darwin as the 'first cultivator' was honoured as the 'genius of the place'.[55] Seward claimed that her verses provided her friend with the idea for a poetic depiction of the Linnaean system, emphasising in a note that the verses 'in their original state, as inscribed here' appeared in Stebbing Shaw's *History of Staffordshire* 'near four years' before Darwin's death and that she as the author 'chose to assert her claim to them' in his lifetime because they had appeared in the periodicals a long time before the *Botanic Garden*'s publication and had 'borne her signature'. She emphasised that, when presented with the poem, Darwin had 'seemed pleased with it', exclaiming that he would send it to a magazine for publication and that 'it ought to form the exordium of a great work'. The Linnaean system was 'unexplored poetic ground' and afforded 'fine scope for poetic landscape', suggesting 'metamorphoses' like those of the Roman poet Publius Ovidius Naso (c. 43 BC–AD 17), 'though reversed'. As Ovid had 'made men and women into flowers, plants, and trees' so Seward could make 'flowers, plants and trees, into men and women', while he would compose the 'scientific' notes. Seward replied that, 'besides her want of botanic knowledge, the plan was not strictly proper for a female pen' and was better 'adapted to the efflorescence of his own fancy'. In response (according to Seward), Darwin cited the problems that 'coming forward an acknowledged poet' might present for his professional reputation, but she retorted that as the subject of the poem was 'connected with pathology', and as he was now in the later stages of his medical career, there was little chance of such problems arising. Darwin therefore 'took his friend's advice, and very soon began his great poetic work'.[56]

Although this may have simply been a misunderstanding, the way in which Darwin subsequently treated Seward's verses became a point of contention for her. A few weeks after their composition Darwin sent them to the *Gentleman's Magazine* in her name and they subsequently appeared in the *Annual Register*, but, as Seward noted, 'without consulting her, he had substituted for the last six lines, eight of his own'. Subsequently, and again without her knowledge, he made them the exordium to *The Economy of Vegetation*, published in 1791 after *The Loves of the Plants* had appeared in 1788/9, and she complained that 'no acknowledgment' had been given that the verses 'were the work of another pen' as 'ought' to have occurred, especially as they had already been previously published in her name. Her annoyance was increased by the fact that they were 'somewhat altered', with eighteen lines of Darwin's own composition being 'interwoven with them'. It seems likely that Darwin intended this as a compliment or tribute to her, as publication in the *Gentleman's Magazine* suggests; and, as we have seen, it reflected the practice of publicly attributing co-authored works to their main author within the Lichfield circle, but Seward remained unhappy and provided a detailed account of her grievances in her memoir.[57]

Darwin intended that his *Botanic Garden* would 'enlist imagination under the banner of science' and 'lead her votaries from the looser analogies' that adorned 'the imagery of poetry' to the 'stricter ones' of the 'ratiocination of philosophy'. He particularly hoped to 'induce the ingenious to cultivate' botanical knowledge by enticing them into 'the vestibule of that delightful science' so they would encounter the 'immortal works of the celebrated' Linnaeus. *The Economy of Vegetation* examined plant physiology and the 'operation of the elements' as they impacted upon vegetable growth.[58] *The Loves of the Plants* explored the Linnaean sexual system and the numerous 'remarkable properties' of many vegetables. In this work, Darwin strove to use his poetical garden like a 'camera obscura' – which was a room or box much utilised by artists in which an inverted image of the view outside was projected upon an opposite wall or screen through a small hole or shutter – in which 'lights and shades' danced on a 'whited canvass … magnified into apparent life'. He urged readers blessed with 'leisure' for 'trivial amusement' to stroll in and experience the delights of his 'enchanted garden' (Figure 9). As discussed with Seward, the premise of the poem was that, just as Ovid had used his craft to transmute 'men, women, and even gods and goddesses, into trees and flowers', so Darwin undertook to employ his pen similarly to 'restore' some of the trees and plants 'to their original animality' as they escaped from the 'vegetable mansions' where they had been long confined. He now exhibited them in a manner akin to 'divers little pictures suspended over the chimney of a lady's dressing room' tenuously connected merely by a 'slight festoon of ribbons' which might amuse even those hitherto unfamiliar with their original qualities by the 'beauty of their persons, their graceful attitudes, or the brilliance of their dress'.[59]

The extensive footnotes and essays in *The Botanic Garden*, which were really a set of philosophical disquisitions in their own right, represented subjects that Darwin considered necessary for a fuller appreciation of Linnaean botany as well as wider

Figure 9. Flora attired by the elements, from E. Darwin, *The Economy of Vegetation*, 4th edn (1799).

observations on natural philosophy and natural history. Despite his fulsome praise for Linnaeus, the notes concerning natural history illustrate the areas that Darwin considered were inadequately explored by the master and his disciples, including vegetable physiology and anatomy. The experience of sojourning in his metaphorical garden, with its mazy paths undertaking philosophical detours, partly explains the charm and appeal of the verses for British readers while underscoring the inspiration provided by creating and experiencing the Lichfield garden, with its admixture of system and tasteful beauties. In his *Philosophia Botanica* (1751) and *Species Plantarum* (1753) Linnaeus aimed to chart the animals and vegetables of the natural world, employing the metaphor of a map and geography of nature. The globe was divided into five climate zones: Australian, oriental, Mediterranean, boreal and, later, the Alpine zone. As Charles Withers has emphasised, the works of Linnaeus and his followers were received differently across Europe, but in Britain there was generally an enthusiastic reception represented by the number of translations of parts of his works and introductions to the system that appeared.[60] Through the efforts of their small Lichfield botanical society, Darwin, Boothby and Jackson aimed to popularise Linnaean botany by translating the original Latin into English to make it more accessible. Darwin shared his love of botany and Linnaeus with his son Charles (1758–1778), introducing him to the 'sweet botanic art' so he could 'in little' read 'great Nature's Laws, from near effects progressive to their cause'.[61]

In the life of his son Charles that Darwin appended to the edition of his son's medical works published in 1780 after Charles' tragically early death, Darwin explained that his boy had from 'infancy' been 'accustomed to examine all natural objects' with an unusual degree of 'attention', and the poem he attached to the work underscores the links between Linnaean botany, herbal medicine and protean richness of place. At the age of nine Charles had accompanied the 'ingenious botanist' Rev. Samuel Dickenson (1733–1823) of Blimhall, Staffordshire, on a French tour, during which he 'acquired a taste' for the subject, and, having 'languished' at Oxford for a year, he 'removed to the robuster exercises of the medical schools of Edinburgh'.[62] According to Darwin, bounteous nature who 'unveils her bosom' and is 'To all a blessing, but supreme to Man' had from her munificence been able to satisfy every need and offer 'cheer' to the heart while charming 'the gazing eyes', having brought pleasure and fascination to father and son alike. Darwin and his son botanised together, Darwin in his writings personifying vegetation as a 'dear enchanting maid' who frequently led the two of them through the fields and woods around Lichfield as the botanic garden was being formed and planted:

> Link'd in arm of each, while smiling she
> Would point each curious grass, flower, shrub and tree;
> Their germination, growth, decline, explain
> How each expires, and bursts to live again!'[63]

The extent of father and son's joint botanising is evident from a wonderful pocket notebook that survives among the Cambridge University Darwin manuscripts. The notebook appears to have been given by Darwin to Charles, as it is inscribed with a printed notice 'C. Darwin, no. 84, Lichfield, 1774'. It primarily consists of 'nature-printed' impressions formed by pressing ink-coated plants upon paper and, like the Lichfield botanical society's later publications, begins with English and Latin plant name indexes. Some of the impressions work well and show the intricate patterns, shapes (and of course sizes) of leaves in fine detail. The representations would not have helped with Linnaean-style identifications of plants, but they would have fostered botanical learning more generally and certainly helped practical identification in the field. Used by Charles with his father's encouragement and help, and probably featuring some of the plants growing around Lichfield referred to by Darwin in the poetical memorial tribute to his son, the notebook was probably taken by Charles to Oxford in 1774 and then to Edinburgh, where his education continued in the medical school. It was reacquired by Darwin after his son's tragically early death and seems to have been subsequently lent out, perhaps, as Anne Secord has suggested, to fellow members of the Lichfield botanical society during their Linnaean translation endeavours. A note at the start states that 'This book is Dr. Darwin's and is to be return'd when done with' and Darwin made some additions and alterations, including the names of the families to which the plants belonged, sometimes changing the genus names provided by Charles if they were incorrect or had been reclassified. Given that at least 35 per cent of the plants listed were also later growing in Darwin's Full Street garden in Derby it seems likely that most of them existed in the Lichfield Beacon Street and botanic gardens as well as the surrounding countryside.[64]

With the importation of so many 'exotic vegetables' into British gardens which according to Darwin and the other members of the Lichfield society were 'gradually' loosing their 'novelty' and being 'mistaken for natives of the soil' and the parallel 'naturalisation' of some Linnaean terminology such as generic names, the time was ripe for new reliable English translations of works providing guides to the system to prevent further confusion. Some Linnaean terms was easily translatable by the Lichfield society members, although other Latin words and phrases required more original English coinage by compounding terms and other means, in which they consulted the 'great master of the English tongue Dr. Samuel Johnson' about the best terminology to use. One principal objective was to bring, or rather return, the Linnaean system democratically from the confines of the study into the world of practical gardening and horticulture and so to improve the botanical education of those with professional or occupational interests in the subject (Figure 10). Although the translations were an important scholarly exercise that in some ways paralleled his progressive political beliefs, Darwin considered that their publication, and particularly that of *The Botanic Garden*, would help to achieve this. Existing translations such as those by Benjamin Stillingfleet (1702–1771) were incomplete, out of date, impractical or expensive. Darwin, Boothby and Jackson noted that, 'like the Bible in Catholic countries', Linnaean botany had been 'locked up in a foreign language, accessible only to the learned few, the priests of Flora'. This meant that gardeners, 'the herb-gatherer, the druggist', farmers and all 'concerned in cultivating' or searching for the vegetable 'tribes' or 'consuming their products', were never able to obtain sufficient knowledge to improve. Equally importantly, reciprocal stimulation between such individuals and scholars in their places of business and study was stymied by such elitism, when in reality all groups were equally able to augment botanical science.[65] However, the cost of the Lichfield botanical society's translations meant they did not circulate widely and never reached a second edition, in contrast to William Withering's *Botanical*

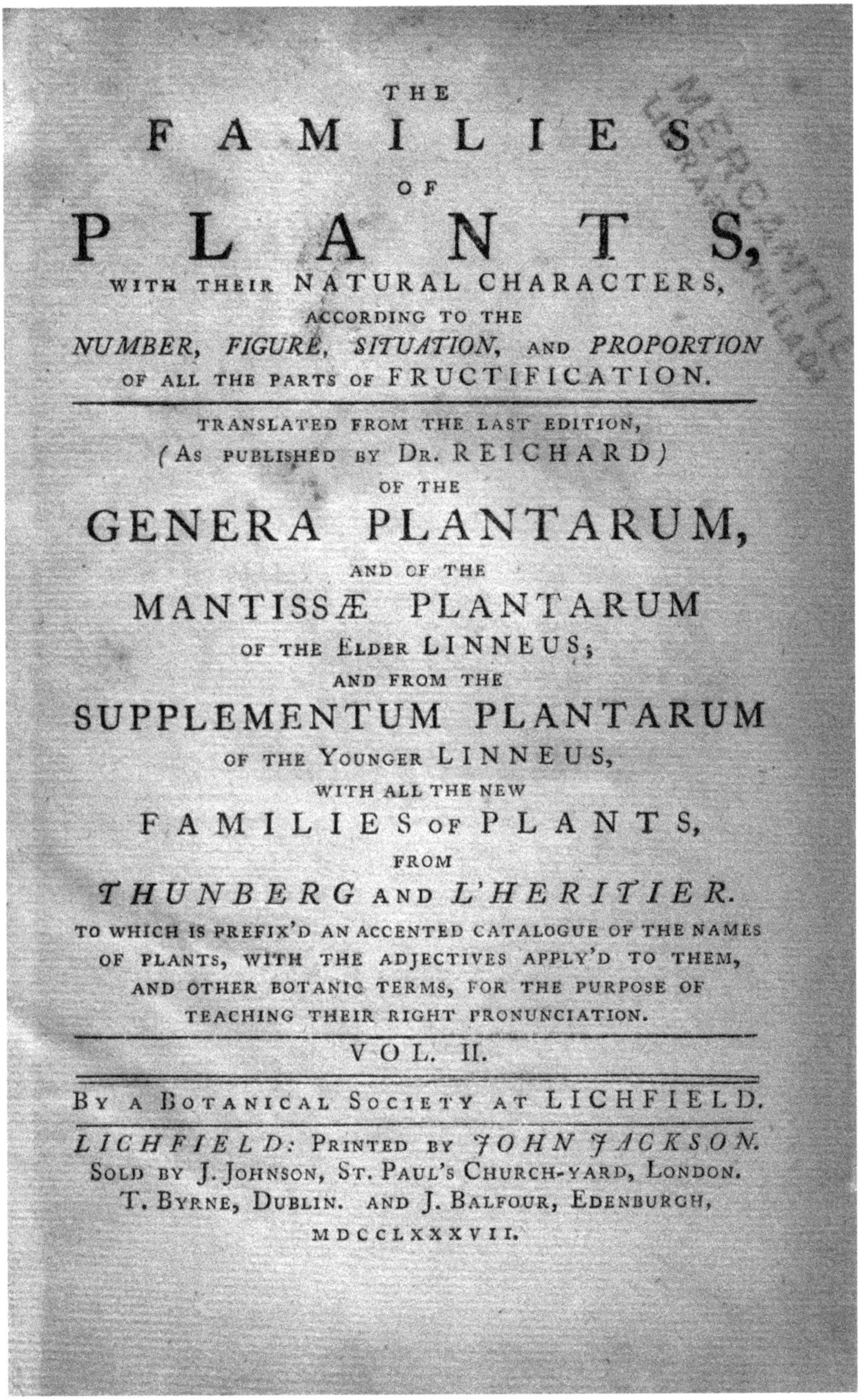

THE

FAMILIES

OF

PLANTS,

WITH THEIR NATURAL CHARACTERS,

ACCORDING TO THE

NUMBER, FIGURE, SITUATION, AND *PROPORTION*
OF ALL THE PARTS OF FRUCTIFICATION.

TRANSLATED FROM THE LAST EDITION,
(AS PUBLISHED BY DR. REICHARD)

OF THE

GENERA PLANTARUM,

AND OF THE

MANTISSÆ PLANTARUM

OF THE ELDER LINNEUS;

AND FROM THE

SUPPLEMENTUM PLANTARUM

OF THE YOUNGER LINNEUS,

WITH ALL THE NEW

FAMILIES OF PLANTS,

FROM

THUNBERG AND *L'HERITIER.*

TO WHICH IS PREFIX'D AN ACCENTED CATALOGUE OF THE NAMES
OF PLANTS, WITH THE ADJECTIVES APPLY'D TO THEM,
AND OTHER BOTANIC TERMS, FOR THE PURPOSE OF
TEACHING THEIR RIGHT PRONUNCIATION.

VOL. II.

BY A BOTANICAL SOCIETY AT LICHFIELD.

LICHFIELD: PRINTED BY *JOHN JACKSON.*
SOLD BY J. JOHNSON, ST. PAUL'S CHURCH-YARD, LONDON.
T. BYRNE, DUBLIN. AND J. BALFOUR, EDENBURGH,

MDCCLXXXVII.

Figure 10. Frontispiece to the Lichfield Botanical Society's *Families of Plants*, 2 vols (1787), vol. 2.

Arrangement (1776), which, as we will see, went through several editions and became the standard British plant reference work. But the publication of the fourth octavo edition of the *Botanic Garden* in 1799 and cheaper North American and nineteenth-century editions with simplified engravings did help Darwin's work to enjoy a wider circulation even as his poetry went out of fashion.[66]

As has often been noted, the emphasis in *The Botanic Garden* on the sexual life of plants and the highly suggestive language employed – aspects widely seen as titillating, provocative and even scandalous, although stimulated by the original Linnaean formulations and Darwin's attitude towards reproduction and carnal enjoyment – were among the reasons Seward declined to pursue the original poetical project idea, as it was, in her view, unbecoming for a female writer. One example is a note explaining that 'the vegetable passion of love is agreeably seen in the flower of the Parnessia, in which the males alternately approach and recede from the female' and, similarly, 'in the flower of the nigella, or devil in the bush, in which the tall females bend down to their dwarf husbands'. Darwin explained that he had been 'surprised' one morning in September 1790 to see amongst Sir Brooke Boothby's valuable Ashbourne plant collection, the 'manifest adultery of several females of the plant Collinsonia, who had bent themselves into contact with the males of other flowers of the same plant' nearby.[67]

DARWIN'S DERBY GARDENS

After acquiring his house in Full Street, Derby, in 1782 and making improvements, Darwin moved in fully with his family in the summer of 1783. The property had 'pleasure and kitchen gardens' to the rear, between the house and the River Derwent.[68] Although Darwin was originally inspired by *materia medica* and the Lichfield botanic garden, the emphasis upon botany and gardening in his poetry also reflected the inspiration that his second wife Elizabeth provided and the centrality that these activities had come to assume in their lives at Derby – as Darwin emphasised to Thomas Day (1748–1789), they both 'loved the country and retirement'. It is also evident in his correspondence between 1780 and 1802, which contains many references to botany, gardening, horticulture and farming. During a trip to London with Elizabeth after their marriage in 1781, Darwin met Joseph Banks, the president of the Royal Society, and on his return to live at Radburn Hall began a series of investigations into vegetable anatomy and physiology, which were detailed in the *Economy of Vegetation* and *Phytologia*. He corresponded with many botanists and publishers concerning the Lichfield society's Linnaean translations, including Banks, Daniel Solander (1733–1782), Anna Blackburne (1726–1793) of Orford near Warrington and William Curtis (1746–1799), whose *Flora Londinensis* he praised and was collecting in parts.[69] The move to Derby had various consequences, and Darwin concentrated more on his books, reduced the number of patients he treated and saw less of his Lunar friends. On the plot around his Full Street house and on the opposite bank of the Derwent he undertook landscape gardening improvements that paralleled and informed the ideas concerning agricultural improvement in *Phytologia*. Within the garden Darwin created a large hothouse and dug an artesian well to supply clean, fresh water, detailing the work and the principles that lay behind it in a paper published in the *Philosophical Transactions* of the Royal Society. The original plaque from the artesian well survives in Derby Museum and it is possible to reconstruct the hothouse using Darwin's correspondence. There was also a summerhouse adjoining the Derwent 'over the water' in which Darwin wrote letters and parts of his books, and it was also used by visitors and guests such as the Rev. John Gisborne (1770–1851), the local poet and author of the Darwinesque *Vales of Wever* (1797), who had married Elizabeth's daughter Millicent Pole (1774–1857), to compose a prayer in memory of his mother in May 1800. Like the Lichfield botanic garden, Darwin's Derby gardens became well known locally and seem to have attracted a lot of visitors, and he complained, probably tongue-in-cheek, to his elder brother Robert Waring (1724–1816), the author of the *Principia Botanica* (1787), that some of his plants 'now waste daily by our gardens being so public and the number of our children'.[70]

The land that Darwin owned on the other side of the Derwent included an orchard and he described the allotment, apparently half-jokingly, as his 'farm' to the artist Samuel James Arnold (1774–1852), who visited in 1799 – but this designation is highly significant, as we will see (Figure 11).[71] Most of the rest of the land in the area belonged to Exeter House. The route to the orchard or 'farm' on its far side, via St Mary's Bridge, was circuitous, and Darwin's solution was to obtain a boat about seven or eight feet wide and twenty-five feet long and then adapt this to be pulled by a rope strung across the river, a fascinating sketch of which was later made by Violetta Harriet Darwin (1826–1880), daughter of Sir Francis Sacheverel Darwin (1786–1859), based upon his recollections (Figure 12). The 'farm' and orchard appear in this sketch, which shows well-established fruit trees, and are clearly visible on Peter Perez Burdett's map of Derbyshire, reissued in 1791, and on George Cole's later map, based upon Burdett.[72]

Darwin's final garden, at Breadsall Priory, where the family moved in 1802, was initially acquired by Darwin's son Erasmus junior, a lawyer. In November 1799 Darwin told his son Robert that Erasmus junior had bought 'a place 5 miles from Derby call'd the Priory' with 'a large old house, and a farm of 80 acres, where he intends to live, but not to occupy the farm; and to sleep away the remainder of his life!' Erasmus had paid £3,500 for the Priory', which was 'thought a very cheap purchase' because it had 'long been exposed to sale'; it was 'a fine situation with 3 fish-ponds descending down a valley', with views of the river Derwent and All Saints Church tower.[73] Erasmus junior died shortly afterwards, however, and, while the Darwins were grief-stricken, the land surrounding the Priory presented a major opportunity for landscape gardening and gardening, and they sold the Full Street house and orchard, the latter as plots for wharfs and building ground, taking advantage of the canal development, in 1802.[74] Darwin described to Richard Edgeworth how they had all moved from town 'about a fortnight' since to the Priory and all liked the 'change of situation', which he described as a 'pleasant house, a good garden, ponds full of fish, and a pleasing valley', rather like Shenstone's Leasowes, as we have noted, with a 'deep, umbrageous' valley and a 'talkative stream'. The house was situated near the valley top and was 'well screened by hills from the east and north, and open to the south', with views of the church tower, while at least four plentiful springs rose nearby that had 'formed the valley' with closed sides which reminded him of the Valchiusa retreat described by the Italian poet Francesco Petrarca (Petrarch, 1304–1374).[75] Just as his gardens had provided him with literary and scientific inspiration so he emphasised to Edgeworth that his daughter Maria (1768–1849), a novelist, might find equal stimulation by visiting. Although Darwin lived at the house for only a short space of time, landscape improvements had begun prior to this and were continued by Elizabeth after Darwin's death for thirty years, visitors to the Priory emphasising how much

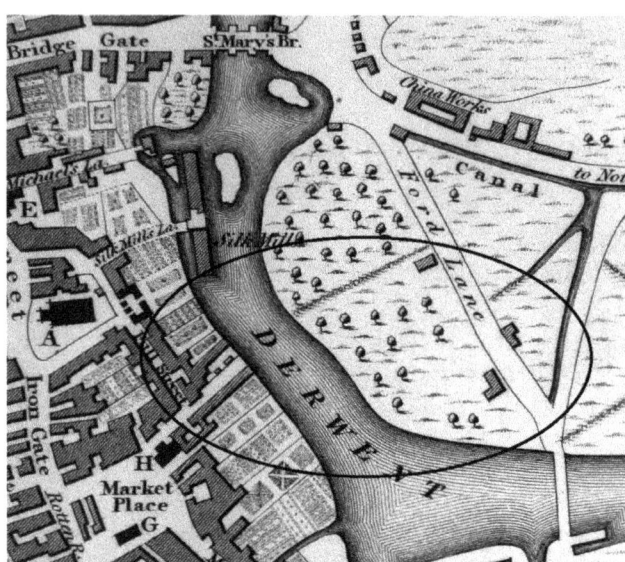

Figure 11. (left)   Detail of map of Derby drawn by G. Cole and engraved by J. Roper (1806), showing the area of Darwin's garden and farm across the Derwent, from W. Hutton, *History of Derby*, 2nd edn (1817).

Figure 12.(below)   Pen and ink drawing by V. H. Darwin of the mechanical ferry designed by Erasmus Darwin for crossing from his house in Full Street, Derby, to his orchard; the child shown in the boat is Francis S. Darwin; 1789, from K. Pearson, *The Life, Labours and Letters of Francis Galton*, vol. 1 (1914).

attention she devoted to the garden. Sending a parcel to Barbara Strutt in summer 1801, Darwin enclosed a branch of a sugar maple, which he contrasted with an ash-leaved maple growing at the Priory, and referred to the purchase of other plants for the gardens there, including a 'double ragged Robin' (*Lychnis floscuculi*) and a 'narrow-leaved Kalmia', the latter being a genus of evergreen shrub from North America.[76]

Is is, however, the Derby garden and orchard that have left the fullest record of plants that grew within. In June 1796 Darwin told Miss Lea that he was compiling a 'new catalogue' of hundreds of plants growing at Full Street and at least some of this is preserved among the Cambridge University library Darwin papers.[77] The notebook lists over 200 trees in the orchard and garden, describing locations and the qualities of fruit produced by each, and the large variety of hardy plants identified with help from William Jackson.[78] Some additional examples referred to as growing in the garden in Darwin's correspondence and publications, especially *Phytologia*, can be added to the list for a fuller picture. As well as his own botanical knowledge, Darwin probably drew upon the list of Derbyshire plants compiled by Rev. James Pilkington for his *History of Derbyshire* (1789), which in turn obtained information from William Withering and a small group of local botanists that included Rev. Dewes Coke (1747–1811) of Brookhill Hall near Mansfield and the medical men William Brookes Johnson (1763–1830: a physician of Coxbench Hall), Jonathan Stokes (1755–1831) of Chesterfield and Thomas Wharton (a surgeon who collected plant information and liaised between Pilkington and Withering).[79] Pilkington's catalogue provided a list of 'the most useful' 'native' plants that grew 'spontaneously' across the county and remarked upon their 'medical virtues, their uses in the arts, and as food, and any other striking or peculiar properties', only including those that had been 'examined by myself' or by Pilkington's colleagues in the endeavour. The 'places of growth' of what he believed were the 'rarer plants' were included, as were the name of the person who had observed them, Linnaean and English nomenclature, times of flowering, age where known and whether trees or shrubs.[80] Like weather diaries and journals, catalogues and lists of plants were a common feature of Enlightenment science and scientific culture and offered opportunities for botanists and gardeners to record plants growing in their gardens or area for their own benefit or sometimes for publication. The motivation was partly taxonomic but more practical, in that enumerating this information was believed to provide a means of making better use of plants for their economic and medical benefits. Like estate records, plant catalogues also provided an opportunity for gardeners to demonstrate the value of their property. However, the vegetable denizens of Darwin's garden seem to have been selected for their beauty as well as their practical medical and domestic uses.

The Darwin family circulated plants and botanical information between one other, especially through Erasmus senior; his son Robert at Shrewsbury; and his elder brother Robert Waring at Elston, from whom, as an authoritative botanist, the family obtained many examples and much advice. Writing to his older brother in 1799, Darwin described how the 'young ladies' had 'filled the vine-house' at Derby with plant pots, mostly 'originally imported from Elston'. Darwin promised to send his son Robert 300 plants 'with their names' in the autumn of 1792, which, he explained, would 'half set you up as a botanist';[81] he also offered advice, and sent vegetable material from the family seat to Robert, promising in November 1799 to despatch 'some grafts ... from the bloody pear from Elston' in the spring and acorns from the oak tree that 'used to be bent down and pruned' by his father Robert when he was a boy.[82] Jackson continued to be involved, too, as he maintained the Lichfield botanical garden, kept the Lichfield botanical society alive and helped Darwin identify trees and plants at Derby. Occasionally there are references in the correspondence to gardeners working for the family. Many plants were probably obtained from Joseph Mason, who, according to John Claudius Loudon, kept a 'good nursery and florist's garden' in Derby. Mason is listed as a dealer and grower of auriculars, polyanthuses, double primroses, carnations and other flowers, and there is a reference in 1752 to Joseph Mason, a 'gardener at Derby', who may have been Mason's father or another relative.[83] The Masons are likely to have been members of a Derby 'Old Friendly Society of Florists' active in the later eighteenth century, who advertised their meetings in local newspapers and awarded prizes. A notice from the stewards, Rev. George Greaves of Stanton and the Derby linen merchant and draper Robert Grayson, calling all 'Gentlemen Florists' to a meeting in April 1789 at the house of Samuel Brackley, the Angel, Cornmarket, Derby on 2 May – just around the corner from Darwin – asked that 'best auriculas and polyanthuses' be brought in pots and vials for judging, prizes being awarded for 'the best and compleatest' examples exhibited on stage and other achievements before dinner from two o'clock. The words of a 'Florist's Song' sung at meetings fuelled by drinks well captures the convivial spirit of these occasions. Like Darwin's *Botanic Garden*, and appealing to 'jolly Gardeners of every degree' from 'the Setter of a Flower, to the Planter of a Tree', the song celebrated the antiquity of horticulture in the post-Edenic world and the pleasures of being in one's 'own enclosure ... as happy as a king', captivated by the 'lowly shrub' and 'lofty spreading trees' forming 'pleasant shade fann'd by th' pleasant breeze'. The florist claimed precedence in the 'delightful art' of nurturing 'his' flowers by 'skilful hand', 'setting each apart', delighting in 'their beauties and all their various hues' and enjoying 'so sweet a life'. While it was 'too tedious' to run through each name, their beloved auriculars claimed precedence as the 'noblest' flowers for their 'smell' and 'beauty'. Carefully managed by their 'impartial judges' and free from 'windy wars', their annual shows inspired each gardener to emulate each other as they spent the day in mirthful feasting and 'pleasure' crowned 'the night'.[84]

In a similar vein, the joy and satisfaction that Darwin's family gained from gardening and horticulture at Derby is evident from two playful poems he composed on 'The Arts of Pruning Melons and Cucumbers' and 'On the Cultivation of Broccoli', the latter of which was a partial translation of a Latin poem by the author and theatre critic Edward Tighe. Although light-hearted, both works present a delightful picture of his gardening in practice intermixed with serious advice and were later included in *Phytologia*, underscoring the associations between his poetical and prose works. Presaging early versions of *The Temple of Nature*, the 'Cultivation of Broccoli' described the operations of the cyclical gardening year in the form of a pastoral poem, associating seasonal gardening operations with the changing nocturnal zodiacal sky as part of the economy of nature, although the detailed practical information in both poems transcends pastoral conventions to some degree. The 'Cultivation of Broccoli' celebrated the bounty of nature, the 'fertile soil' and 'grateful toil', as the 'watch'ful gardener' scatters seed and rakes with 'iron teeth' the 'yielding glebe', the 'vegetable birth' in the 'womb of the earth', carefully watering, dividing and nurturing his 'swelling 'stalks' to 'gigantic size', 'rich viands' in their 'ample beds'. So:

> 'Oft in each month, poetic Tighe! be thine
> To dish green Broccoli with savoury chine;
> Oft down thy tuneful throat be thine to cram
> The snow-white cauliflower with fowl and ham!'

In the 'Art of Pruning' Darwin explained when and how the operation should be undertaken, 'Arm'd with fine knife or scissors good' to stimulate the growth of lateral branches 'like Hydra's fabled head'. The second part of the poem is a strong reaffirmation of the sexualised style and imagery of *The Loves of the Plants*, portraying the flowers as 'fair Belles in gaudy rows' and saluting their 'vegetable beaux' as they lost their 'virgin bloom'. Then, as the 'pregnant womb' 'swells' after cutting off the 'barren flowers' if the plants were 'confin'd' within 'frames' and could not be impregnated on the 'breezy' winds, so 'tall males' had to be artificially bent towards the 'fair' or pollen had to be plucked to shed the 'genial pollen' over their bed so 'each happier plant' would 'unfold, prolific germs, and fruits of gold'.[85]

The Derby garden was enclosed, bounded by the house, walls and the Derwent, and it is clear from the notebook that, apart from the prospect across the river towards the orchard, its most attractive features were the individual plants, and there would have been a real mixture of colours, primarily reds, whites and yellows. Most were presumably planted by Darwin, Elizabeth, their children or their gardeners, although some, such as the fruit trees, must have pre-dated their arrival in the 1780s. Catalogue alterations and crossings-out reveal which plants died and, usually, their replacements. The verdant variety probably made the garden appear unkempt and many plants were not conventional garden cultivars, but native species more usually found in meadows, fields and woodlands. We can therefore see why, just as visitors to the Lichfield botanic garden thought it was merely an 'extensive wilderness', some who came to the Derby garden likewise dismissively described it as a collection of weeds; it must have been difficult, too, for the gardeners to distinguish between the specially planted specimens and plants considered weeds.[86] In 1794 Darwin complained to his son Robert that his '*Polygonium proliferum*', having been 'nearly destroyed by the dry season', was 'almost hoe'd up by an ignorant gardener' – but perhaps the gardener deserves some sympathy for not recognising the significance of this frazzled clump of knotgrass![87] Darwin regarded at least some of his Derby garden plants as a systematic collection akin to the Lichfield garden and described it as a 'botanical garden' in 1787 when requesting '2 or 3 or a few plants' of the mangelwurzel or root of scarcity from Dr John Coakley Lettsom.[88]

Only hardy plants are listed in the Derby notebook, but there would probably have been many annuals in the garden each year, too, such as nasturtiums (*Tropaeolum majus*), of which Darwin believed the flowers to possess 'an agreeable acrimony'; they could be eaten raw, 'shred with the fresh leaves of lettuce, young mustard plants, or red cabbage'. Some were utilised in experiments, such as madder (*Rubia tinctorum*), sun spurge (*Euphorbia helioscopia*) and Christmas rose (*Helleborus niger*), which were employed in his studies of vegetable physiology and anatomy undertaken at Radburn Hall in 1781 and possibly brought from there to Derby (or separately planted).[89] The sections in the list are not defined according to botanical categories, but follow a walk from the back of the house to the fountain, then circling and passing the fishpond and continuing beyond the stone table. Returning from the Derwent, Darwin lists the plants moving towards his summerhouse overlooking the river, before describing those plants between there and the well.[90]

Darwin marked some plants in the notebook with a 'B', indicating those regarded as especially beautiful, most of which produced particularly large or colourful flowers. Beyond the fishpond he highlighted for their beauty the blue and white *Aconitum* (monk's-hood or wolf's-bane); *Achillea millefolium* (yarrow); *Ranunculus gramineus*, grass-leaved buttercup; *Kalmia angustifolia*, narrow-leaved sheep laurel; a dwarf *Narcissus* or daffodil; *Spirea filipendula* (syn. *Filipendula vulgaris*), drop-wort (Figure 13); and an orange-coloured *Cistus*. He had ordinary tulips growing and a double yellow variety, remarking in *Phytologia* that while, with its delightful 'beauty ... and its greater longevity', this had 'much to recommend it to common eyes', yet the 'endless variety in the colours of single tulips has long deservedly been the admiration of florists'.[91] When at least one of Darwin's double yellow tulips unfortunately died he chose an iris to replace it, unquestionably one of his favourite flowers, with thirty-nine different kinds scattered across the garden, according to the notebook, although not all grew simultaneously and riverside location and soil conditions of course, must have played a role in plant choices. These included yellow, Spanish, yellow-vein leaved and large blue irises, and another beyond the stone table described as 'tall with twisted stem'. Dead plants such as common speedwell beyond the stone table and a burning bush beyond the fishpond were replaced with irises and efforts were made, unsurprisingly, to select plants appropriate for the habitat. Many of those growing around the fishpond, for instance, such as aromatic or sweet flag (*Acorus calamus*), brooklime (*Veronica beccabunga*), purple loosestrife (*Lythrum salicaria*), river birch (*Betula nigra*), ragged robin (*Lychnis flos-cuculi*) and white water lilies (*Nymphaea alba*), were known to thrive near water or be tolerant of damp conditions.[92]

As we have seen, it is striking that more than a third of the plants listed by Charles Darwin in his nature-printed notebook between 1774 and 1778 were later growing in the Full Street family garden. This was probably partly because some had practical culinary and medical applications, but also, as the number labelled as 'beautiful' demonstrates, because they were some of Darwin's favourite plants. It may also be that, in choosing many of the plants depicted by Charles during their botanical excursions for his Derby garden, they served as a memorial to Darwin's son, whose death had hit him hard. These included some of Darwin senior's favourite plants, such as meadowsweet (*Filipendula ulmaria*), of which there were at least seven examples growing at Full Street, including a double-flowered kind and another that replaced

Figure 13. Dropwort (*Filipendula vulgaris*), one of Darwin's favourite plants, from J. Hill, *The Family Herbal* (1812).

Dropwort

a defunct silver-veined geranium. Other Full Street plants that had previously been nature-printed by Charles included peonies (*Paeonia*) and broad-leaved plantain (*Plantago major*), which also grew at Derby, different kinds of saxifrage (*Saxifraga*), silver weed (*Potentilla anserina*) or wild tansy (*Tanacetum vulgare*), the aquatic plant brooklime, sea wormwood (*Artemisia maritima*), St John's wort (*Hypericum perforatum*), wallflowers (*Erysimum*) and yarrow (*Achillea millefolium*).[93]

CONCLUSION

Through his two great mock-heroic but serious poems, *The Loves of the Plants* and *The Economy of Vegetation*, Darwin made what Anna Seward called his 'wild' and 'swampy' Lichfield botanical garden the most famous garden of its kind in late Georgian society. This chapter has demonstrated the extent to which these poetical and prose creations originated from their physical place, in terms of both a literary and philosophical midlands community and a specially designed space created to blend picturesque planting and landscaping with systematic collecting. Darwin's Lichfield creation sought not only to underline the scientific benefits of Linnaean botany but also to celebrate the pleasures of plants as individual entities and as parts of a configured and managed garden. It has also been argued that, while the Lichfield garden remained well known long after its demise, living on through its creator's poetry, Darwin's later Derby garden has been comparatively neglected. Yet, as we have seen, there are considerable similarities between the two and, in many ways, the Full Street garden and orchard were a continuation of the gardening and horticultural endeavours begun at Lichfield, serving equally as places for entertainment of family and friends as well as for literary and epistolary enquiry and celebration.

In some ways Darwin's gardens were places of enjoyment and retreat as he began to reduce his medical activity, as a surviving sketch of Edward, Emma and Violetta Darwin playing together at Full Street demonstrates (Figure 14). Another of Violetta Darwin's marvellous sketches based upon the memories of her father Francis growing up at Full Street shows the children having fun, with him caught on the spike of the railings adjoining the Derwent around 1795.[94] However, these green spaces were also at the same time closely linked to Darwin's medicine in many important ways, as we will see in the next chapter. The Lichfield and Derby gardens were places of scientific observation and experimentation, spaces for the study of plant physiology and taxonomy, horticulture, agriculture and other sciences, as well as providing him with opportunities to consider the dynamic operation of the economy of nature and landscape aesthetics. They were shaped by suggestions and plants provided by family and friends and were not just embodiments of Darwin's intentions, as his adoption of ideas from his wife Elizabeth, his brother Robert Waring and friends Seward and Boothby demonstrate. Of these, it is clear that Seward and Elizabeth were the most crucial figures who stimulated and challenged Darwin in creating and managing his Lichfield and Derby gardens respectively, but, whereas the former's role is evident from her *Memoirs of Dr. Darwin* and other sources, we get only glimpses of the latter's input from surviving correspondence and a few other sources. However, it is striking

Figure 14. Sketch of Edward, Emma and Violetta Darwin playing together as children at Full Street, Derby, from K. Pearson, *The Life, Labours and Letters of Francis Galton*, vol. 1 (1914).

that one of the last conversations Darwin had on the day before his death, with Elizabeth and one of her female friends, concerned the future effects of improvements on the house and gardens at the Priory.[95] Darwin's garden plants provided himself and his family with important culinary, household and medical resources and were therefore, in a very real sense, an important component of his daily life and medical practice and an extension of his social, domestic and professional life. With their inspiration from landscape gardening as well as Linnaean botany, Darwin's Lichfield botanical garden and *Botanic Garden* helped to inspire the development of the semi-public subscription botanical gardens and arboretums that emerged during the later eighteenth and early nineteenth centuries at Liverpool (1802), Hull (1808), Whitby (1812), Birmingham (1836) and elsewhere. Needing to retain patrons and subscribers and attract paying visitors, these institutions sought to combine labelled systematic collections for scientific, medical and recreational purposes with picturesque features such as plantations, hills and lakes, with published catalogues and guidebooks being an important part of their strategy.[96]

NOTES

[1] E. Darwin, *Phytologia; or the Philosophy of Agriculture and Gardening* (London, 1800), 483.

[2] S. Bowerbank, *Speaking for Nature: Women and Ecologies of Early-Modern England* (Baltimore, 2004), 170–80.

[3] S. Coleridge, *Biographia Literaria* (1817), edited by G. Watson (London, 1975), 9–10, quoted in D. King-Hele, *Erasmus Darwin and the Romantic Poets* (Basingstoke, 1986), 90; Bowersock, *Speaking for Nature*, 175.

[4] Studies of Darwin's botany and the *Botanic Garden* include: A. Bewell, '"Jacobin plants": botany as social theory in the 1790s', *Wordsworth Circle*, 20 (1989), 132–9; J. Browne, 'Botany for gentlemen: Erasmus Darwin and the *Loves of the Plants*', *Isis*, 80 (1989), 593–620; D. Coffey, 'Protecting the botanic garden: Seward, Darwin, and Coalbrookdale', *Women's Studies* 31 (2002), 141–64; King-Hele, *Erasmus Darwin and the Romantic Poets*; J. V. Logan, 'The poetry and aesthetics of Erasmus Darwin', *Princeton Studies in English*, 15 (1936), 46–92; J. McGann, *The Poetics of Sensibility: A Revolution in Literary Style* (Oxford, 1996); M. McNeil, *Under the Banner of Science: Erasmus Darwin and His Age* (Manchester, 1987); M. McNeil, 'The scientific muse: the poetry of Erasmus Darwin', in L. J. Jordanova ed., *Languages of Nature: Critical Essays on Science and Literature* (London, 1986); J. Browne, 'Botany in the boudoir and garden the Banksian context', in D. P. Miller and P. H. Reill eds, *Visions of Empire: Voyages, Botany*

and *Representations of Nature* (Cambridge, 1996), 153–72; D. King-Hele, *Erasmus Darwin: A Life of Unequalled Achievement* (London, 1999); J. Uglow, *The Lunar Men: The Friends who Made the Future, 1730–1810* (London, 2002); C. Packham, 'The science and poetry of animation: personification, analogy, and Erasmus Darwin's *Loves of the Plants*', *Romanticism*, 10 (2004), 191–208; M. Page, 'The Darwin before Darwin: Erasmus Darwin, visionary science, and romantic poetry', *Papers on Language and Literature*, 41 (2005), 146–69; L. Schiebinger, 'The private life of plants: sexual politics in Carl Linnaeus and Erasmus Darwin', in M. Benjamin ed., *Science and Sensibility: Gender and Scientific Enquiry, 1780–1945* (Oxford, 1991); A. B. Shteir, *Cultivating Women, Cultivating Science: Flora's Daughters and Botany in England 1760 to 1860* (Baltimore, 1996); F. J. Teute, '*The Loves of the Plants*; or, the cross-fertilization of science and desire at the end of the eighteenth century', *Huntington Library Quarterly*, 63 (2000), 319–45; P. Fara, *Erasmus Darwin: Sex, Science and Serendipity* (Oxford, 2012); M. Priestman, *The Poetry of Erasmus Darwin: Enlightened Spaces, Romantic Times* (Farnham, 2013).

5   P. A. Elliott, *The Derby Philosophers: Science and Culture in English Urban Society, 1700–1850* (Manchester, 2009); Uglow, *The Lunar Men*; C. U. M. Smith and R. Arnott eds, *The Genius of Erasmus Darwin* (Aldershot, 2005); D. Coffey, 'Protecting the botanic garden: Seward, Darwin, and Coalbrookdale', *Women's Studies*, 31 (2002), 141–64.

6   A. Seward, *Memoirs of the Life of Dr. Darwin* (London, 1804); 'Lichfield: The Cathedral Close', in M. W. Greenslade ed., *A History of the County of Stafford, Vol. 14, Lichfield* (London, 1990), 57–67.

7   S. Whatley, *England's Gazetteer; or an Accurate Description of all the Cities, Towns and Villages of the Kingdom*, 3 vols (London, 1751), vol. 1, unpaginated.

8   Seward, *Memoirs of Dr. Darwin*, 14–15.

9   Bowerbank, *Speaking for Nature*, 161–2, 170–80.

10 T. Barnard, *Anna Seward: A Constructed Life: A Critical Biography* (Farnham, 2009), 122; Bowerbank, *Speaking for Nature*, 186–7.

11 A. Seward, letter to M. Powys, Lichfield, 27 February, 1777, Anna Seward correspondence, Samuel Johnson Birthplace Museum, Lichfield, 2001.76.6.

12 L. Schiebinger, *Plants and Empire: Colonial Bioprospecting in the Atlantic World* (Cambridge MA, 2004); P. Wallis, 'Exotic drugs and English medicine: England's drug trade, c1550–c1800', *Social History of Medicine*, 25 (2012), 20–46; Z. Baber, 'The plants of empire: botanic gardens, colonial power and botanical knowledge', *Journal of Contemporary Asia*, 46 (2016), 1–21.

13 R. Porter, 'Erasmus Darwin: doctor of evolution?' in J. R. Moore ed., *History, Humanity, and Evolution: Essays for John C. Greene* (Cambridge, 1989), 39–69, at 40–2, 58–60.

14 E. Darwin, *Zoonomia; or the Laws of Organic Life*, 3rd edn, 4 vols (London, 1801), vol. 2, 399, 400, 426.

15 Darwin, *Zoonomia*, vol. 2, 427, 450.

16 Darwin, *Zoonomia*, vol. 2, 451–2, 469–72; for copaifera and olibanum see J. Lindley, *Flora Medica* (London, 1838), 171 and 278 respectively.

17 Darwin, *Zoonomia*, vol. 2, 472–5, 515–18.

18 Darwin, *Zoonomia*, vol. 2, 519, 528–9.

19 Darwin, *Zoonomia*, vol. 2, 529–30, 535–6.

20 Darwin, *Zoonomia*, vol. 2, 536–7, 550–1.

21 P. A. Elliott, *Enlightenment, Modernity and Science: Geographies of Scientific Culture and Improvement in Georgian England* (London, 2010), 125–66; P. A. Elliott, C. Watkins and S. Daniels, *The British Arboretum: Trees, Science and Culture in the Nineteenth Century* (London, 2011), 11–36; J. Evelyn, *Sylva or a Discourse of Forest Trees*, 4th edn [1706] edited by J. Nisbet, 2 vols (London, 1908); J. Evelyn, *Directions for the Gardener and other Horticultural Advice*, edited by M. Campbell-Culver (Oxford, 2009); K. Taylor and R. Peel, *Passion, Plants and Patronage: 300 Years of the Bute Family Landscapes* (London, 2012).

22 F. D. Drewitt, *The Romance of the Apothecaries' Garden at Chelsea*, 3rd edn (Cambridge, 1928), 19–28, 31, 46; S. Minter, *The Apothecaries' Garden: A History of the Chelsea Physic Garden* (London, 2003), 1–2, 11.

23 W. B. Coyte, *Hortus Botanicus Gippovicensis, or, A systematical enumeration of the plants cultivated in Dr Coyte's botanic garden at Ipswich* (Ipswich, 1796); *Index Plantarum* (Ipswich, 1807); J. M. Blatchley and J. James, 'The Beeston-Coyte *Hortus Botanicus Gippovicensis* and its printed catalogue', *Proceedings of the Suffolk Institute for Archaeology and History*, 39 (1999), 339–52; W. B. Coyte, letter to J. E. Smith, 27 Feb 1805, Linnean Society Library, GB110/JES/COR/21/87; M. Walker, *Sir James Edward Smith MD, FRS, PLS, 1759–1828: First President of the Linnean Society of London* (London, 1988), 7–10, 18–21; A. T. Gage and W. T. Stearn, *A Bicentenary History of the Linnean Society of London* (London, 2001), 5–21.

24 King-Hele, *Erasmus Darwin*; D. King-Hele ed., *The Collected Letters of Erasmus Darwin* (Cambridge, 2007), 133–5; C. Linnaeus, *Systema Naturae*, 10th edn (1758); G. L. Buffon, *Histoire Naturelle*, 44 vols (Paris, 1749–1804).

25 W. Withering, *A Botanical Arrangement of all the Vegetables Naturally growing in Great Britain, with descriptions of all the Genera and Species according to the celebrated system of Linnaeus*, 1st edn, 2 vols (Birmingham, 1775); R. Schofield, *The Lunar Society of Birmingham: A Social History of Provincial Science and Industry in Eighteenth-century England* (Oxford, 1963),122–7; King-Hele ed., *Collected Letters*, 130–7; Uglow, *Lunar Men*, 270–91, 381–3, 423–4, 477.

26 Schofield, *Lunar Society*, 306–13.

27 J. J. Rousseau, *Rousseau Juge de Jean Jacques: Dialogues: Premier Dialogue; d'Après le Manuscrit de M. Rousseau, laissé entre les mains de M. Brooke Boothby* (Lichfield, 1780); A. Seward, letter to W. Hayley 15 March 1785, in *The Letters of Anna Seward Written between the years 1784 and 1807*, edited by A. Constable, 5 vols (Edinburgh, 1811), vol. 1, p. 20; M. A. Hopkins, *Dr. Johnson's Lichfield* (London, 1957), 186–8, 230; B. Nicolson, *Joseph Wright Painter of Light*, 2 vols (London, 1968), vol. 1, p. 129; J. Zonneveld, *Sir Brooke Boothby* (Den Haag, 2003); Elliott, *Derby Philosophers*, 102–5.

28 Obituary of W. Jackson, *Gentleman's Magazine*, 68 (1798), 730.

29 Seward, *Memoirs of Dr. Darwin*, 98–100.

30 Lichfield Botanical Society, *A System of Vegetables … translated from the thirteenth edition of the Systema Vegetabilium*, 2 vols (Lichfield, 1783); Lichfield Botanical Society, *The Families of Plants … translated from the Last Edition of the Genera Plantarum*, 2 vols (Lichfield, 1787).

31 A. Seward, *The Complete Works of Anna Seward*, edited by W. Scott, 3 vols (Edinburgh, 1810), vol. 1, preface, v; M. Ashmun, *The Singing Swan* (New Haven, 1931), 1–2, 4; Bowerbank, *Speaking for Nature*, 161–87; Barnard, *Anna Seward*, 21–5. For women, nature and gardens generally see Bowerbank, *Speaking for Nature*; S. Bending, *Green Retreats: Women, Gardens and Eighteenth-Century Culture* (Cambridge, 2013).

32 Seward, *Memoirs of Dr. Darwin*, 101; Barnard, *Anna Seward*, 117.

33 Letter from A. Seward to Right Hon. Lady E. Butler and Miss Ponsonby, Lichfield, 4 June 1798 from H. Pearson, *The Swan of Lichfield* (London, 1936), 222–4; J. Thomson, *The Seasons, to which is prefixed the life of the author by Patrick Murdoch DD FRS … and an essay on the plan and manner of the poem by John Aikin, MD* (London, 1803); J. Barrell, *The Idea of Landscape and the Sense of Place, 1730–1840* (Cambridge, 1972), 12–63.

34 Ashmun, *The Singing Swan*, 178–87, 239–42; Hopkins, *Dr. Johnson's Lichfield*, 105–21; Barnard, *Anna Seward*, 69–71, 74–6, 82–7.

35 A. Seward, letter to Mrs Blore, 17 May 1804, letter to L. Philips Esq., 1 June, 1804 in Pearson, *Swan of Lichfield*, 275–8.

36 S. Shaw, *The History and Antiquities of Staffordshire*, 2 vols (London, 1802), vol. 1, 346; Seward, *Letters*, vol. 2, 181; vol. 5, 79; Hopkins, *Dr. Johnson's Lichfield*, 186–9.

[37]   W. Shenstone, 'Unconnected thoughts on gardening', 'A description of the Leasowes, the seat of the late William Shenstone, Esquire' and 'verses to Mr. Shenstone in W. Shenstone, *Works in Verse and Prose*, 2 vols (London, 1764), vol. 2, 125–47, 333–72; J. Dixon Hunt, *The Picturesque Garden in Europe* (London, 2002), 49–50; T. Richardson, *The Arcadian Friends: Inventing the English Landscape Garden* (London, 2008), 444–67.

[38]   E. Darwin, letter to R. L. Edgeworth, 17 April 1802, in King-Hele ed., *Collected Letters*, 579–81.

[39]   J. Floyer and E. Baynard, *The Ancient Psychrolusia Revived, or, an Essay to Prove Cold Bathing Both Safe and Useful* (London, 1702); J. Jackson, *History and Antiquities of the Cathedral Church of Lichfield* (Lichfield, 1796), 71–2; D. D. Gibbs, 'Sir John Floyer', *British Medical Journal*, 1 (1969), 242–5. An inscription on the baths commemorated their restoration at the close of the nineteenth century: 'This stone marks the site of the Ancient Bathhouse purchased by Dr Erasmus Darwin of Lichfield and his son Erasmus Darwin the Younger from Thomas Weld of Lulworth Castle, Dorset, Esq. in the 20th year of the reign of King George III ... the Bath was restored (with the original materials) by Albert Octavius Worthington at Maple Hayes, in the 53d year of the reign of Queen Victoria.'

[40]   Darwin, *Phytologia*, 234–5.

[41]   E. Darwin, *The Economy of Vegetation*, 4th edn (London, 1799), 2–4; Shaw, *Staffordshire*, vol. 1, 347.

[42]   C. E. Stringer, *A Short Account of the Ancient and Modern State of the City and Close of Lichfield* (Lichfield, 1819), 126, quoted in King-Hele, *Erasmus Darwin*, 150.

[43]   'Lichfield: economic history', from Greenslade ed., *Lichfield*, 109–31, 167, 169; P. Borsay, *The English Urban Renaissance: Culture and Society in the Provincial Town, 1660–1770* (Oxford, 1991), 162–72; Bowerbank, *Speaking for Nature*, 170–80.

[44]   Account book of the Lichfield Conduit Lands Trust, 1741–1856, 7 January 1773, 17 January 1776, Staffordshire County Record Office, Lichfield, D126/2/1; J. Snape, *A Plan of the City and Close of Lichfield* (Lichfield, 1781); J. Rawson, *An Enquiry into the History and Influence of the Lichfield Waters* (Lichfield, 1840), 14–15, 55; P. Laithwaite, *The History of the Lichfield Conduit Lands Trust, 1546–1946* (Lichfield, 1947), 43; Greenslade ed., *Lichfield*, 159–70; Bowerbank, *Speaking for Nature*, 178–80.

[45]   Coyte, *Hortus Botanicus Gippovicensis*, 122–32; Blatchley and James, 'The Beeston-Coyte *Hortus Botanicus Gippovicensis*', 339–52.

[46]   Jackson, *History and Antiquities*, 71–2; E. Darwin, letter to J. Johnson, 27 August 1787 in King-Hele ed., *Collected Letters*, 290–1.

[47]   E. Darwin, letter to R. J. Thornton, Autumn 1798? in King-Hele ed., *Collected Letters*, 525–6.

[48]   E. Darwin, letter to Robert Darwin 8 August 1799, in King-Hele, *Collected Letters*, 530–1. The Bond involved was probably either Thomas, a mercer of Lichfield, or William, agent to the Royal Exchange Assurance Company.

[49]   E. Darwin, letter to R. Darwin 8 February 1800, in King-Hele ed., *Collected Letters*, 540.

[50]   E. Darwin, letter to R. Darwin 2 June 1800, in King-Hele ed., *Collected Letters*, 546.

[51]   *Gardener's Magazine*, 14 (1838), 345–6. By 1838 the garden had been purchased by John Atkinson and became part of the farm adjoining his estate of Maple Hayes.

[52]   C. Darwin, *Life of Erasmus Darwin*, edited by D. King-Hele (Cambridge, 2003), 31.

[53]   Ordnance Survey 25-inch map, Staffordshire, LII.14 (1884).

[54]   84 R. Dodsley, 'A description of the Leasowes' in Shenstone, *Works in Verse and Prose*, vol. 2, 331–92; Lichfield Botanical Society, *Families of Plants*, vol. 1, i; W. Withering, *A Systematic Arrangement of British Plants*, 5th edn, corrected and enlarged by W. Withering, 4 vols (1812), vol. 1, 5–6.

[55]   Seward, *Memoirs of Dr. Darwin*, 127–30.

[56]   Seward, *Memoirs of Dr. Darwin*, 130–2; Shaw, *Staffordshire*, vol. 1, 347.

[57]   Seward, *Memoirs of Dr. Darwin*, 130–2.

[58] Darwin, *Economy of Vegetation*, advertisement, iii–iv.

[59] E. Darwin, *The Loves of the Plants*, 4th edn (London, 1799), proem; H. and A. Gernsheim, *A Concise History of Photography* (London, 1971), 9–15.

[60] C. Withers, *Placing the Enlightenment: Thinking Geographically about the Age of Reason* (Chicago, 2007), 122–5.

[61] E. Darwin, 'Elegy on the much lamented death of a most ingenious young gentleman', in *The Prince, My Son, A hero: Three Elegies*, edited by S. Harris (Sheffield, 2009), 30–1.

[62] E. Darwin, 'Life of Charles Darwin' in C. Darwin, *Experiments Establishing a Criterion between Mucaginous and Purulent Matter* (Lichfield, 1780), 127, 128, 131; King-Hele, *Erasmus Darwin*, 73–5.

[63] Darwin, 'Elegy', 32.

[64] C. Darwin, manuscript nature printed notebook, Cambridge University Library, MS.AD 101.41 reproduced with notes by A. Secord at: https://cudl.lib.cam.ac.uk/view/MS-ADD-10141

[65] C. Linnaeus, *Miscellaneous Tracts Relating to Natural History, Husbandry and Physick*, translated by B. Stillingfleet (London, 1759); Lichfield Botanical Society, *System of Vegetables*, vol. 1, p. xi; Lichfield Botanical Society, *Families of Plants*, vol. 1, x

[66] E. Darwin, letter to Miss S. Lea, 6 June, 1796 in King-Hele ed., *Collected Letters*, 499–500.

[67] Darwin, *Economy of Vegetation*, 220, Note XXXVIII: Vegetable Impregnation, 457–61; T. Connolly, 'Flowery porn: form and desire in Erasmus Darwin's *The Loves of the Plants*', *Literature Compass*, 13 (2016), 604–16.

[68] King-Hele ed., *Collected Letters*, 211–18. A painting showing the Darwin garden and Exeter House on the left attributed to Robert Bradley of Nottingham (but may be by Robert Rigby of Madeley, Staffs, or Henry Lark Pratt). It is dated 1838 and hangs in Derby Museum and Art Gallery (DMAG200240).

[69] Darwin, letter to T. Cadell, 5 April 1781, letter to J. Banks, 13 September 1781 in King-Hele ed., *Collected Letters*, 183, 186–8.

[70] E. Darwin, letters to R. Darwin, 17 June 1788 and to R. W. Darwin, 7 October 1799 in King-Hele ed., *Collected Letters*, 317–19, 531–3; E. Nixon ed., *A Brief Memoir of the Life of John Gisborne, to which are added Extracts from His Diary* (London and Derby, 1852), 101.

[71] *Derby Mercury*, 25 February 1802. The land almost became an island after the opening of the Derby Canal in 1796 with the river and Chaddesden Brook on the other sides. Darwin's friend William Strutt (1756–1830) acquired the land adjoining his friend's orchard.

[72] *Derby Mercury*, 2 October 1783; K. Pearson, *The Life, Letters and Labours of Francis Galton*, 3 vols (London, 1914–30), vol. 1, 16; M. Craven, 'Canary Island, Derby: an area rich in heritage', *Bygone Derbyshire* website (no longer available). Encouraged by the canal, the area later became built up for housing and industry and the gardens and orchard disappeared during the nineteenth century.

[73] E. Darwin, letter to R. Darwin, 23 November 1799, King-Hele ed., *Collected Letters*, 535.

[74] *Derby Mercury*, 25 February 1802.

[75] E. Darwin, letter to R. L. Edgeworth, 1802, in King-Hele, ed., *Collected Letters*, 579.

[76] King-Hele, *Erasmus Darwin*; D. and S. Lysons, *Magda Britannia*, vol. 5, *Derbyshire* (London, 1817), 67–8; N. Redman, *Illustrated History of Breadsall Priory* (Derby, 1998); M. Craven and M. Stanley, *Derbyshire Country House* (Ashbourne, 2001), 51–3; King-Hele ed., *Collected Letters*, 574, 578–9.

[77] E. Darwin, letter to Miss S. Lea, 6 June 1796 in King-Hele ed., *Collected Letters*, 499–500.

[78] E. Darwin, Notebook, Cambridge University Library (DAR 227.2:11); Darwin, *Phytologia*, 449–54; King-Hele ed., *Collected Letters*; F. C. Laird, *Topographical and Historical Description of the County of Nottingham* (London, 1820), 48–76, 349–52; Darwin, *Life of Erasmus Darwin*, 31; J. C. Brown, *The Forests of England and the Management of Them in Bye-Gone Times* (London, 1883); S. Daniels, 'The political iconography of woodland in later Georgian England', in D. Cosgrove and S. Daniels eds, *The Iconography of Landscape* (Cambridge, 1988), 43–82; Zonneveld, *Sir*

Brooke Boothby; C. Jarvis, *Order out of Chaos: Linnaean Plant Names and their Types* (London, 2007); Elliott, Watkins and Daniels, *The British Arboretum*, 1–26.

[79] J. Pilkington, *A View of the Present State of Derbyshire*, 2 vols (Derby, 1789), vol. 1, vi, 319–479.

[80] Pilkington, *Derbyshire*, vol. 1, 320–2.

[81] E. Darwin, letter to R. Darwin, 30 May 1792 in King-Hele ed., *Collected Letters*, 402–4.

[82] E. Darwin, letter to R. W. Darwin, 7 October 1799 and letter to R. Darwin, 23 November 1799 in King-Hele ed., *Collected Letters*, 531–3, 535.

[83] *Derby Mercury*, 5 June 1752; J. C. Loudon, *Encyclopaedia of Gardening*, 1st edn (London, 1822), 1240; I. Emmerton, *A Plain and Practical Treatise on the Culture and Management of the Auricula, Polyanthus, Carnation, Pink and the Ranunculus*, 2nd edn (London, 1819), 224.

[84] *Derby Mercury*, 9 April 1789; L. Jewitt ed., *The Ballads and Songs of Derbyshire* (London, 1867), 184–7. For Stanton, R. S. Fitton and A. P. Wadsworth, *The Strutts and the Arkwrights, 1758–1830: A Study of the Early Factory System* (Manchester, 1958), 123, 126, 130; for Greaves and his gardening interests, J. Baker, 'The Rev. George Greaves (1746–1828) and Stanton-by-Bridge during his incumbency', *Derbyshire Miscellany*, 17 (2004), 11–15.

[85] Darwin, *Phytologia*, 392.

[86] Stringer, *Short Account*, 126.

[87] E. Darwin, letter to R. Darwin, 18 August 1794 in King-Hele ed., *Collected Letters*, 451–3.

[88] E. Darwin, letter to J. Coakley Lettsom, 8 October 1787 in King-Hele ed., *Collected Letters*, 295–6.

[89] Darwin, *Economy of Vegetation*, 443–8, 450–7.

[90] E. Darwin, manuscript notebook, Cambridge University Library, DAR 227.2.11; Uglow, *Lunar Society*, 383.

[91] Darwin, *Phytologia*, 494.

[92] W. Withering, *A Systematic Arrangement of British Plants*, 5th edn, 4 vols (London, 1812), vol. 2, 427, 17, 541, 427, vol. 3, 599–600.

[93] C. Darwin, manuscript album of nature-printed plants, Cambridge University Library, MS Add. 10141; E. Darwin, manuscript catalogue of hardy plants (1796), Cambridge University Library, DAR 227.2.11.

[94] Pen and ink drawing by Violetta Harriet Darwin, '[Francis Darwin] Caught on the spikes in the bottom of the garden in the Full Street in the year of the great flood 1795', Galton Papers, University College London Library, Special Collections, GALTON/1/1/12/1/6.

[95] Seward, *Memoirs of Dr. Darwin*, 423–4.

[96] Elliott, *Enlightenment, Modernity and Science*, 124–66; N. Johnson, *Nature Displaced, Nature Displayed: Order and Beauty in Botanical Gardens* (London, 2011).

CHAPTER 2

MEDICINAL PLANTS AND
THEIR PLACES

Georgian medicine depended heavily upon plant-based remedies, which were utilised by a range of healers and practitioners, including apothecaries and druggists, as well as in domestic medicine, and there was a widespread belief in the efficacy of what were designated the 'virtues' of some vegetables. The teaching of botany therefore remained a crucial component of the education of physicians, surgeons and apothecaries. Many Latin and English-language herbals or books listing and describing medicinal plants were available to guide the application of vegetable parts for healing during the seventeenth and eighteenth centuries. Some were new editions of long-established works or translations of French texts and others were marketed as 'new' books, although in practice there was much recycling of information. Some herbals were written or translated into other languages of the British Isles and Ireland – such as a Welsh translation of Nicholas Culpeper's herbal (1818) – while others were written by or for women, such as Leonard Sowerby's *Ladies Dispensatory* (1651) and Elizabeth Blackwell's *Curious Herbal* (1737). Some, such as Henry Barham's *Hortus Americanus* (1794), were aimed at – or derived knowledge from – the colonies and plantations in the Americas or East Indies, where colonists and transplanted populations faced new forms of disease and difficult climate conditions. Medico-botanical knowledge and plant materials were obtained from indigenous peoples (who had lived there a long time before the Europeans came) such as the Amerindians or those moved by trade and slavery, such as African slaves forced to Caribbean plantations in the Atlantic world, as Judith Carney, Richard Rosomoff, Londa Schiebinger and Suman Seth have shown.[1]

Numerous healers, nostrum mongers, gardeners, publishers and other professionals advertised herbal remedies, plants and plant products through word of mouth and in shop windows, newspapers and pamphlets, often with fulsome testimonials from apparently satisfied customers restored to health. In the face of the polypharmaceutical (multi-drug) practices prevalent in the seventeenth and early eighteenth centuries and the growing importation of new plant products from around the world that were utilised in medical treatments, some apothecaries, healers and druggists strongly advocated what they idealised as a return to the homespun wisdom of local plant lore, which became a marketing point for some new herbals. Sir John Hill (1714–1775), for example, author of the *Family Herbal* (1754) – one of the most widely read such Georgian texts, complained about the large number of 'foreign plants brought into our stoves' from across the earth at considerable 'expense' by nurserymen and collectors to fill 'the eye with empty wonder', when everyone ought to be making best use of local common plants and getting to know thoroughly the properties of the 'meanest herb which grows in the next ditch'.[2] According to Hill, medicine had literally lost its roots and become 'entirely chemical', at the cost of thousands of lives, as 'curious botanists' became obsessed with flowers rather than whole plants, coining 'unintelligible terms' and obsessing about seeming novelties that confounded the memories of students. What they actually needed was more understanding of vegetable 'virtues' 'confirmed by practice' and utilising simple arrangements based upon the 'English alphabet' for the 'English reader' with general plant descriptions and instructions on locations that could be 'understood by all'.[3] 'Nature' in Britain had provided 'herbs of its own growth' and remedies for the kind of diseases that its inhabitants were most likely to be subject to, while the 'medicine of nature' and the 'druggist's shop' provided examples from nearby fields and gardens that were 'more efficacious', safer and gentler than 'foreign' 'chemical' remedies.[4]

Although inquiring into the 'virtues of herbs' was often dismissed as a women's domain, this should perhaps be regarded as 'an honour' to the female sex, who studiously strove to pursue their botanical 'studies' for practical uses rather than obsessively scouring the world yet remaining 'ignorant of all they left at home'.[5] For Hill, it was of much greater 'consequence' and more satisfying to have discovered the virtues of a single new herb than to have reclassified 16,000 plants, because the former had genuine practical benefits rather than being an exercise in 'mere curiosity'. Of course, correct identification of plants was essential, but men of 'public spirit and humanity' clearly saw that 'a [Nicholas] Culpepper' was a much 'more respectable person than a [Carl] Linnaeus or a [Johann Jacob] Dillenius' (1687–1747) because searching for disease remedies was one of the most 'honourable sciences in the world' and far above mere taxonomy.[6]

This belief that there was a growing division between practical herbalists using traditional plant lore and systematic Enlightenment botany has been accepted by some historians of herbal medicine. Barbara Griggs has maintained that the medical systems taught in major British and European institutions such as Edinburgh University during the eighteenth century discouraged the use of plant-based remedies and reinforced the division between medical professionals such as physicians and surgeons, other healers and domestic medicine. The methods employed by William Cullen (1710–1790) at Edinburgh, for instance, which followed those of his Leiden teacher Herman Boerhaave, placed the emphasis upon system but employed fairly traditional humoural treatments such as vomiting, purging, bleeding and blistering, as did those of his successor James Gregory. Although challenged in some respects by supporters of John Brown, with his emphasis upon attaining balance in degrees of excitability, the result was reliance upon a limited range of fairly standard treatments and drugs.[7]

In her stimulating study of the history of domestic plant medicine, which argues for continuity in vegetable-based domestic rural remedies and blends oral historical accounts with printed and manuscript herbal sources, especially from the early modern period, Gabrielle Hatfield sees the eighteenth century as a watershed in the strong and growing 'divergence' between elite 'written herbal medicine' and 'official medical thinking', which included 'ideas gathered from abroad', and the 'ordinary grassroots medicine of the majority of country people'. In Hatfield's view, while there was 'occasional input from domestic medicine into "official" medicine', there was 'very little input' from 'official to traditional' medicine and a growing 'divergence' between the two. With the 'so-called "rationalisation" of medicine in the eighteenth century' medical professionals became 'reluctant to acknowledge' their 'debt' to 'popular' medicine or 'take plant medicine seriously enough', seeking to 'distance' themselves from the 'non-scientific remedies of the past', while herbalists were 'forced either underground or abroad'.[8] This chapter argues, however, that Erasmus Darwin's use and understanding of medico-botany suggests that the division between herbal and 'official' medicine in the period between 1730 and 1830 should not be exaggerated and, furthermore, that the division between urban and rural medicine is not very useful for this period. In fact, through their botanical studies, omnivorous collecting of plant information, strong advocacy of the Linnaean taxonomical system and use of vegetable-based remedies – all the while claiming to reject the methods of 'ancient herbalists' in favour of 'modern practice' (the rhetoric that was required to make a living) – medical practitioners such as Darwin and his Lunar friend William Withering determinedly sought to push the boundaries of herbal medicine, which was, for them, not distinct from mainstream practice.

On the face of it, as one of the most successful advocates of plant collecting and Linnaean botany in late Georgian society through the Lichfield Society translations and *Botanic Garden* volumes, Darwin had little in common with Sir John Hill, and seems to have had a low opinion of Hill's herbal because he had heard it used plant illustrations recycled from older herbals with 'one branch or leaf' added to 'disguise' the source.[9] However, as a doctor he too had been led towards the botanical study of plants by his medical interests and authors such as Peter Ayres, who described Georgian botany as 'moribund' and 'enslaved by medicine', and as progressing only after its 'liberation' from this, under-appreciate the many ways in which Darwin's botany and natural history were informed by his medical practice.[10] Unlike Hill, who claimed that Linnaean botany was a threat and barrier to wider understanding of herbal medicine, Darwin and Withering strongly believed that it provided an approachable key to the subject for medical professionals, healers, gardeners and the general public alike, which would help facilitate the discovery – and better usage – of a large number of novel and effective vegetable-based treatments. Before Darwin catastrophically fell out with Withering he enthusiastically suggested names for his friend's major new English translation of the Linnaean system *The Botanical Arrangement* (1776), including 'The Scientific Herbal', 'Linnean Herbal' or 'English Botany', which underlined his belief in the major benefits to medicine of a combination of a new botanical system with practical plant growing and collecting. In practice, however, the Linnaean approach to taxonomy was not necessarily that helpful to apothecaries and those seeking to benefit from plant medicine because of its emphasis on one aspect of floral parts rather than the many other dimensions of vegetable physiology and the context in which plants grew.

This chapter argues that Darwin's experience of growing and closely observing plants in his gardens, seeing their usage in culinary and domestic life, employing vegetable products in medicine and working towards a 'Linnean Herbal' drew him towards a critique of Linnaean botany and a more 'natural' system that sought to recognise the rich and dynamic variations that occurred in practice rather than in textbooks or dried herbaria. When Darwin's living and printed gardens are considered in relation to each other, we can see how much he employed both to drive medical, domestic, agricultural and manufacturing improvements, blending plant-collecting with classification, paying equal regard to nomenclature and taxonomy in order to forge a more effective medico-botany.[11]

PLANTS AND PRESCRIBING IN GEORGIAN MEDICINE

Healers utilising traditional plant remedies competed with professionals and quacks in the Georgian medical marketplace, jostling for business and patient clients. Apothecaries and chemists such as John Mason Good (1764–1827), who later became a physician, sought to be taken seriously as professionals by formalising their education and qualifications and employing new drugs and the latest treatments to fill the gap left by a shortage of physicians, worrying about differentiating themselves from druggists. Pills and potions became marketed and manufactured in new ways, using handbills, newspaper advertisements and lists of recommendations, and by the 1790s hundreds of London apothecaries organised to defend their profession. Branded nostrums with what were claimed to be special formulas distinguishable by their own bottles and labels included Godfrey's Cordial, a soothing syrup marketed at children by the 1770s that included anise, coriander, caraway, sassafras and opium.[12]

Apothecaries and chemists sought to secure their position as traders by joining in the establishment of guilds and trading organisations, which were designed to regulate training and apprenticeships, govern conduct and petition rulers and governments (cooperating with other traders outside London). The activities of apothecaries were also partly governed by the production of pharmacopoeias such as the *Pharmacopoeia Londinensis*, produced by the College of Physicians with 'official' sanctioned lists of recommended treatment formulas, while some apothecaries established physic gardens for themselves or helped facilitate the formation of institutional botanical gardens. Physicians were concerned to keep control over the process of prescribing and dispensing medicines, and resisted encroachment by apothecaries upon this territory for fear that it would undermine their income and professional status – and, given the importance of plant-based medicines, the emphasis upon practical botanical knowledge in university medical education was one means of achieving this. The publication of the first *Pharmacopoeia Londinensis* in 1618 was partly driven by the fact that the London apothecaries had split from the Grocer's Company the previous year to create their own society and the physicians were keen to retain control over the process of prescribing medicines rather than risk this passing to the new Society of Apothecaries. In 1704, however, the College of Physicians eventually lost an action against the apothecary William Rose, who successfully defended the right to practise physic and prepare medicines independently of physicians if desired by patients after appealing to the House of Lords, although this did lead to a bifurcation between apothecaries and druggists (who prepared and dispensed medicines but did not practise physic and generally had less training) during the eighteenth century.[13]

In most provincial towns and villages, however, surgeons and physicians needed to cooperate closely with local apothecaries and druggists and there were also many practitioners, such as apothecary–druggists or apothecary–surgeons, who sought to combine elements of both occupations. Through decades of medical practice and after his own initial failure in Nottingham, Darwin gained much understanding of how to succeed as a provincial physician–surgeon or apothecary and strove to keep good relations with the midlands chemists and druggists that he used or recommended to patients and their families. When framing his proposals for a Derby dispensary in 1784, where local physicians and surgeons might give their services freely, for example, he proposed that the poor be directed to take their prescriptions from all local apothecaries in turn to forestall potential opposition that might arise from them.[14] When discussing helping one of his son Robert's friends, who was considering starting as an apothecary or surgeon–apothecary in the Lichfield area, Darwin explained how he had long been acquainted with all the apothecaries of this vicinity, some 'from their infancy', and as a result could not provide any particular letters of recommendation because they would 'both feel and resent it'. For this reason, he had been careful to take 'no part' in supporting Samuel Mellor, who settled in Lichfield from Derby as a surgeon–apothecary.

Darwin recommended that budding apothecaries mix with 'all ranks' and put on colourful window displays to attract business, explaining how Richard Green's well-presented retail business in Lichfield had provided him with £100 per year. Darwin recollected how a 'foolish garrulous' Cannock apothecary had obtained 'great business without any knowledge, or even art' by 'persuading people that he kept good drugs' and commenting in great detail upon their qualities: 'here's a fine piece of assfoedita, smell of this valerian, taste this album grecum' – which was so fine that 'Dr. Fungus says he never saw such a fine piece in his life'. Other 'arts of the Pharmacapol' included dining out on market days and taking part in dancing and card assemblies to become known in society, and getting letters of recommendation to the proctors or lawyers in the ecclesiastical courts, who would 'forward his getting acquaintance'. 'Journeymen apothecaries' had wages little better than many servants, which made the early years difficult, and it was thus essential for apothecaries or surgeon–apothecaries, in order to set themselves 'well up for life', to project an image of success early in their careers, so that they 'appear[ed] well' and could not be 'better laid out'. Darwin asserted to his son – presumably at least half jokingly – that if all else failed medical men could support their business by 'perpetual boasting like a charlatan', which 'suited a black-guard character' but not 'a more polish'd or modest man'.[15] In 1792 Darwin recommended to Robert that another prospective surgeon–apothecary considering how best to establish a practice locate in a 'large village ... about five or six miles from a town [such] as Abbot's Bromley', where, although no apothecary had hitherto 'succeeded', this was probably owing to having no 'pretensions to success, from ignorance or drunkenness'. Some midlands apothecary–surgeons known to Darwin, however, such as John Power (1730–1791), at Polesworth near Tamworth, and Lightwood of Yoxall, had 'both gain'd fortunes in villages', while another at Appleby Magna, Leicestershire, was 'now doing well, where no surgeon had settled before'; he suggested, therefore, that a surgeon–apothecary might succeed at Childs Ercall, Leicestershire, or other places away from market towns.[16]

Although Hatfield has argued that these should not be exaggerated and (convincingly) that much sound practical workmanlike knowledge and experience underlay apparent irrational belief and custom, aspects of humoural medicine remained influential in Georgian society, as did related astrological classifications of remedies based upon different mineral, animal and herbal substances and the 'doctrine of signatures' – the idea that some obvious characteristic of natural substances revealed their medicinal virtues and the diseases for which they were most efficacious (such as red-flowered plants being useful in treating blood disorders and yellow-flowered plants for jaundice, or the holes, or stoma, in St John's wort leaves meaning that it was good for skin diseases).[17] Many traditional organic remedies continued to be utilised by the sick, healers and medical professionals during the eighteenth century and one historian has estimated that 80 per cent of medicines before the development of the nineteenth-century pharmaceutical industry were plant-based.[18] Different parts of plants were used, including roots, leaves, flowers, seeds, bark, stems and exudates such as pitches or resins. As William Withering emphasised, these various parts 'often manifest very different properties', with wormwood leaves being bitter while the roots were aromatic and poppy seedheads being 'narcotic' while the seeds themselves had 'no such quality'.[19] According to Darwin, flowers employed for 'medicinal purposes' included those of hops (*Humulus lupulus*), chamomile (*Anthemis nobilis*), roses (*Rosa*), violets (*Viola*), cardamine (probably lady's smock, *Cardamine pratensis*, or perhaps a reference to various plants in that genus) and nasturtiums (*Tropaeolum* sp.). As it was normally the petals that contained vegetable virtues, their effectiveness and 'quality' would be much increased if double or multi-flowered examples were bred, as was already practised in the cases of roses and chamomile (Figure 15). Darwin emphasised that 'many acres' of land near Chesterfield in Derbyshire were already devoted to growing chamomile, which was sold to mix with hops for medical and brewing purposes, and considered that this was an industry that might be expanded to nurture double or multi-floral versions of other medicinal herbs.[20] Plant leaves employed as medical treatments included those of 'blessed thistle' (*Cnicus acarna*), used as an emetic, and foxglove (*Digitalis*), used 'as an absorbent' in cases of anasarca or oedema. The bitter leaves of wood sage (*Teucrium scordium*), which had the 'odour of garlic', worked 'with success' to cure agues, as Darwin had witnessed, and, as it grew relatively easily upon 'dry barren soils', might perhaps be cultivated for use 'in some diseases' instead of Peruvian bark.[21]

The manner in which vegetable substances were prepared in apothecaries' shops during the eighteenth century also varied, from direct consumption or topical application to infusions (steeping in very hot or boiling water), decoctions (boiling and straining) and tinctures (dissolved in wine, vinegar or spirits). Medicines were

Figure 15. Chamomile, from J. Hill, *Family Herbal* (1812).

frequently prepared by grinding plants using a mortar and pestle, and the resulting material was dispensed as powders, mixtures or compounds, in the form of hand-rolled pills if necessary. Other methods intended to obtain the plant's most active or essential principles included burning to ash (or ashing) and fermentation or distillation to produce substances including organic salts, such as natron, or acids, such as vinegar. Some healing herbs were dried; according to Darwin, this needed to be undertaken in the shade, just as dried flowers were produced, otherwise the plants became 'bleached' and lost their 'colour' and 'odour' through 'too great insolation [irradiance] and exhalation', which risked the potency of the herbs' virtues and 'nutritive' qualities.[22] Herbs were gathered during sunshine and after dry weather ideally before being dried further in shade, then pressed by weights or a press and enclosed between paper sheets. They were sometimes stored hanging up in 'loose bundles' but were better retained in the paper in a dry place.[23]

There was a change from polypharmacy towards simpler formulations in the two centuries from 1600, especially from the mid-eighteenth century, and a reduced emphasis upon the idea that virtues or active plant principles were best sought at particular times of the annual or astrological calendar when stellar alignments were most favourable, although these notions continued to have some 'influence' and were reflected in some Georgian herbals. Some of the preparations listed in the original *London Pharmacopoeia* (1618) contained twenty, thirty or even fifty or more ingredients, for example, whereas the edition of 1746 saw a large reduction in article numbers and a simplification of preparation formulae. This was no simple progressive tale, however, as some plants, such as deadly nightshade (*Atropa belladonna*) and henbane (*Hyoscyamus niger*), went in and out of favour during the period.[24]

As intimated above, there were various reprints and editions of well-established herbals, such as those by John Parkinson (1567–1650), Nicholas Culpeper (1616–1654) and John Gerard (c. 1545–1612), alongside those marketed as new works by Hill and others, although there was much recycling of medicinal plant information throughout the seventeenth and eighteenth centuries.[25] Popular domestic medical works such as John Wesley's deliberately inexpensive *Primitive Physic* (1747), which went through twenty-three editions in his lifetime, were designed to help the poorer sick who could not afford physicians or surgeons, and placed much reliance upon herbal remedies while recognising that some were becoming hard to find growing in urban areas. It was well known that many advocates of plant medicines exaggerated the list of 'virtues' associated with each example, which was something that the authors of new herbals, such as Hill, Robert John Thornton (1768–1837) (Figure 16) and William Meyrick (c. 1755–1794), complained about; this situation was exacerbated by the frequent reprinting and circulation of older family favourites and the recycling of information. A reliance upon tried staples of plant medicine was being overlaid by the increased usage of a range of exotic plants and knowledge imported from around the globe, which included accounts of the wisdom and experience of indigenous peoples across the Americas, for example. As a result, treatments based upon, for example, spices such as hemp, sandalwood, cardamom, nutmeg and cinnamon, and plants of American origin, such as cocoa, sassafras, sarsaparilla and tobacco, began to appear in European pharmacopias from the sixteenth century, accelerating markedly in the eighteenth century as a mass market developed and prices dropped. The East India Company, for instance, began to grow the cultivated species of opium poppy (*Papaver somniferum*) in its Indian territory and sell it to China and elsewhere from the 1770s, combining the trade with exports of tea and silk and exchanges of precious metals.[26]

re 16.   Portrait of Robert John Thornton with his herbal, from R. J. Thornton, *A Family Herbal*, 2nd edn
4).

Many herb and physic gardens were established on the edges of towns and cities, often around market and seed gardens, by herbalists, nurserymen, gardeners and seedsmen to supply what John Claudius Loudon described as 'the demand of certain classes of medical men' (although many healers and apothecaries were women too), including 'self-doctors … quacks and irregular practitioners'. These were still common in the eighteenth century, when apothecaries 'generally grew' most 'of their own herbs' and often collected 'the rest in the fields' – which was why so many were eminent botanists. Herb and physic gardens grew widely used plants such as peppermint, lavender, chamomile, rosemary and wormwood, as well as licorice and rhubarb, which needed a 'deep free soil'. Some of these gardens, such as the herb garden of the Scottish nurseryman James Dickson (1738–1822) of Croyden, survived throughout the period and into the nineteenth century. Messrs. Dickson and Anderson's 'long-established and respectable shop' at Covent Garden during the 1820s still stocked more than 500 species, including 'all the varieties' mentioned by Nicholas Culpepper and other seventeenth-century herbalists.[27]

Aided by their practical interest in plants, as Loudon emphasised, apothecaries formed plant collections and physic gardens and developed a sophisticated botanical knowledge that was utilised by other medical professionals, such as surgeons and physicians. Although they contributed more to botany than any other of the 'developing sciences', according to Burnby, they were hampered by the fact that, despite improvements in data collection and the systematic arrangements of ideas, as well as 'advances in the description and classification of plants', by 1760, there were few 'theoretical principles or laws of biology' that could really help.[28] Medicinal plants were arranged and classified in physic or botanical garden collections in a number of ways. Non-medical arrangements included alphabetical systems similar to those in many pre-Linnaean herbals, such as Hill's *English Herbal*, geographical ones, according to their actual or supposed place of origin, or systems based on botanical systems, such as Linnaean or Jussieuian arrangements. Arrangements based on plants' medical qualities might emphasis the ailment they were believed to be the most effective at treating or their effects on the body (such as diuretics, expectorants, purgatives and so on).[29]

The Chelsea Physic Garden, which was formed by the London Company of Apothecaries in 1673 for practical, medical and professional purposes and to facilitate easy access to materials for visitors, served as an important institution for the apprenticeship, education and training of apothecaries. A 'Botanical Demonstrator' taught the apprentices names, classes of plants and medical uses and led them on spring and summer plant-hunting expeditions in the metropolitan vicinity; during the eighteenth century the apprentice who learnt to identify and study the largest number of plants received a prize that included William Hudson's *Flora Anglica* (1762).[30] Major support came from the Irish naturalist and former physician in Jamaica, Sir Hans Sloane (1660–1753), president of the Royal Society, who presented the Company of Apothecaries with the freehold of the land, which had previously been rented to them, on condition that they provided annual instalments of plants to the Royal Society for study; this resulted in thousands of specimens being sent during the eighteenth century.[31] As John Haynes' plan of 1751 demonstrates, externally the 'physical plants' at Chelsea were placed in alphabetical order in a large bed at the front of the greenhouse, 'bulbous rooted flowers', 'annual and biennial plants' and 'perennial plants' each had their own separate areas, and most trees and shrubs were positioned in two 'wilderness' zones towards the Thames. A library, herbarium and seed cabinets also facilitated study and a series of ever more comprehensive catalogues and other works was published, including a catalogue of the physic plants by the long-serving gardener Philip Miller (1691–1771), author of the *Gardener's and Florist's Dictionary* (1731), and a more comprehensive version in Latin for the sciences of 'botany and especially Materia Medica' by the Praefectus Horti and Demonstrator Isaac Rand (1739), laid out alphabetically with references to John Ray (1627–1705) and the French botanist Joseph Pitton de Tournefort (1658–1708).[32]

The advent of Linnaean botany in Britain provided physicians with another opportunity to assert their position in the face of the challenge from apothecaries with prescribing aspirations, which helps to explain the enthusiasm with which Darwin and Withering took up the Linnaean system. Withering's *Botanical Arrangement* (1776) concentrated on 'vegetables naturally growing in Great Britain' and utilised the Linnaean system with additional information on the preparation and storage of plants and methods of undertaking microscopic analysis, expressly claiming to reject the methods of 'ancient herbalists' and replacing them with what were claimed to be more 'accurate' and experimental observations of 'modern practice' (Figure 17). In his *Account of the Foxglove* (1785) he emphatically asserted that older forms of 'chemical' analysis by burning had little benefit, while the 'more obvious and sensible

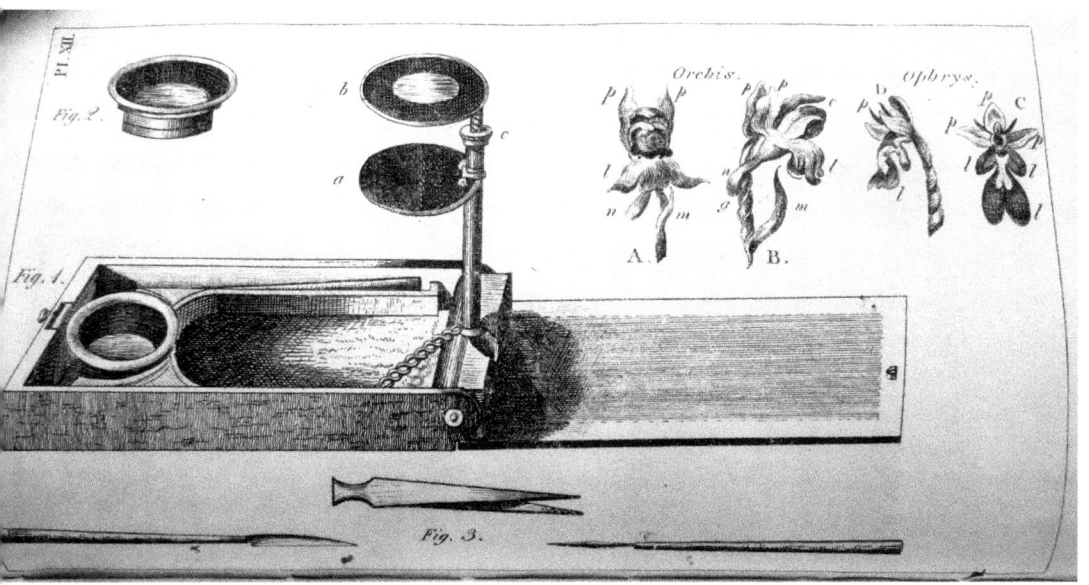

ıre 17.   Botanical microscope, magnifying glass and dissecting instruments, from W. Withering, ematic Arrangement of British Plants, 5th edn, 4 vols (1812), vol. 1, plate xii.

properties of plants', such as colour, taste and smell, had 'little connexion with the diseases they are adapted to cure' and therefore their 'peculiar qualities' had 'no certain dependence upon their external configuration'. Although Withering argued that Linnaeus and his followers had emphasised that the 'medical properties of plants' were only one aspect of their botanical significance, by concentrating on plants 'naturally growing' in Britain and providing some details of places where they thrived and an account of medicinal virtues in his *Botanical Arrangement*, and by increasing the information concerning their locations, qualities and characteristics in successive editions with assistance from botanical associates on the edge of the Lunar circle (such as Jonathan Stokes (1755–1831), who later published his own comprehensive medico-botanical treatise), he sought to attract apothecaries and chemists as well as botanical collectors and readers. This also explains why he included the plant-name synonyms of the Swiss physician and botanist Gaspard Bauhin (1560–1624) 'with a view to medical students', because he believed that these were still generally employed by 'writers on the materia medica'.

Withering maintained that medico-botany had been confused by the assumption that 'every common plant' might be a 'cure for almost every disease' and traditional polypharmacy, which was the 'fashion' of combining multiple 'ingredients' to answer particular purposes. This, however, masked 'the real efficacy of any of them', while a fear of poisons led to the rejection of many plants that might otherwise be medically efficacious. Withering believed that close observations of the behaviour of 'the meanest of mankind' (by which he meant many of the indigenous peoples from around the globe) should be undertaken to see how they employed 'their remedies', especially in 'uncompounded form' through 'long experience', as Linnaeus and his supporters had done – this had revealed the medical benefits of '[*Cephaelis*] ipecacuanha, [*Dorstenia*] contrayerva and [*Smilax*] sarsaparilla'.[33] In his work on the foxglove Withering argued that researching it demonstrated how little progress had been made in determining potential medicinal plant virtues that might be most efficacious against particular diseases, how best to apply them, which plant parts to employ, how to undertake clinical tests and other matters. He also emphasised the potential benefits of taking a Linnaean approach to medico-botany using analogies between plants of similar properties (which was similar to Hill's argument, but lacking the taxonomy), observing that the virtues of foxglove might have been guessed from the fact that its 'congenera' within the Linnaean natural order of Luridae included the genuses *Nicotiana* (including tobacco), *Atropa* (including deadly nightshade), *Hyoscyamus* (including henbane), *Datura* (including thornapple) and *Solanum* (which included the eggplant or aubergine).[34]

DARWIN'S MEDICO-BOTANY

Darwin intended that the new system of medicine detailed in *Zoonomia* would help the sick public at the mercy of quacks and nostrum mongers, including the many competing advocates of different vegetable virtues and new and old herbals. With his fellow members of the Lichfield botanical society, Lunar Society associates such as Jonathan Stokes and Withering and some of the other Derby Philosophical Society members,

Darwin believed that more widespread use of Linnaean botany would facilitate medical improvements by better exploitation of both familiar plants and the numerous new exotic examples being imported into the country, making plant identification easier and leading to new discoveries and the exploitation of their virtues. He agreed with Withering that plant experimentation, when combined with 'empirical usages and experience', would reveal more about vegetable virtues and was an inclusive process 'within the reach of everyone who is open to information regardless of the source from which it springs'. When combined with the benefits of systematic Linnaean botany Withering's call potentially made information from everyday botanical amateurs, plant collectors, gardeners, healers and botanists as useful for medicine as research and observations by physicians, surgeons, apothecaries and chemists. Taking the lead in utilising the new botanical system also offered the opportunity for physicians such as Withering, Stokes and Darwin to reinforce their professional position and take maximum advantage of opportunities for new treatments that might be uncovered as a result, which helps to explain why Darwin enthusiastically encouraged his physician son Robert to set 'up as a botanist', as detailed above.[35]

An interest in understanding more about the medicinal uses of plants and extending the available range of vegetable products is evident from some of the books ordered by Darwin with the approval of members for the Derby Philosophical Society library between the 1780s and early 1800s.[36] These included Charles Bryant's *Flora Dietica* (1783), on edible plants, the Scottish physician William Fordyce's study of the cultivation and medicinal uses of rhubarb (1784), Samuel Crump and the physician Martin Wall's studies of opium and its current and historical applications (1793 and 1786) and the Coventry surgeon Bradford Wilmer's book on poisonous plants (1781),[37] as well as editions of the *Pharmacopia Londonensis*, catalogues of botanical collections such as William Aiton's *Hortus Kewensis* (1789) and transactions of medical, botanical and natural historical societies.[38]

The Philosophical Society also obtained a copy of a study of the effects of tobacco (*Nicotiana*) on 'dropsies', 'dysuries' (difficulty of passing urine owing to kidney or bladder stones) and 'cases of pain and difficulty of passing urine' (1785) by Thomas Fowler (1736–1801), a physician at Stafford General Infirmary, where Darwin had served as consulting 'Physician Extraordinary'. Here Fowler, modelling his work on Withering and Darwin's analyses of foxglove, reported on a series of cases of in- and out-patients at the Stafford hospital in which the 'celebrated Indian [Amerindian] plant' tobacco was taken, usually as an infusion in a 'small cup full of water', as a diuretic, sometimes anally as a clyster and sometimes with foxglove too. These demonstrated that it succeeded better as an 'evacuant' with dropsical patients suffering from swelling of legs, feet, stomach and other body parts than diaphoretics, sudorifics and sialagogues, emetics or purgatives because it produced no ill effects and apparently prevented a recurrence of the problems. Fowler emphasised that because tobacco was a 'vegetable', and because such 'productions of nature' were 'generally constant and uniform', it had an advantage over the 'often variable and uncertain' operation of the 'powerful chemical remedies' frequently employed. His work was supported by letters from other practitioners who had used it successfully, including Thomas Arnold (1741–1816), physician to the Leicester General Infirmary and a member of the Derby Society.[39]

Another of Fowler's works acquired by the Philosophical Society was a study of the effects of mineral arsenic in the 'cure of agues, remitting fevers and periodic headaches', which sought to encourage its usage and dispel the common assumption that it usually had a 'deleterious' effect upon the body. Published by Joseph Johnson with supporting letters from Arnold (again) and Withering, and quoting the observation in the latter's *Botanical Arrangement* that 'poisons in small doses are the best medicines, and the best medicines in too large doses are poisonous', Fowler again presented a series of 'trials' with patients, claiming that arsenic 'bids fair to hold a place among the best and most valuable medicines' to 'rank with Peruvian bark in the cure of agues, remitting fevers and periodic headaches'.[40] The close relationship that Fowler had with Darwin and the members of the Derby Society is underlined by his presentation of a copy of another of his studies on animal electricity, which was usually done only when an author had attended a meeting and given a talk.[41]

Many treatments recommended in *Zoonomia*'s materia medica and in Darwin's surviving prescriptions and correspondence were vegetable-based and intended, as outlined above, to restore the body to its 'natural state' with effective and balanced operation of all 'irritative motions'. It is striking that some of the prescribed plants for each section of the *Zoonomia*'s materia medica grew at Full Street and on the allotment or 'farm', although others growing in the garden and widely seen as medicinally efficacious, such as cat mint (*Nepeta cataria*), horehound (*Marrubium album*), dyer's broom (*Genista tinctoria*), common bugloss or ox-tongue (*Anchusa officianalis*), wormwood (*Artemisia absinthium*), wood avens (*Geum urbanum*) and sea wormwood (*Artemisia maritima*), do not appear to have been employed by Darwin, while other plant-based medical mainstays, such as Peruvian bark (*Cinchona officianalis*), were imported from abroad.[42] By the same token, gaps in Darwin's plant knowledge can be explained by the plants concerned lacking apparent medical applications, and he admitted to his son Robert (1766–1848), on being asked about grasses, for instance, that he had 'never studied' them 'with any accuracy', recommending instead that he consult a work on the subject by the German physician–botanist Christian Gottlieb Ludwig (1709–1773) or even (sarcastically, given their strained relations) 'your friend' Withering on these.[43]

Darwin frequently discussed plants with his patients as both medicines and potential additions to their gardens. In 1796 he enquired of Miss S. Lea whether she could manage opium pills to prevent the return of her 'paroxysms of asthma' and recommended she try breathing oxygen, but the letter also discussed a parcel of Full Street plants that he had sent her by carrier. This 'hamper of roots' wrapped in paper, which he hoped had arrived 'before they were spoil'd', included meadia (or American cowslip, *Dodecatheon meadia*) 'with white foliage', white lychnidea (*Phlox lychnidea*), great bulbous-rooted fumitory (*Fumaria bulbosa*), feather grass (*Stipa tenuissima*) and grape hyacinth (*Hyacinthus muscaris*) – the first of which Darwin advised should be 'kept set in a pot and moistish' and be 'shaded from the sun' when flowering. Furthermore, he offered to send more plants in the autumn if desired, subsequently a further '3 or 400 hardy' examples – though 'not all at once' – and the Lichfield society's *Families of Plants* (1787) and *System of Vegetables* (1783).[44] In such engagements with patients Darwin combined general discussion about plants with promotion of botanical knowledge and education, supporting his medical practice and publications certainly, but in this case the motivation appears to have been the shared pleasures and sociability of gardening and collecting.

Darwin emphasised the importance of different environments in shaping the character of plants for medicinal use which took advantage of their adaptation to cope with these, and here his medico-botany reciprocally supported and stimulated his developmental or evolutionary emphasis upon the protean qualities of living beings operating in the natural economy. The medicinal power of vegetables from around the world was a major theme of *The Loves of the Plants* and Darwin believed that, shaped by the opportunities and challenges of different climates, plants developed characteristics to defend themselves. In a note, Darwin emphasised that many flower smells that were 'so delightful' to the senses, along with the 'disagreeable scents of others', were caused by the 'exhalation of their essential oils'. Some of these were volatile and 'all inflammable'; while some were poisonous, such as laurel and tobacco, others had 'narcotic' qualities, such as oil of cloves, oil of cinnamon and balsam of Peru, which could relieve toothache, hiccups and some ulcers respectively. All these plants had acquired the power to produce such essential oils in the 'vegetable economy' to 'protect them from the depredations of their voracious enemies', such as various insects. The qualities of these oils could sometimes be harnessed by humans for their own purposes, and apothecary's shops depended upon the 'resins, balsams and essential oils' produced by plants growing in Britain and from around the world.[45]

The qualities acquired by plants as a result of their local environments was emphasised in *Loves of the Plants*: 'Round the vex'd isles where fierce tornados roar, Or tropic breezes sooth the sultry shore', for example, 'With gloomy dignity DICTAMNA' stalked, emitting 'sulphurous eddies' that kindled 'into flame' as a result of an 'inflammable air or gas', which was especially potent on 'still dry evenings of dry seasons'.[46] The dangerous qualities of the caustic, toxic juice produced by the 'Grim MANCINELLA' or Hippomane tree (*Hippomane mancinella*) were equally evident and it was employed by Amerindians as a milky poison for their arrows, while other 'noxious plants' abounded globally, including, in Britain, deadly nightshade (Atropa belladonna), henbane (*Hyoscyamus niger*), hound's-tongue (*Cynoglossum officinale*), among others – which were, therefore, largely left unmolested by animals.[47] The provision of thorns, stinging barbs, hooks and noxious fluids in plants such as nettles and brambles had similar origins and purposes, while the 'deleterious exhalations' of the white-flowered *Lobelia longiflora* (known today as *Hippobroma longiflora*, or horse madness) of the West Indies were so strong that humans felt unwell 'at many feet distance' from it when growing in a hothouse or room.[48]

Darwin's favourite example of phytotoxicity – which he believed to be probably the most powerful in the world – was the Upas or 'HYDRA-TREE of death' (now known as *Antiaris toxicaria*), which grew on the upland plains of the island of Java in the East Indies. This 'vegetative serpent' was so deadly that its roots envenomed the soil across 'ten square leagues', killing humans, animals and other plants for up to fourteen miles around by its 'deadly effluvia'. This paragon of vegetable potency absorbed Darwin so much that he devoted thirty-eight lines of poetry in *The Loves of the Plants* to a dramatic depiction of its deadly destructiveness, along with a footnote and the book's longest additional note, while in the fourth edition he included more material on it from Dutch and Swedish authorities, providing a source of inspiration for the poet William Blake, among others. Certainly, the Javanese tree offered Darwin an opportunity for an arresting and macabre narrative but also demonstrated his belief in the fundamental relationship between the healing and hazardous qualities of plants that arose from the same imperative in the economy of nature, the fight to survive against a sea of threatening predators.[49]

Less extreme or deadly examples of plants whose characteristics had long been utilised by humans were the many aquatic species that had long featured in the materia medica because of the usefulness of their 'bulbous and palmated' roots, which were adapted to cope with inundations or boggy soil. These included squill or sea-onion (*Drimia maritima*), yellow water flag (*Iris pseudacorus*) and aromatic flag (*Acorus calamus*).[50] Bog-bean (*Menyanthes trifoliata*) was employed as a 'corroborant' in medicine and might be 'plentifully cultivated on boggy grounds', helping the 'public' save 'more fertile soil' for cultivating corn or other 'valuable vegetables'.[51]

The most effective plant-based treatments for the various ailments were listed in *Zoonomia*, which categorised them according to condition type and most common symptoms. Many of the Full Street plants appear among these. Those useful as 'Nutrientia', for example, which fostered 'growth' and the restoration of bodily systems, included common garden plants and vegetables such as peas, carrots, cabbage and artichoke and fruits such as apples, pears, and plums, along with the 'abundance' of cucumbers, melons and grapes growing in the hothouse.[52] Those that acted as 'Incitantia' in Darwin's system, stimulating the 'exertions of all the irritative motions', included black cherry and deadly nightshade, while plants that were 'Secerentia', stimulating glands and secretions, included the sialogogue cowhage (*Mucuna pruriens*), the expectorants garlic, leek, onion, marshmallow (*Althaea officinalis*), coltsfoot (*Tussilago farfara*), liquorice (*Glycyrrhiza glabra*) and honey, the cathartic rhubarb (*Rheum*) and the diuretic asparagus. Darwin's employment of coltsfoot is significant, as it was a plant that, according to Withering in 1776, was 'formerly much used in coughs and consumptive complaints' and 'perhaps not without reason', but which had fallen from favour (Figure 18). Darwin's use of the berries of deadly nightshade as a painkiller is likewise striking given that it was normally administered as an infusion of the leaves and that, although favoured in the sixteenth and seventeenth centuries, for much of the ensuing century its employment 'fell into complete desuetude'; indeed, Darwin admitted that its effects were 'less known' and the best doses 'not ascertained'.[53] Those plants listed for 'Sorbentia', which increased the 'irritative motions', included apple for the skin, sea wormwood (*Artemisia maritima*), tansy (*Tanacetum vulgare*), gentian (*Gentiana*), artichoke leaves and wood avens (*Geum urbanum*) for the cellular membranes, cuckoo-flower (*Cardamine pratensis*), celery and cabbage for the veins, and rhubarb, red roses and silverweed (*Argentina anserina*) for the intestines.[54]

Figure 18. Coltsfoot, *Tussilago farfara*, from J. Hill, *Family Herbal* (1812).

Plants that served to invert the 'natural order' of the vascular or irritative motions, or 'Inverentia' in Darwin's scheme, included, as we have seen, foxglove as an emetic and diuretic and the 'violent' errhines and sialogogues European wild ginger (*Asarum europaeum*), spurges (*Euphorbia*) and apparently both white and black hellebore (*Veratrum*, false hellebore).[55] Plants that worked as 'Reverentia' by restoring the 'natural order' of 'inverted irritative motions' – combatting hysteria and working on particular bodily parts such as the stomach, intestines or urinary tract – included valerian (*Valeriana*) for the 'inverted motions' accompanying the 'hysteric disease' and rhubarb to 'reclaim' the 'inverted motions of the intestinal lymphatics'.[56] Finally, Full Street examples that Darwin recommended as 'Torpentia', which diminished the exertions of 'irritative motions', included fruits such as gooseberry, apple, melon and grape, cherry-tree gum, peas, marshmallow – Darwin had shrub marshmallow (*Althaea frutescens*), growing in Derby – and snowdrop root.[57] Of course, some of these may pre-date the Darwin family's arrival at Full Street and many were common culinary plants or orchard trees, but the fact that there was a range of plants available for each part of his treatment regime suggests that the garden functioned as a family physic and kitchen resource as well as place of retreat, entertainment and study.

Three kinds of plants growing at Derby, valerian, rhubarb and foxglove, were favourites of later Georgian medicine, and Darwin played a major role in promoting use of the latter. He had both common valerian (*Valeriana officinalis*) and the more potent red valerian (*Valeriana rubra*) growing at Full Street, the latter between the summerhouse and well (Figure 19). He recommended it for various conditions, including for 'inirritative debility', as well as an emetic and for 'hysteric affections'; for the latter, it acted 'in the usual dose without heating the body' by reclaiming the 'inverted motions which attend the hysteric disease'. Following the Scottish physician John Fordyce (1716–1760), Darwin prescribed 'two drams' of powdered valerian root three or four times daily with bark, opium and other treatments for 'cold paroxysm or intermittents' with headaches and 'general inaction' of the entire 'system', along with 'gentle exercise in the open air, flesh diet, small beer, wine' and the alternation of 'regular hours of sleep' with periods awake.[58] Between 1786 and 1788 he recommended valerian to the Loughborough surgeon and Derby Philosophical Society member Thomas Hunt to treat his patient John Hood, who suffered from hemiplagia (paralysis of one side) and a feeble pulse, as well as to Henrietta Sneyd (with the bark) for pain, stomach problems, fever and indigestion and Mary Thorneycroft and Josiah Wedgwood for eye problems. Darwin's approach to valerian usage was measured and careful. Thorneycroft, Sneyd and Wedgwood appear to have been already taking it and Darwin either stopped this altogether, modified how it was administered or added other treatments. Thorneycroft had been taking it with bitter-tasting powdered quassia wood (amargo, *Quassia amara*) and columbo (*Frasera caroliniensis*, yellow gentian) for 'nervous rheumatism', which Darwin believed particularly affected those with 'bilious constitutions'; he suggested she discontinue this and take rhubarb and aloe pills instead. He suggested that Sneyd continue with bark and valerian but in a different form, while he advised Wedgwood to try electrification while continuing with the plant. In support of using valerian, Darwin noted that Fordyce had described taking half an ounce a day for two or three years with 'great advantage'. After relocating his copy of Fordyce's book *Historia Febris Miliaris et de Hemicrania* (1758), which he had mislaid, Darwin translated for Wedgwood the relevant passage from Latin in which wild valerian root was advocated in 'larger doses' as the 'principal remedy', which would 'alone' cure 'pains whether acute or chronical'. Having been 'daily afflicted so violently for four years' – so much so that life had become a burden – Darwin emphasised how much relief Fordyce obtained

OFFICINAL VALERIAN, E. P.
VALERIANA OFFICINALIS, P. E.
WILD VALERIAN, L. P.
VALERIANA SYLVESTRIS, P. L.

Figure 19. Valerian, *Valeriana officinalis*, from R. J. Thornton, *A Family Herbal*, 2nd edn (1814).

from the plant, the 'whole virtue' of which he believed lay 'in the root'. Growing 'spontaneously' or naturally, the roots needed to be 'dug up before the stem shoots' and administered in doses of 'a dram three or four times' daily, or even two drams if the stomach would 'bear it'.[59]

Despite being considered hard to grow and propagate by seed initially, rhubarb was a frequently used medicine that was first acquired in Europe through international trade and which became an important commodity for the East India Company (Figure 20). It was frequently recommended by Darwin as a cathartic or 'Secerentia', which would drive gland stimulation and secretion, and a 'Sorbentia', which would increase the 'irrititative motions' of the intestines. As we have seen, he ordered the treatise on the subject by William Fordyce (1724–1792) for the Philosophical Society library; that paper urged that the plant be much more widely cultivated in Britain and demonstrated how best to transplant and propagate it for medical usage.[60] Darwin had two types at Full Street, common rhubarb (*Rheum rhaphonticum*) and palmate rhubarb (*Rheum palmatum*); however, by closely observing these he noticed a cross, which he called *Rheum hybridum*, or mule rhubarb, which had appeared both in his garden and in that of his neighbours 'without being previously placed or sown there', so demonstrating that it had originated spontaneously. The leaves of this were 'very large and pointed, without being palmated' and it grew more quickly than its cousins, being 'a week or two forwarder in the spring', while its 'peeled stalks' were 'asserted by connoisseurs in eating' to make superb tarts, much better than those of the 'palmated or raphontic' kinds, being even more valuable 'as a luxury' because they preceded the ripening of gooseberries and 'early apples' by a month. They did not 'produce seed in all summers', but it appeared that they could be propagated by root division. Having been established in the relatively damp and mild conditions beside the Derwent, all three kinds of rhubarb grew vigorously and furnished the Darwin family dinner table as much as their medicine cabinet.[61]

Following Withering's pioneering work, Darwin helped encourage the systematic medical use of foxglove (*digitalis*), publishing two accounts of its use in patients (Figure 21). The plant had, however, long been employed medically despite remaining controversial. In 1775 Withering found that foxglove, when mixed together with other herbs, was traditionally used by the 'common people, in Shropshire and elsewhere for dropsy', and, after finding little information in herbals, he experimented with different plant parts from examples growing at varied times of the year to find the most effective preparation. In the first edition of his *Botanical Arrangement* (1776) Withering commented that foxglove was 'certainly a very active medicine' that excited 'violet vomiting' and merited 'more attention than modern practice bestows on it'.[62] As the royal physician Sir George Baker (1722–1809) emphasised in 1785, digitalis usage had experienced varied fortunes, sometimes being seen as a 'powerful remedy' and at other times 'utterly rejected' because of the degree of nausea and sickness it could induce and uncertainty concerning preparation methods and dosage levels. It was not listed in the materia medica of the *London Pharmacopoeia* of 1650 or in the 1746 edition, for instance, but was included in the 1721 edition. Equally it was in the 1744 and 1783 editions of the *Edinburgh Pharmacopoeia*, but was omitted in 1756 and 1774.

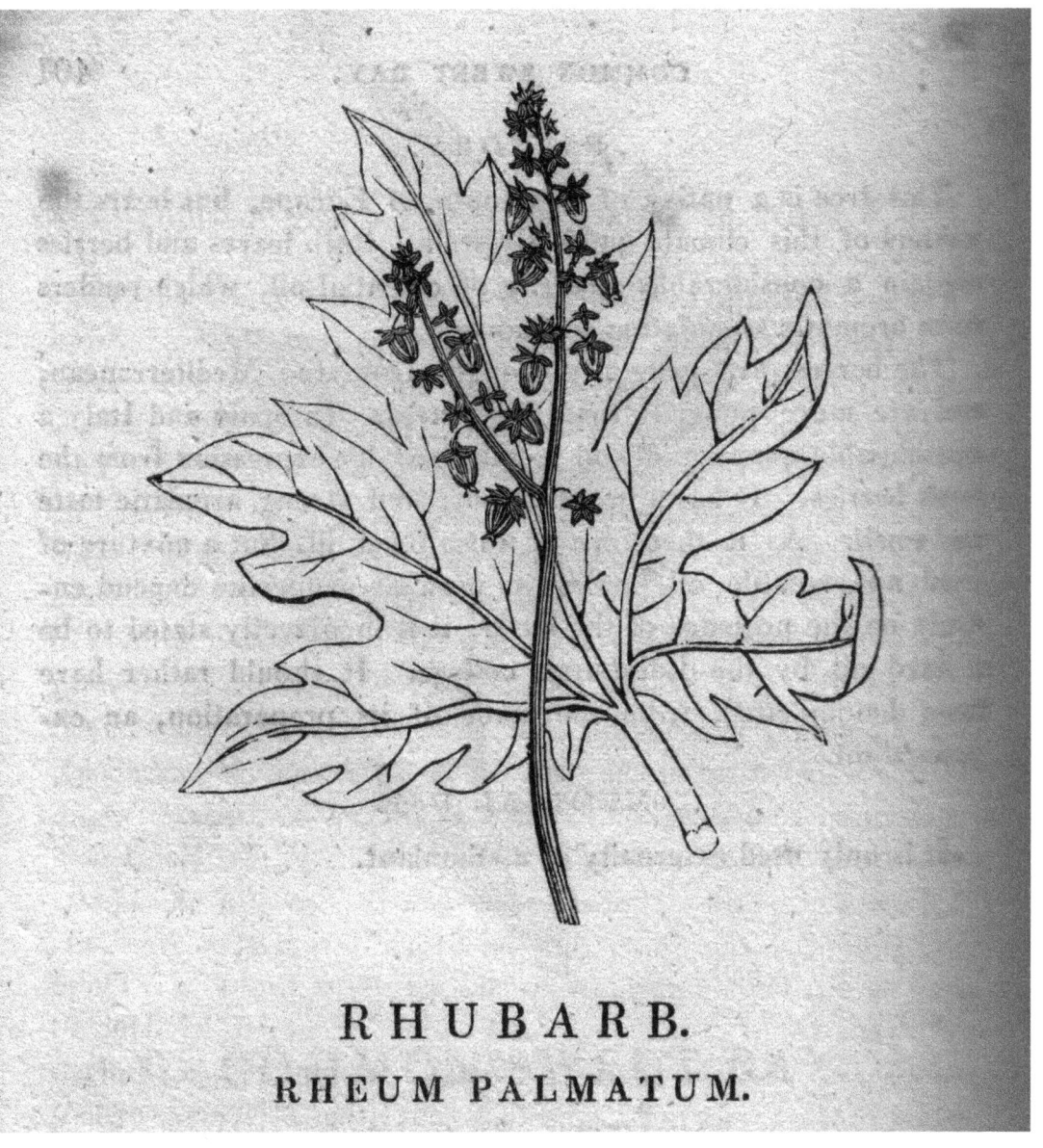

RHUBARB.

RHEUM PALMATUM.

ure 20.   Rhubarb, *Rheum palmatum*, from R. J. Thornton, *A Family Herbal*, 2nd edn (1814).

Some French and German pharmacopoeias also included it.[63] According to John Hill, writing in 1755, foxglove was a 'very beautiful wild plant in our pastures and about our woodsides' which was useful for 'quartan agues', the 'falling sickness' (epilepsy), scrophulous sores (as an ointment), rheumatism and 'other stubborn complaints', although it worked 'violently upwards and downwards'.[64]

Figure 21. Foxglove, *Digitalis*, from J. Hill, *Family Herbal* (1812).

With Withering, Darwin advocated a foxglove 'decoction' (concentrated in fluid) for oedema or heart failure and described how it was 'given to dropsical patients' with 'considerable success', being preferable to squill (*Drimia maritima*, known to Darwin as *Scilla maritima*) or other 'evacuants' used as diuretics to facilitate urination.[65] Having, along with Lunar friends, encouraged Withering with his *Systematic Arrangement* (1775), Darwin unfortunately fell out with him over who deserved greatest credit for successfully promoting the effective medical employment of foxglove. His sons Robert and Charles were drawn into the disagreement and Darwin failed to acknowledge Withering's use of the plant both in Charles's posthumously published pamphlet on the subject (1780) and in his own paper concerning it (1785).[66]

The quarrel obscured the cooperative role that Withering and Darwin had played in encouraging the medical use of digitalis, until the print war and the involvement of the sons of both men made reconciliation impossible. A pivotal case on which they worked closely together from June 1776, which promoted 'very general use' of foxglove in Shropshire, was that of Miss Hill of Aston, near Newport, a lady in her forties suffering from 'a severe cold shivering fit' fever, pains, shortness of breath, coughing and 'copious expectoration' (coughing up phlegm). The treatment illustrates clearly how a range of largely plant-based substances was utilised by Darwin and contemporary physicians for dropsy or serious fluid retention. After Miss Hill's condition declined, with a weakening of the pulse, serious breathing difficulties, coldness, severe swelling and problems passing urine, with only vomiting, induced by ipecacuanha (a preparation from the dried root of *Cephaelis ipecacuanha*), providing some small relief. Withering proposed digitalis, to which Darwin agreed. Their use of ipecacuanha, which was native to highland areas of what is now Brazil and had been employed by Amerindians to combat diarrhoea before the coming of the Portuguese (who also adapted the name from them), provides a good example of their readiness to utilise relatively new imported plant preparations. In this case ipecacuanha, brought to Europe in the later seventeenth century, was added to the *Pharmacopoeia Londonensis* in 1788, although its usage had been advocated by Hans Sloane and others and its sweat-inducing nauseating effects appeared to chime with residual humoural medical notions that it was successfully eliminating injurious bodily substances. Subsequently, after digitalis draughts induced more vomiting, Miss Hill urinated more freely, her swelling decreased markedly and her pulse rate and breathing improved. Darwin gave her *Pareira brava* (virgin vine), guaiacum shavings (*Lignum vitae*), a pill of myrrh (gum resin from the *Commiphora myrrha* tree), white vitriol (zinc sulphate) and a calomel (mercury chloride) and aloe (vera) pill to prevent constipation. Other medicines employed by both doctors for Hill included pills of soap, rhubarb, tartar of vitriol and more calomel with a saline draft to treat her suspected 'diseased liver', tincture of bark (cinchona), and more pills of aloe, guaiacum and sal martis or green vitriol (ferrous sulphate). Finally, after Withering took over fully, continuation of her dropsical symptoms necessitated repeat prescriptions of dried digitalis leaves infused in water to facilitate urination.[67]

Subsequently Darwin used foxglove fairly frequently, publishing an account of its applications for oedema and pulmonary consumption in the *Medical Transactions* of the

College of Physicians, which included an appendix by Sir George Baker. A postscript to the appendix noted the publication of Withering's book while the sheets were with the printers.[68] Both Darwin and Withering were aware of the potency of digitalis and the latter remarked that its incorporation into the *Edinburgh Pharmacopia* at the behest of Dr John Hope (1725–1786), professor of botany at the university, might initially kill more people than it benefited. In his commonplace book for 1776 Darwin described how a lady who was 'asthmatic and dropsical' but otherwise appeared in reasonable health nevertheless quickly died 'upon the close-stool' after taking 'four drafts' of foxglove decoction, having vomited and been purged twice.[69] Most of Withering's experiments were designed to determine which conditions it was most useful for and optimum dosage levels. It is perhaps not surprising that as well as the common foxgloves growing at Full Street between the summerhouse and the well, when a marsh marigold or meadow bolt ('*Caltha palustris*') died beyond the fishpond it was replaced by an 'Iron foxglove' or '*Digitalis ferruginea*'.[70]

Darwin emphasised that the foxglove he employed medically was *Digitalis purpurea* (Linnaeus), which was 'found growing plentifully' in all British sandy soils but he thought not in the clay lands. Being a biennial plant, it could be 'procured fresh' in all seasons, a useful advantage for medicine. Darwin used foxglove decoction generally to induce nausea, vomiting and urination, which he took as signs that it was working effectively,after experimentation using 'four ounces' of 'fresh green leaves' rather than other plant parts, boiled down from two to one pints of water and combined when strained with 'two ounces of vinous spirit'. In most cases a large half-ounce spoonful was given in the early morning and repeated hourly in dropsical cases until patients had received three to nine spoonfuls or until sickness or 'disagreeable sensations' occurred. There was considerable variation, but the oedema usually disappeared within a day or two, often without increased urine production but sometimes with vomiting, 'a large flow of urine' or 'purging stools'. With some conditions, such as consumption (tuberculosis) or scropholous ulcers, Darwin varied the dose or combined with peppermint water or bark decoction, and if no nausea occurred then some patients were given higher doses, although he usually began with an 'under' rather than an 'excessive' dose, given the drug's potency and patient vulnerability.[71]

Darwin focused upon exploring which types of oedema foxglove worked best for and the most effective means of administration, finding that dropsies of the thorax and abdomen accompanied by anasarca (severe oedema) of the limbs benefited most. Most such patients – such as Mr Saizoe, an Ashbourne music teacher, and 'Mr. Y', aged about sixty, of Derby – were middle-aged or older and were too partial to drinking 'fermented or spirituous liquors', which had caused gout, corpulency, serious 'oedematous' leg and thigh swellings, breathing problems and irregular pulses. For some patients, more treatment was not required; however, many had to repeat it in ensuing years. A day or two after initial treatment Darwin administered 'an infusion of artichoke stem leaves or bark decoction' with 'chalybeate (iron-salt water)' and enough rhubarb and aloe to induce daily stools with a grain of opium for 'some weeks'. He also encouraged patients to have as healthy and varied a diet as possible to build up general strength by eating as much meat, shell-fish, eggs, spice and 'acrid vegetables', such as celery, water cress and raw red cabbage 'shred fine and eaten as salad', as they could manage and, of course, to reduce, dilute or preferably cease their intake of 'so concentrated a poison' as gin, rum and brandy, which it was believed had often caused the problems in the first place.[72]

Darwin continued to investigate whether the foxglove decoction benefited patients suffering from different dropsies, pulmonary consumption, scropholous ulcers, asthma and even melancholy. He believed that successful treatment of a Shropshire farmer's wife for abdominal oedema with the digitalis, opium, bark and calomel, and treatment of Sarah Morris and 'Mr. W' of Derby for abdominal oedema after 'excessive use of fermented and spirituous liquors', underlined its effectiveness.[73] Other successful cases, such as a twenty-year-old shoemaker from Church Broughton, Derbyshire, suffering from pulmonary consumption and two young women with scropholous ulcers, suggested that, combined with treatments such as plaster bandages, digitalis formulated as 'powder, poultice, or fomentation' strongly merited further trials for internal and external use. The use of foxglove to treat asthma and melancholia appeared to bring few benefits, and Darwin concluded that a young lady suffering from the latter was so healthy that her 'dejection and despair' arose from 'powers of the mind', making 'corroborant medicines' more useful.[74]

In 1786 Darwin treated Samuel Johnson's lifelong friend Dr John Taylor of Ashbourne with foxglove decoction for hydrothorax following debilitation caused by 'unnecessary blood-letting' by the physician Dr Joseph Denman (1731–1812). After administering 'half an ounce every two or three hours' up to seven doses, the most Darwin had hitherto given, Taylor was not surprisingly 'violently sick', but then he regained mental and physical strength and his leg oedema reduced for more than 'several years'. Darwin was 'certain' that without foxglove death would have followed within two days.[75] For the rest of his medical career he continued to advocate foxglove decoction, believing that it cured two old men of abdominal and leg fluid retention in 1787, recommending it for Jessie (Janet Watt) to James Watt in 1794 because of its 'miraculous' powers of absorption from the lungs and prescribing it for whooping cough to his son Robert in 1788. He likewise advocated it for other members of his family, including his mother with opium and bark in 1796 and his cousin, John Bromhead (1742–1818), manager of the Kedleston estate and husband of Ann Darwin (1748–1837), who suffered from fluid retention in the lungs and legs.[76]

Overall, Darwin believed that digitalis was most efficacious for chest dropsies, where it reversed the 'retrograde motions of the lymphatic vessels', rather than other conditions, such as other bodily oedemas or ulcers, and that more trials were needed to gain a fuller understanding. He was pleased that 'the faculty' had gained 'so valuable an addition to the materia medica' and recalled previous patients with oedemas who had to endure 'squills, emetic tartar, and drastic purges' and might have lived if digitalis decoction had been administered. The new understanding of treatments utilising foxglove was therefore, he believed, one of the greatest achievements of modern medicine, which had much reduced the grievous suffering of many sick people, increased their comfort and extended their lives – all through the use of a widely available garden and woodland plant.[77]

GARDENING AND TAXONOMY

With the discovery, import and breeding of new vegetable varieties and a greater appreciation of their medical benefits, a consistent and reliable system of naming and identifying plants (taxonomy) became, if anything, all the more crucial. While John Ray's system remained influential, it was superseded during the second half of the eighteenth century by the Linnaean system, which employed differences in the sexual organs of plants to place them into categories. Naturalists strove to formulate or discover a more 'natural' system of botanical classification, working 'upward' from observations of a range of plant characteristics, rather than 'downward' towards an 'artificial' system, using what were defined as the most essential general qualities, but the simplicity of Linnaeus's system based upon identification using the sexual organs was attractive.[78]

Darwin was, of course, an enthusiastic Linnaean, but, having utilised so many plant-based products in medicine, which required knowledge of the virtues of all the plants' parts, having created his botanic garden and having observed plants growing elsewhere during the 1770s before he helped to complete the Lichfield society translations, he believed that improvements could be made in the system to better reflect differences in growth patterns according to localities and environments. Linnaeaus recognised the value of a more 'natural' taxonomy alongside his 'artificial' one and made an initial effort to explore how this might be undertaken, while Withering accepted that it was 'confessedly imperfect' and that exceptions to 'these rules' were 'numerous', as they served as 'rude and imperfect out-lines, which the industry of future ages must correct and compleat'.[79] Many Enlightenment naturalists and botanists relied upon textbooks and dried herbariums for their taxonomic studies, but Darwin's Linnaean translations and *Botanic Garden* were primarily stimulated by his first-hand experiences planting and designing gardens, growing plants and observing those kept by friends, which gave him an appreciation of the strengths and weaknesses of Linnaean taxonomy. He observed discrepancies, anomalies, inconsistencies and other characteristics in growing plants which complicated their relationship to Linnaean genera, families and orders, the most glaring of which were problems placing grasses and cryptogamia and difficulties caused by effacing ancient and widely recognised divisions between trees and plants. According to Withering, it was 'well known that the attention of Linnaeus was much less engaged by the Class Cryptogamia' – which he divided into four orders, mosses (*Musci*), ferns (*Filices*), funguses (*Fungi*) and flags (*Algae*) (Figure 22) – than by others classes with more obvious 'fructifications'.[80]

In their translation, the Lichfield botanical society rendered cryptogamia as 'clandestine marriages' whose 'fructifications seclude themselves from our sight', with 'structures' different from those of other plants. They had some similarities with grasses, also seen to be at the lower end of the plant hierarchy, which were characterised as 'plebeians' and 'tributary ... to all herbivorous animals', while funguses were 'vagabonds ... barbarous, naked, putrescent, rapacious, voracious'. However, both were 'tenacious of life', with grasses constituting the 'multitude and strength of the vegetable kingdom', rising up the more they were 'trod upon'. Fungi

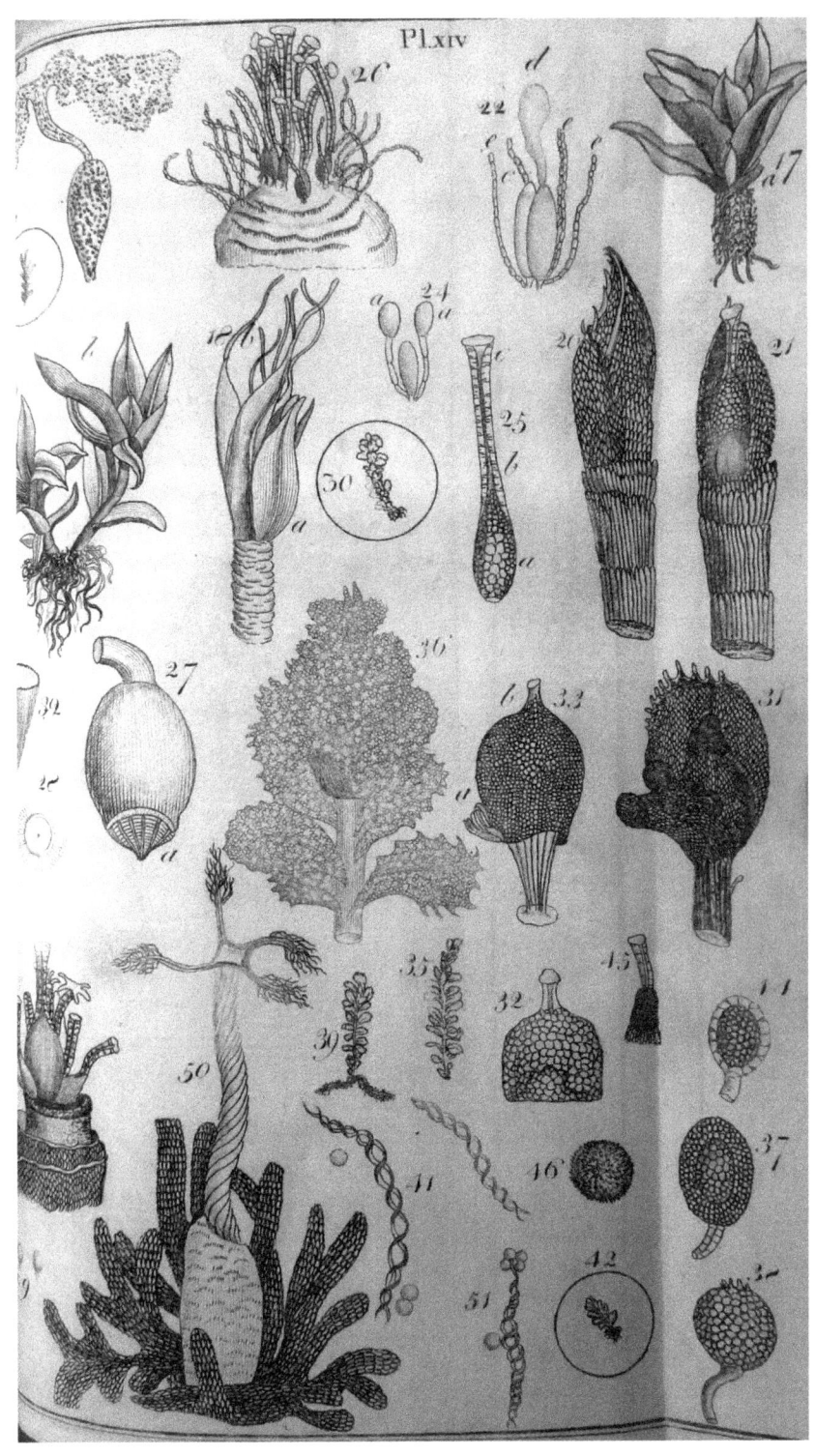

were 'supported by the recrements of other plants', mosses protected the 'roots of other plants' and collected 'for the benefit of others, the daedal soil', while flags, or algae, though 'slaves', were 'the first vegetation on uncultivated places'.[81] Darwin's efforts to encourage the study of cryptogamia and his fascination with fungi are evident in the interest taken in these by the Derby Philosophical Society members, who ordered the *Historiae Naturalis Muscorum Frondosorum* (1782) of German physician and botanist Johann Hedwig (1730–99) and the first volume of *Fasciculus Plantarum Cryptogamicarum Britannicae* (1785–1801) by James Dixon (1738–1822) for their library; the latter included hundreds of species of algae and fungi from across the British Isles.[82]

The grasses, most of which were comprised within the class Triandria, were difficult to interpret for similar reasons to the cryptogamia, usually lacking striking flowers, and Withering admitted that originally 'no part of botany appeared to me more difficult than their study'.[83] While undertaking the Lichfield Society's Linnaean translation Darwin wrote to Joseph Banks explaining how James Crowe of Tuckswood, near Norwich, sent him 'specimens of two rare grasses, the *Agrostis littoralis* and the *Phalaris phleoides*' and suggested to him that a 'character' should 'be introduced' in their translation of Linnaean taxonomy to demarcate British natives. Crowe also provided information on their 'habitations' and the context in which they thrived, while the botanist John Lightfoot FRS (1735–1788) supplied additional information. Subsequently Darwin tried growing some grasses in his Derby garden, such as saltmarsh grass (*Puccinellia maritima*) but with mixed success.[84]

Darwin hoped that the efforts of the Linnean Society of London, founded by James Edward Smith (1758–1828), aided by his acquisition of the the master's own materials and especially the evidence for original or 'type' specimens, would produce a 'plan for disposing part' of the 'vegetable system' into 'more natural classes and orders' during the 1780s and 1790s. Taxonomy was of vital importance to scientific, economic and social progress and should not be considered mere distant, dry, abstract scholarly study; Darwin thus believed that more 'natural classes' would facilitate identification and discovery and enable plants to be better exploited for culinary, artistic, medical and manufacturing purposes, such as 'dying, tanning, architecture, ship-building'. This had 'already been happily experienced' through application of the knowledge of 'genera or families of plants', which were 'all natural distributions' because the 'same virtues or qualities generally exist among all the species of the same genus, though perhaps in different degrees'.[85]

In *The Temple of Nature* Darwin noted how the 'anthers and stigmas of flowers' were 'probably nourished' by honey, which was 'secreted by the honey-gland' or 'nectary' and possessed 'greater sensibility or animation' than other plant parts. Darwin believed that the flower's corol (or petals) was probably a 'respiratory organ' associated with the anthers and stigmas, which 'oxygenat[ed] the vegetable blood' to produce 'anther dust' and honey, which was 'exposed to the air in its receptacle or honey-cup'. Darwin thought that petals were needed for 'further oxygenation' because many flowers, such as monk's hood, delphinium, larkspur, honeysuckle and woodbine, had a complex 'apparatus' apparently especially designed for protection against insects, while the petals and 'nectar' drooped with the anthers and stigmas when the 'pericarp' was 'impregnated'.[86] Although reproduction by other means was possible, Darwin believed, encouraged by his observations of gardening and horticulture, that sexual reproduction was the healthiest and only proper means of reproduction. Thomas Andrew Knight (1759–1838), the Herefordshire landowner and naturalist, had observed that apple and pear trees propagated for over a century by grafting were 'now so unhealthy' that there was little point in cultivating them. Similarly, Darwin suspected that potato curl leaf and diseases of strawberry plants which caused 'barren flowers' occurred because such plants were 'too long raised from roots' or 'solitary reproduction', occasioning these 'hereditary' ailments, and not from seeds or sexual reproduction.[87] In *The Temple of Nature* he explained how 'feeble births acquired diseases chase, Till death extinguish the degenerate race' and grafted trees became a 'waning lineage, verging to decay' unless 'amended by connubial powers' from which 'seedling progenies' could arise from 'sexual flowers'.[88]

Darwin was well aware that some botanists faced hostility from Linnaeus and his followers for allegedly tampering with the Linnaean system. Indeed, he, Boothby and Jackson distinguished between 'true' and 'false' botanists in their English translation, as well as reproducing the master's summary of his principles of classification. In this Linnaeus emphasised how God had proceeded from the 'simple to the compound, the few to the many', forming as many different plants 'as there are natural orders' at the creation. He had then 'so intermixed' plants from these orders through 'marriages with each other' that 'distinct genera' had appeared. Although the timeframe involved was not specified, 'Nature' had then 'intermixed' such generic vegetables by 'reciprocal marriages' that did not alter the floral structure and 'multiplied them into all possible existing species'. 'Mule plants', were, however, excluded from being new species because they were 'barren'. Each genus was therefore 'natural', with nature 'assenting to it, if not making it'. Linnaeus believed that the defining characteristics were never to constitute the genus but had to be 'diligently … constructed according to the genus of nature'. The 'diagnosis' of plants, by which they were identified, named and placed within their taxonomic group using the binomial system, was made using 'the affinity of the genus' and 'difference of the species', which was reflected in plain nomenclature providing a 'generic family name' and a 'specific trivial name'. 'True' botanists, according to Linnaeus, moved from genus, determined by the 'characters of the displayed plant or flower', to designating species and their 'synonymies', founded upon the work of generations of naturalists who imparted the august botanical wisdom of the ages.

Linnaeus forcefully contrasted 'true botanists', who primarily laboured to describe 'obscure, rare and new plants' according to these rules, always prefixing genus to species and defining plants by their 'natural character from the situation, figure, proportion' of all fructification parts, from 'False botanists'. These tried to proclaim 'the laws of the art before they have learned them', extolling 'absurd authors', appropriating ideas from others, boasting of 'little knowledge', pretending they had 'discovered a natural method' and asserting 'the genera to be arbitrary'. The way in which his system ignored many plant features and homed in upon the most essential and fundamental was taken by Darwin and the Lichfield botanical society members as a proof of its philosophical vigour and scientific efficacy.[89]

However, in the final section of *Phytologia*, underscoring the relevance of its subject matter for agriculture and gardening, Darwin presented a plan for adapting part of the Linnaean system into 'more natural classes and orders' which utilised his observations of many different plants growing in his gardens and elsewhere, his experiences undertaking the translations and his knowledge of the virtues and characteristics of plant parts from medical practice and experiment. While admitting that he had 'often admired' the Linnaean sexual taxonomy, Darwin contended that some classes seemed to him to be 'more excellent than others' because they approached 'nearer to natural ones'. Although Darwin's fascination with sexual reproduction has often been emphasised, particularly in relation to *The Loves of the Plants*, his critique of Linnaean botany, founded upon careful observations of growing plants in different contexts, sought to reduce the emphasis – and reliance – upon vegetable reproductive organs by botanists. He believed that the Linnaean system might have been framed differently if Linnaeus and his followers had chosen to make the filaments, anthers or other features the bedrock of the system. There was, therefore, scope for botanists to utilise different characters to alter or augment the Linnaean taxonomy. Darwin's gardening experiences encouraged him to regard changes in environmental conditions as major determining factors in the mutability of sexual organ characters. The number of sexual organs within flowers was 'more liable' to be altered by the 'influence of soil or climate', or the 'progress of time', than by 'their situations or proportions or forms', and could therefore be usefully employed to distinguish between classes and orders as well as making these designations 'more natural'. Using a greater range of characteristics should therefore produce a botanical taxonomy that more accurately and sensitively reflected the realities of how plants grew and changed on the ground.[90]

The different ways in which plants grew in varied contexts meant that inexperienced, 'young' and even veteran botanists might find difficulties making identifications if dependent upon Linnaean textbook descriptions only or if faced with unusual features. Darwin therefore focused upon situations where his plants grew in an unusual manner, just as his grandson Charles was to do more systematically. For instance, Darwin had various snapdragons (*Antirrhinum*) growing in his Derby garden, and in later summer 1799 he obtained an 'antirrhinum peloria'. Normally snapdragons produce zygomorphic, or bilaterally symmetrical, flowers; however, occasionally they produce peloric, or actinomorphic, flowers. Darwin remarked to his brother Robert Waring that the 'antirrhinum peloria' was 'a most uncommon flower, but I suppose a hardy one', as it survived the winter.[91] As we have seen, Darwin emphasised that, as the medical virtues of many plants were often contained in the flowers and their petals, so double or multi-flowered examples had great worth. His grandson Charles later explored instances of peloria in snapdragons while investigating the inheritance of floral characteristics for his *The Variation of Animals and Plants under Domestication* (1868), providing another example of the similarities in their approaches to botany and the parallel role that gardening and horticulture played in supporting their evolutionary theories.[92]

In his essay in *Phytologia* on adapting the Linnaean system, Darwin emphasised that the 'luxuriant growth of many cultivated flowers' or 'the duplicature or multiplication' of petals or nectaries meant that various plant species had just 'half the number of stamina which other species of the same genus possess'. This occurred 'so frequently' that the 'defect of number' was often 'expressed as an essential character of the species'. The 'vernal flowers' of the tropical shrub slippery burr (*Corchorus siliquosus*), for instance, had 'but four stamina', while the autumnal ones had numerous; the common flax, growing in Britain, had 'but five perfect stamina and five without anthers on their summits', whereas Portuguese flax (*Linum lusitanicum*) possessed ten complete ones. Similarly, vervain (*Verbena officinalis*) in Britain had four stamens while in Sweden it had just two. These example demonstrate the 'uncertainty of depending upon numbers alone for distinguishing the classes of plants'. Nor did the number of pistils offer a 'more certain criterion' for the orders, which also demonstrated the 'great confusion' that was occasioned by a reliance upon pistil numbers alone for defining plant orders.[93] Darwin contended that classes based upon proportions, situations and sometimes number of stamens would produce 'more natural distributions of vegetables' than those based upon number alone, and that it would have been 'more fortunate' for botanical science if Linnaeus had based all of his classes upon proportions, situations and number of stamens combined. Likewise, orders might have been better identified by the proportions, situations, forms and perhaps, 'conjointly', numbers of the pistils.[94] Founding plant classes upon the 'situations, proportions or forms' of sexual organs rather than the number also had the 'great advantage' that such classes and orders would be 'much less subject to variation'.

Despite the high praise he had for Linnaeus, Darwin thus sought to improve upon his system by making suggestions for more 'natural' taxonomies using information from the works of the 'great master' naturalist and his disciples using his own plant observations. He pointedly warned, however, that 'great general inconvenience' would result by altering a comprehensive and well-established system and criticised 'idle efforts' by botanists to add classes 'deduced from situation or proportion' to those 'simply numerical'. He also believed that if too many changes were made to the basis of Linnaean taxonomy by interweaving classes based upon the situation and proportion of the sexual organs with those distinguished simply by number, this would damage the 'great system' that naturalists were seeking 'to amend'. Furthermore, Darwin admitted he was 'incapable' of fully executing the plan for improvements that he had propounded because it would require a 'genius' and a profound knowledge of botany, which came only through 'many years of unremitted application', taking 'every opportunity' to visit botanical gardens, examine dry collections and inspect 'prints and drawings of vegetables', to refine 'the complex and intricate to the simple and explicit'. However, the state of botanical collecting and study provided a wonderful opportunity for the Linnaean system to be 'intrinsically improved' using the 'situation, proportions, or forms', with or without the numbers of sexual organs to distinguish between orders and classes, which would lay a 'foundation' upon which a 'great architect' would 'erect' a new 'superstructure.' For Darwin this also had the major advantage that it would better utilise knowledge gained in medico-botanical practice and help to place this, too, on a stronger scientific footing by facilitating the identification of apparently anomalous plants and providing greater information concerning all vegetable parts and their virtues, enabling the production of a new Linnaean herbal of the kind he had originally urged Withering to create.[95]

CONCLUSION

While Darwin was always an enthusiastic Linnaean, his experiences utilising, growing and experimenting with plants and observing them in town and countryside convinced him that improvements could be made to Linnaean botany to render it more 'natural' and medically efficacious. He utilised his garden plants to help produce a 'plan for disposing part of the vegetable system of Linnaeus into more natural classes and orders' which exploited his experience designing the botanical garden, translating Linnaeus and composing the *Botanic Garden*. This provided him with an unrivalled knowledge of Linnaean taxonomy as it related to growing garden plants rather than using the dried herbariums that dominated eighteenth-century botany. Discrepancies, anomalies, inconsistencies and other problems were highlighted by observing growing plants in the botanic garden and comparing them with Linnaean genera, orders and families of plants. Some of the benefits of this are most fully evident when we consider how Darwin approached the medical usage of foxglove. Long known to have medical benefits, foxglove had failed to gain a consistent place in the materia medica because of continued uncertainty about how it should be administered and the strength of dosage. Darwin's understanding of plant physiology and anatomy as well as taxonomy and medicine allowed him, along with Withering, to start determining the illnesses that foxglove most effectively combated, the plant parts to be employed, the level of dosage and the means in which it was to be administered.

Darwin hoped that improvements of the Linnaean system encouraged by differences between theory and the reality of growing plants would render it more 'natural', which would make the kind of 'Linnaean herbal' he had originally advocated to Withering a reality by encouraging the usage of all vegetable parts found to be useful. Observations of plants in his various gardens and growing wild, along with the stimulation provided by reading and involvement in gardening and scientific networks, would be combined to produce what he hoped would eventually be a more natural, stable, practically useful and consistent medico-botanical system. Although reaffirming the significance of the sexual organs, which were believed to be the most fundamental and important in the life of the plant, Darwin contended that it was necessary to pay attention to other characters within the sexual organs rather than just number of stamens, which he found to be less fundamental and more unstable than previously supposed.

NOTES

[1] E. S. Rohde, *The Old English Herbals* (London, 1922); W. S. C. Copeman, *The Worshipful Society of Apothecaries of London: A History* (London, 1967); J. G. L. Burnby, *A Study of the English Apothecary from 1660 to 1760, Medical History*, supplement no. 3 (1983); C. Stockwell, *Nature's Pharmacy: a History of Plants and Healing* (London, 1989); B. Griggs, *New Green Herbal: the Story of Western Herbal Medicine*, 2nd edn (London, 1997); S. Minter, *The Apothecaries' Garden: A History of the Chelsea Physic Garden* (Stroud, 2000); G. Hatfield, *Memory, Wisdom and Healing: the History of Domestic Plant Medicine* (Stroud, 2005); L. H. Curth ed., *From Physick to Pharmacology: Five Hundred Years of British Drug Retailing* (Aldershot, 2006); L. Schiebinger, *Plants and Empire: Colonial Bioprospecting in the Atlantic World* (Cambridge MA, 2004); P. Chakrabarti, *Materials and Medicine: Trade, Conquest and Therapeutics in the Eighteenth Century* (Manchester, 2010); J. Carney and R. N. Rosomoff, *In the Shadow of Slavery: Africa's Botanical Legacy in the Atlantic World* (Berkeley, 2011); L. Schiebinger, *Secret Cures of Slaves: People, Plants and Medicine in the Eighteenth-century Atlantic World* (Chicago, 2017); S. Seth, *Difference and Disease: Medicine, Race and the Eighteenth-Century British Empire* (Cambridge, 2018).

[2] J. Hill, *Family Herbal* (London, 1812 [1755]), preface, v.

[3] Hill, *Family Herbal*, preface, vi–vii.

[4] Hill, *Family Herbal*, iii–iv.

[5] Hill, *Family Herbal*, preface, v.

[6] Hill, *Family Herbal*, preface, iii–v.

[7] Griggs, *New Green Herbal*, 141–2.

[8] Hatfield, *Memory, Wisdom and Healing*, 7, 9, 11.

[9] E. Darwin, letter to A. Reimarus (July? 1756) in D. King-Hele ed., *The Collected Letters of Erasmus Darwin* (Cambridge, 2007), 31–3.

[10] P. Ayres, *The Aliveness of Plants: The Darwins at the Dawn of Plant Science* (London, 2008), 25–6.

[11] E. Darwin, letter to W. Withering, 13 May 1775 in King-Hele ed., *Collected Letters*, 136–7; W. Withering, *A Botanical Arrangement of all the Vegetables Growing Naturally in Great Britain*, 1st edn, 2 vols (Birmingham, 1776); D. King-Hele, *Erasmus Darwin: A Life of Unequalled Achievement* (London, 1999), 123–4.

[12] J. M. Good, *The History of Medicine in so far as it Relates to the Profession of the Apothecary* (London, 1795), 145–55; W. G. Black, *Folk Medicine: A Chapter in the History of Culture* (London, 1883); J. Grier, *A History of Pharmacy* (London, 1937), 262; Griggs, *New Green Pharmacy*, 143–4; R. Porter, *Health for Sale: Quackery in England, 1660–1850* (Manchester, 1989); F. W. Robertson, *Early Scottish Gardeners and their Plants, 1650–1750* (East Linton, 2000), 129–40; P. J. Corfield, 'From poison peddlers to civic worthies: the reputation of the apothecaries in Georgian England', *Social History of Medicine*, 22 (2009), 1–21.

[13] Grier, *History of Pharmacy*, 45–7; Burnby, *The English Apothecary*, 4–23, 92–116; Stockwell, *Nature's Pharmacy*, 89–90, 95–6; A. W. Sloan, *English Medicine in the Seventeenth Century* (Durham, 1996), 91–101.

[14] C. Darwin, *The Life of Erasmus Darwin*, edited by D. King-Hele (Cambridge, 2003), 83.

[15] E. Darwin, letter to R. Darwin, 17 December 1790 in King-Hele ed., *Collected Letters*, 372–4.

[16] E. Darwin, letter to R. Darwin, 30 May 1792, in King-Hele ed., *Collected Letters*, 402–3. J. Power, 'Of the use of fermenting cataplasms in mortifications', *Medical Transactions of the College of Physicians*, 3 (1785), 47–53; S. Shaw, *The History and Antiquities of Staffordshire*, vol. 1 (London, 1798), 99.

[17] N. Culpeper, *The Complete Herbal and English Physician Enlarged* (Ware, 1995 [1653]), v–viii; Grier, *History of Pharmacy*, 51–2; Griggs, *New Green Pharmacy*, 46–7; Hatfield, *Memory, Wisdom and Healing*, 5–7, 11–12, 19–20, 27, 73–81, 87–8, 135–8.

[18] Stockwell, *Nature's Pharmacy*, 18.

[19]   Withering, *Botanical Arrangement*, 1st edn, vol. 1, xii–xiii.

[20]   E. Darwin, *Phytologia; or the Philosophy of Agriculture and Gardening* (London, 1800), 498–9; Lichfield Botanical Society, *The Families of Plants ... Translated from the Last Edition of the 'Genera Plantarum'*, 2 vols (Lichfield, 1787), vol. 2, 691, 572, 450, vol. 1, 252.

[21]   Darwin, *Phytologia*, 466; Lichfield Botanical Society, *A System of Vegetables ... translated from the thirteenth edition of the Systema Vegetabilium of the late Professor Linneus*, 2 vols (Lichfield, 1783), vol. 2, 596, vol. 1, 60; Lichfield Botanical Society, *Families of Plants*, vol. 2, 540, 390; C. Jarvis, *Order out of Chaos: Linnaean Plants Names and their Types* (London, 2007), 882.

[22]   Darwin, *Phytologia*, 461.

[23]   J. C. Loudon, *Encyclopaedia of Gardening*, 2nd edn (London, 1824), 1054.

[24]   Culpeper, *Complete Herbal*, 393–553; Grier, *History of Pharmacy*, 23–30, 50–6, 58–9, 61–2, 169–70, 172–3, 179–81; Stockwell, *Nature's Pharmacy*, 24, 103.

[25]   J. Parkinson, *Theatre of Plants or a Universal and Complete Herbal* (London, 1640); Culpeper, *Complete Herbal*; Stockwell, *Nature's Pharmacy*, 88–93.

[26]   J. Wesley, *Primitive Physic: or an Easy and Natural Way of Curing Most Diseases* (London, 1747); Hill, *Family Herbal*; W. Meyrick, *The New Family Herbal; or Domestic Physician* (Birmingham, 1790); R. J. Thornton, *A Family Herbal: or Familiar Account of the Medical Properties of British and Foreign Plants* (London, 1814); Grier, *History of Pharmacy*, 46, 85–92; Stockwell, *Nature's Pharmacy*, 77–87, 109; Griggs, *New Green Pharmacy*, 54–5, 91–5; R. Waller, *John Wesley: A Personal Portrait* (New York, 2003), 113–15; S. King, 'Accessing drugs in the eighteenth-century regions', in L. H. Curth ed., *From Physick to Pharmacology: Five Hundred Years of British Drug Retailing* (Aldershot, 2006), 49–78; P. Wallis, 'Consumption and cooperation in the early modern medical economy', in M. S. R. Jenner and P. Wallis eds, *Medicine and the Market in England and its Colonies, c1450–1850* (Basingstoke, 2007), 47–68; Schiebinger, *Plants and Empire*; P. Wallis, 'Consumption, retailing and medicine in early-modern London', *Economic History Review*, 61 (2008), 26–53; H. Barker, 'Medical advertising and trust in late-Georgian England', *Urban History*, 36 (2009), 379–98; Chakrabarti, *Materials and Medicine*; Carney and Rosomoff, *In the Shadow of Slavery*; P. Wallis, 'Exotic drugs and English medicine: England's drug trade, c1550–c1800', *Social History of Medicine*, 25 (2012), 20–46; Schiebinger, *Secret Cures of Slaves*; Seth, *Difference and Disease*.

[27]   Loudon, *Encyclopaedia of Gardening*, 2nd edn (London, 1824), 1054.

[28]   C. R. B. Barrett, *History of the Society of Apothecaries of London* (London, 1905); Burnby, *English Apothecary*, 62–8.

[29]   P. A. Elliott, *Enlightenment, Modernity and Science: Geographies of Scientific Culture and Improvement in Georgian England* (London, 2010), 124–66; N. C. Johnson, *Nature Displaced, Nature Displayed: Order and Beauty in Botanical Gardens* (London, 2011), 1–50.

[30]   T. Faulkner, *An Historical and Topographical Description of Chelsea and its Environs* (London, 1810), 18–27; H. Field, *Memoirs of the Botanic Garden at Chelsea belonging to the Society of Apothecaries at London*, continued by R. H. Semple (London, 1878); F. D. Drewitt, *The Romance of the Apothecaries' Garden at Chelsea*, 3rd edn (Cambridge, 1928); Copeman, *The Worshipful Society of Apothecaries of London*; H. Le Rougetel, *The Chelsea Gardener: Philip Miller, 1691–1771* (London, 1990); Burnby, *English Apothecary*, 63–6; Minter, *The Apothecaries' Garden*; R. Drayton, *Nature's Government: Science, Imperial Britain and the 'Improvement' of the World* (New Haven, 2000), 35–7.

[31]   Renewed attention to the wealth Sloane gained through marriage from sugar plantations worked by slaves has recently clouded his reputation.

[32]   J. Haynes, *An Accurate Survey of the Botanic Garden at Chelsea with the Elevation and Ichnography of the Green House and Stoves* (London, 1751); I. Rand, *Horti Medici Chelseiani Index Compendiarius* (London, 1739); Faulkner, *Historical and Topographical Description*, 18–29; Field, *Memoirs of the Botanic Garden at Chelsea*, 60–2; Drewitt, *Romance of the Apothecaries' Garden*, 62–93; Minter, *Apothecaries' Garden*, 11–28.

33 Withering, *Botanical Arrangement*, 1st edn, vol. 1, iii, ix, xiv, xxiv, 2nd edn in 3 vols (Birmingham, 1792); J. Stokes, *A Botanical Materia Medica*, 4 vols (London, 1812); R. Schofield, *The Lunar Society of Birmingham: a Social History of Provincial Science and Industry in Eighteenth-Century England* (London, 1963), 223–6, 231–2, 307–10, 316–17, 417–18.

34 Withering, *Botanical Arrangement*, 1st edn; W. Withering, *An Account of the Foxglove and some of its Medical Uses* (Birmingham, 1785), xiii–xx, 1–3; Griggs, *New Green Herbal*, 136–41.

35 Withering, *Account of the Foxglove*, xiii–xx, 1–3; E. Darwin, letter to R. Darwin, 30 May 1792, in King-Hele ed., *Collected Letters*, 402–3.

36 Derby Philosophical Society, *Rules and Catalogue of the Library of the Derby Philosophical Society*, with supplements of 1795 and 1798 (Derby, 1793–8); Derby Philosophical Society, *Rules and Catalogue of the Library of the Derby Philosophical Society* (Derby, 1815).

37 B. Wilmer, *Observations on the Poisonous Vegetables which are either Indigenous in Great Britain or Cultivated for Ornament* (London, 1781); C. Bryant, *Flora Dietica; or History of Esculent Plants* (London, 1783); W. Fordyce, *The Great Importance and Proper Method of Cultivating and Curing Rhubarb in Britain for Medical Uses* (London, 1784); M. Wall, *Clinical Observations on the use of Opium in Low Fevers and in the Synochus* (Oxford, 1786); S. Crump, *Inquiry into the Nature and Properties of Opium* (London, 1793); J. Lane, 'Eighteenth-Century medical practice: a case study of Bradford Wilmer, surgeon of Coventry, 1737–1818', *Social History of Medicine*, 3 (1990), 369–86.

38 Derby Philosophical Society, *Rules and Catalogue of the Library of the Derby Philosophical Society*, 16, 19, 20, 24, 28; W. Aiton, *Hortus Kewensis*, 3 vols (London, 1789).

39 T. Fowler, *Medical Reports of the Effects of Tobacco in the Cure of Dropsies and Dysuries*, 2nd edn (Stafford, 1788), iii–v, xviii, ix, 1–74, 84–6, 96; Derby Philosophical Society, *Rules and Catalogue of the Library of the Derby Philosophical Society* with supplements of 1795 and 1798 (Derby, 1793–8), 11.

40 T. Fowler, *Medical Reports of the Effects of Arsenic in the Cure of Agues, Remitting Fevers and Periodic Headaches* (London, 1786).

41 Derby Philosophical Society, *Rules and Catalogue of the Library of the Derby Philosophical Society* (1793–8).

42 For the use of dyer's broom, common bugloss, wormwood, wood avens, sea wormwood and deadly nightshade see Thornton, *New Family Herbal*; on cinchona, Grier, *History of Pharmacy*, 94–104; Wallis, 'Exotic drugs and English medicine', 32–4, 36.

43 E. Darwin, letter to R. Darwin, 18 August 1794 in King-Hele ed., *Collected Letters*, 451–3; it is not clear to which of Christian Gottlieb Ludwig's works Darwin was referring.

44 E. Darwin, letters to Miss S. Lea, April/May 1796? and 6 June 1796, in King-Hele ed., *Collected Letters*, 495–6, 499–500; Lichfield Botanical Society, *System of Vegetables*.

45 Darwin, *Loves of the Plants*, 137–8.

46 Darwin, *Loves of the Plants*, 137–8.

47 Darwin, *Loves of the Plants*, 138–9; J. Lindley, *Flora Medica* (London, 1838), 189.

48 Darwin, *Loves of the Plants*, 140–1; Lichfield Botanical Society, *System of Vegetables*, vol. 2, p. 653; Jarvis, *Order out of Chaos*, 638.

49 Darwin, *Loves of the Plants*, 143–5, 246–65; King-Hele, *Erasmus Darwin and the Romantic Poets*, 47, 52–4. The latex of *Antiaris toxicaria* contains a highly poisonous cardiac glycoside that increases the contraction of heart muscles. The incredible nature of Darwin's account later drew criticism from Mary Anne Schimmelpenninck, who was related to him by marriage and suggested in her memoirs that he had not been sufficiently concerned about the accuracy of his sources; however, Charles Darwin rebutted the charge in his memoir, emphasising that his grandfather had took care to give his authorities 'in full detail' (Darwin, *Life of Erasmus Darwin*, 77).

50 Darwin, *Phytologia*, 439.

[51] Darwin, *Phytologia*, 466.

[52] E. Darwin, *Zoonomia; or the Laws of Organic Life*, 3rd edn, 4 vols (London, 1801), vol. 2, 399, 400, 426; E. Darwin letter to R. L. Edgeworth, 20 February 1788 in King-Hele ed., *Collected Letters*, 304–5.

[53] Darwin, *Zoonomia*, vol. 2, 427, 433, 450, 451–2, 469–72; Withering, *Botanical Arrangement*, 1st edn, vol. 1, p. 349; for copaifera and olibanum, Lindley, *Flora Medica*, 171 and 278 respectively; Grier, *History of Pharmacy*, 58–9.

[54] Darwin, *Zoonomia*, vol. 2, 472–5, 515–18.

[55] Darwin, *Zoonomia*, vol. 2, 519, 525–6, 528–9.

[56] Darwin, *Zoonomia*, vol. 2, 529–30, 535–6.

[57] Darwin, *Zoonomia*, vol. 2, 536–7, 550–1.

[58] Darwin, *Zoonomia*, vol. 2, 530–1, 535, vol. 3, 99, 210–12; J. Fordyce, *Historia Febris Miliaris, et de Hemicrania Dissertatio* (London, 1758); Lindley, *Flora Medica*, 471.

[59] E. Darwin, letter to Mr Hunt, 5 December 1786, in King-Hele ed., *Collected Letters*, 267–8; E. Darwin, letter to Mrs [Mary?] Thorneycroft, 11 March 1787 in King-Hele ed., *Collected Letters*, 274–5; letter to Henrietta Sneyd, 29 July 1787 in King-Hele ed., *Collected Letters*, 288–9; E. Darwin letter to J. Wedgwood, 5 January 1788, in King-Hele ed., *Collected Letters*, 299–300; letter to J. Wedgwood, 29 January 1788, in King-Hele ed., *Collected Letters*, 301–3.

[60] Fordyce, *The Great Importance and Proper Method of Cultivating and Curing Rhubarb*; Lichfield Botanical Society, *System of Vegetables*, vol. 1, 311; Thornton, *Family Herbal*, 402–16; Lindley, *Flora Medica*, 353–9; C. M. Foust, *Rhubarb, the Wondrous Drug* (Princeton, 1992).

[61] Darwin, *Zoonomia*, vol. 2, 516; Darwin, *Phytologia*, 525–6.

[62] Withering, *Account of the Foxglove*; Withering, *Botanical Arrangement*, 1st edn, vol. 1, 376; W. Withering, *The Miscellaneous Tracts of the late William Withering MD FRS to which is prefixed a Memoir*, edited by W. Withering junior, 2 vols (London, 1822), vol. 1; Grier, *History of Pharmacy*, 64–8; Griggs, *New Green Pharmacy*, 137–41.

[63] E. Darwin, 'An account of the successful use of Foxglove in some dropsies and in the pulmonary consumption', *Medical Transactions*, 3 (1785), 255–86, G. Baker, appendix and postscript to the appendix, 301–8; see also Withering, *Account of the Foxglove*, xiv–xx.

[64] Hill, *Family Herbal*, 144.

[65] C. Darwin, *Experiments Establishing a Criterion between Mucaginous and Purulent Matter* (Lichfield, 1780); Withering, *Account of the Foxglove*; for use of squill see Darwin, *Zoonomia*, vol. 2, 524–8, 484–5, 470.

[66] Withering, *Account of the Foxglove*; W. Withering, *A Systematic Arrangement of British Plants*, 5th edn, edited and enlarged by W. Withering junior, 4 vols (Birmingham, 1812), vol. 3, 684–5; J. F. Fulton, 'Charles Darwin (1758–1778) and the history of the early use of digitalis', *Journal of Urban Health*, 76 (1934), 533–41; P. Sheldon, *The Life and Times of William Withering: His Work, His Legacy* (Studley, 2004).

[67] Withering, *Account of the Foxglove*, 12–16; Grier, *History of Pharmacy*, 80–2.

[68] Darwin, 'An account of the successful use of Foxglove', 255–86, Baker, appendix and postscript to the appendix, 448.

[69] E. Darwin, Commonplace Book, 8, quoted in King-Hele, *Erasmus Darwin*, 133.

[70] Withering, *Account of the Foxglove*; Cambridge University Library, E. Darwin, manuscript catalogue of hardy plants (1796), DAR 227.2.11; Darwin, *Zoonomia*, vol. 2, 484–7, 520–9.

[71] Darwin, 'An account of the successful use of Foxglove', 55–7.

[72] Darwin, 'An account of the successful use of Foxglove', 258–68.

[73] Darwin, 'An account of the successful use of Foxglove', 268–82.

[74] Darwin, 'An account of the successful use of Foxglove', 283–4.

[75] E. Darwin, Letter to R. Clive, 4 October 1786 in King-Hele ed., *Collected Letters*, 260–262.

[76] E. Darwin, Letter to J. C. Lettsom, 8 October 1787 in King-Hele ed., *Collected Letters*, 295–6; Darwin, letters to J. Watt, 1 January, 25 May, 1794 in King-Hele ed., *Collected Letters*, 426–7; Darwin, letters to R. Darwin, 8 March 1796, 8 June 1798 in King-Hele, *Collected Letters*, 493–5, 519–20.

[77] Darwin, 'An account of the successful use of Foxglove', 285, 277, 281–2.

[78] C. E. Raven, *John Ray, Naturalist: His Life and Works* (Cambridge, 1950), 106–8, 197–8, 283–8, 288–94; D. Knight, *Ordering the World: A History of Classifying Man* (London, 1981); A. Pavord, *The Naming of Names: the Search for Order in the World of Plants* (New York, 2005), 372–400; Jarvis, *Order out of Chaos*; P. A. Elliott, C. Watkins and S. Daniels, *The British Arboretum: Trees, Science and Culture in the Nineteenth Century* (London, 2011), 37–53.

[79] Withering, *Botanical Arrangement*, 1st edn, vol. 1, p. xviii.

[80] Withering, *Systematic Arrangement of British Plants*, vol. 1, 346–95, quotation at 346; vol. 3, 920–1037; vol. 4, 1–440.

[81] Lichfield Botanical Society, *System of Vegetables*, vol. 1, quotations, vol. 1, 4–5; vol. 2, 789–842, quotation, 789; Lichfield Botanical Society, *Families of Plants*, vol. 2, 737–51, quotation at 737.

[82] Derby Philosophical Society, *Rules and Catalogue* with supplements of 1795 and 1798, 22; A. T. Gage and W. T. Stearn, *A Bicentenary History of the Linnean Society of London* (London, 2001), 9, 28.

[83] Withering, *Systematic Arrangement of British Plants*, vol. 1, 131.

[84] E. Darwin, letter to J. Banks, 29 September 1781 in King-Hele ed., *Collected Letters*, 189–92.

[85] Darwin, *Phytologia*, 513–14; Gage and Stearn, *Bicentenary History*, 3–21, 175–81, 183; M. Walker, *Sir James Edward Smith (1759–1828): First President of the Linnean Society of London* (London, 1988); Jarvis, *Order out of Chaos*, 13–101; Elliott, Watkins and Daniels, *British Arboretum*, 37–53.

[86] E. Darwin, *The Temple of Nature; or the Origin of Society* (London, 1803), 63.

[87] Darwin, *Temple of Nature*, 57.

[88] Darwin, *Temple of Nature*, 56–7.

[89] Lichfield Botanical Society, *System of Vegetables*, vol. 1, 11–12.

[90] Darwin, *Phytologia*, 515.

[91] Cambridge University Library, Darwin, catalogue of hardy plants; E. Darwin, letter to R. W. Darwin, 7 October 1799 in King-Hele ed., *Collected Letters*, 531–3.

[92] C. Darwin, *The Variation of Animals and Plants Under Domestication* (London, 1868), 33–4.

[93] Darwin, *Phytologia*, 514–15.

[94] Darwin, *Phytologia*, 513–14.

[95] Darwin, *Phytologia*, 524.

CHAPTER 3

AGRICULTURAL IMPROVEMENT: ENCLOSURE AND THE APPLICATION OF SCIENCE AND TECHNOLOGY

Erasmus Darwin's most important contribution to the science of farming was the publication *Phytologia*, a major treatise on agriculture and gardening that attracted some contemporary praise.[1] More recently, Robert Schofield was puzzled that an 'elderly, established doctor' with 'little public indication of an interest in agriculture' should have published a treatise on the subject. And while Maureen McNeil argued that Darwin's inspiration lay primarily in Scottish writings concerning agricultural improvement, the Agricultural Revolution and the 'scarcity crisis of the 1790s', along with some aspects of Linnaean botany and the creation of the Lichfield botanic garden, she still found it hard to explain how these provided 'sufficient stimulus for a six-hundred page treatise on agriculture'. While some interest in the technologies of agricultural improvement and plant physiology were already evident in Darwin's commonplace book from the 1770s, for McNeil 'explanations of Darwin's agrarian interest' founded only upon his 'personal situation' are 'inadequate' and she saw 'no necessary transition' from prose and poetical 'nature studies' to agriculture. His medical interests did not provide a full explanation either, given the small part of agriculture that was devoted to generating 'tools of medicine'.[2] However, this underestimates the extent to which he mixed with farmers and landowners throughout his career, the extent of his experiences combating animal and plant diseases and the impact of his medical ideas and experiences upon his analyses of agriculture and horticulture.

Darwin was a strong supporter of the Georgian ideology of improvement, which, as Raymond Williams argued, like the notion of 'cultivation', contained meanings that were 'historically linked but in practice so often contradictory': 'working agriculture' existed alongside the costly 'improvement of houses, parks, artificial landscapes', interweaving the desire for increased wealth and productivity with landscape aesthetics reinforced by moral judgement.[3] This chapter argues that Darwin's belief that the sciences and medicine could be used to better harness nature and help realise a more productive countryside is evident in *Phytologia*, his medical treatise *Zoonomia* and the references to agriculture and gardening in his poetry and correspondence. McNeil has argued that, as such, his emphasis was on the 'intellectual, rather than on the manual aspect' of farming and that he was 'eager to celebrate' both agricultural as much as industrial 'capitalists' as 'social heroes' as well as those who helped encourage farmers to better organise their 'space, capital and time'.[4] Darwin came from a Nottinghamshire and Lincolnshire landowning family, as did his second wife Elizabeth, and most of his friends and patients were either landowners or farmers, or made a living from agriculture in some way. Like Richard Weston (1733–1806), the prolific agricultural writer and secretary of the Leicestershire Agricultural Society, Darwin believed that when farmers 'in the low class of life', the 'gentleman farmer', nurserymen and 'men of opulence and learning' all worked together, they brought the 'lights and assistance of philosophy' and the 'new husbandry' and 'garden-culture' together to foster agricultural improvement.[5]

After examining Georgian attitudes towards agricultural improvement and enclosure and the way in which these were promoted in the midlands and especially Staffordshire, Derbyshire, Nottinghamshire and Leicestershire, the counties that Darwin knew best, this chapter will explore the ways in which Darwin sought to promote a more scientific form of farming in *Phytologia*, arguing that his efforts were shaped by his medical practice and landowning family background. Using the examples of his ideas for applying technological and meteorological innovations to agriculture, it then demonstrates how he believed that technology and the sciences would facilitate improvements in farming and horticulture and thereby society in general, by increasing productivity and the range of crops and animals nurtured.

AGRICULTURAL IMPROVEMENT AND ENCLOSURE

The Georgian notion of progress in the arts and sciences, known to contemporaries by the term 'improvement', was a concept that originated in agriculture and particularly involved the application of novel ideas and techniques to make farms and estates more manageable, valuable and productive, as well as more symbolically important both for the lineage of established landed families and for those with newer claims to such status. As Alistair M. Duckworth emphasised in his study of their role in Jane Austen's novels, Georgian estates were both 'ordered physical structure' but also a 'metonym for other inherited structures – society as a whole, a code of morality, a body of manners, a system of language'. Improvement was the mode in which people related to their 'cultural inheritance', a 'means of distinguishing responsible from irresponsible action' and a way of 'defining a proper attitude to social change'. Political power, of course, remained largely in the hands of the landowning classes, connected through networks of family and friends, while farming dominated society, as the effects of food price fluctuations and their relationship to crime, food riots and patterns of public contentment and discontent demonstrates (Figure 23).[6]

Farming was the mainstay of the economy and an endeavour that impacted upon the lives of everyone, although the drivers and meanings of agricultural improvement varied according to social class, occupation, geographical location and other factors. There was a growing elite consensus about what constituted agricultural improvement by the final decades of the eighteenth century, but disputes continued about the real efficacy of particular enclosure schemes and increases in farm sizes, the consequences of loss of access to common lands and rights and the impact of depopulation and movement of settlements, and could conflict with time-sanctioned customs and established

ure 23. Markeaton Hall from the lake, drawn by J. P. Neale and engraved by T. Barber (1824).

landed paternalism. While there was some agreement on the need for – and nature of – agricultural development and rural improvement before the French Revolution among the landed elites, there were growing divisions between Tory and Whig notions of rural improvement. There were also divergent views about the comparative efficacy of more traditional farm management methods and those promoted as new, more rational and 'modern' by agriculturists such as Arthur Young (1741–1820) and Sir John Sinclair (1754–1835). Meanwhile, the social and political importance of land was reinforced by representations of nature in the arts, philosophy and natural philosophy. While the pace of change varied considerably across the British Isles, the agricultural revolution and improvements in the arts and sciences brought new ideas, techniques, materials and technologies that could be applied to significantly enlarge the extent of land in cultivation and degrees of productivity. From its agricultural origins, by the seventeenth century the notion of improvement was increasingly transferred to other dimensions of society and culture following the work of Francis Bacon (1561–1626) and others. Bacon's vision of agricultural improvement and its utilitarian applications was reaffirmed in various influential works.[7]

By the eighteenth century, measures that were believed to do the most to increase productivity included enclosures, which enabled the augmentation of estates; land drainage and reclamation; the introduction of the Norfolk, or four-course rotation, system; animal and plant breeding; improved methods of fertilising; tree planting; the creation of new farm buildings and other structures; and the application of new or improved technologies, such as better ploughs and seed drills. Farmers following the Norfolk system, for example, grew wheat in the first year, turnips and barley in the second and clover and ryegrass in the third, with clover and ryegrass being grazed or harvested for animal feed in the ensuing year. Turnips were used to feed cattle and sheep during the winter and the method meant that it was unnecessary to leave land fallow for a year, as had been the practice previously, because it was enriched by successive spreadings of manure.[8] According to Simon Schaffer, the drive for improvement and higher productivity, exemplified by Jethro Tull (1674–1741) and his *New Horse-Hoeing Husbandry* (1731), saw 'traditional custom' as a 'set of obstacles to the … rationalities of new systems of agricultural production', the 'transmutation of 'nature's capacities' into 'market values' and the naturalising of the 'laws of political economy'.[9] As Charles Withers has emphasised in the Scottish context, through the activities of agricultural societies, the universities and philosophers such as John Walker (1731–1803) and James Hutton (1726–1797) the notion of improvement privileged what were claimed to be 'modern' methods of rural land management, surveying and landowning linked to civic virtue and patriotic endeavour over 'traditional' methods, especially in the north-east and central lowlands, although the 'receptiveness of local landowners' varied.[10] Efforts were made to make places perceived by improvers to be 'redundant wastes', such as bogs, dunes and common lands, more productive (although they had traditionally been useful resources for local populations), replacing what Arthur Young called the 'old barbarous story' of the open-field system of fallow, wheat, beans or peas and a lack of hoeing ('you might as well recommend an Orrery to their inspection as a hand-hoe').[11]

The productivity of some farms and estates thus increased through the applications of 'modern' agricultural methods, timber plantations, mining (for coal, ironstone, copper, lead and other materials) and changes in management and administration, such as increases in farm size. Georgian improvements in river, road and canal networks also had some impact upon agriculture, facilitating the movement of products and materials. The parks around the houses of the elite were transformed to suit the demands for sports such as foxhunting and shooting, and by fashionable 'natural' landscape gardening methods associated with Lancelot 'Capability' Brown (1715/16–1783), William Emes (1729/30–1803), Humphry Repton (1752–1818) and others. The oppressive aspects of agricultural improvement and subsequently industrialisation were masked by Enlightenment conceptions of the purposeful, productive, utilitarian aspects of the natural economy and pastoral portrayals of bucolic bounty and contended countryside. Writers and farmers such as Henry Home, Lord Kames (1696–1782) and Arthur Young promoted their vision of agricultural improvement focused upon optimising production for the national interest, although the emphasis was primarily upon the benefits for larger landowners and landlords with farms run as effective capitalist units, rather than upon smaller farms and tenant farmers.[12]

These efforts were supported by scientific and agricultural societies, including the Royal Society, the Society of Arts – which offered prizes for agricultural and horticultural improvement – and the *Museum Rusticum et Commerciale* (1763–6), which published essays on agriculture, arts and manufactures. Inspired by Arthur Young, Joseph Banks, John Sinclair and others during the 1790s, county surveys were undertaken under the auspices of the government's Board of Agriculture, although the quality of these varied considerably and the surveyors sometimes faced suspicion and hostility for social and economic reasons (when highlighting the burden of tithes, for instance). Nonetheless, they helped to publicise a vision of wartime patriotic productivity with the aim of fostering national self-sufficiency in foodstuffs and timber plantations to supply the British navy. The publication of agricultural treatises including the Board of Agriculture's county studies and a series of countryside tours examining the state of farming by authorities such as Young provided one means for disseminating the ideology and methods of improvement and helped to transform the countryside, although there was considerable variation in how these ideas were applied and practised.[13] Nevertheless, innovations associated with improvement 'heroes' such as Thomas William Coke, first earl of Leicester ('Coke of Norfolk', 1752–1842), and Robert Bakewell (1725–1795) were detailed and celebrated by the Society of Arts, the Board of Agriculture, county farming societies, gentlemen's clubs and other associations.[14]

The varied regional impact of improvement is evident across the north midlands and Derbyshire and was partly determined by geographical factors as well as the behaviour and attitudes of landowners, estate managers and farmers. A distinction was (and is) usually drawn between the high ground of the Peak and the fertile soils of southern Derbyshire and Staffordshire, but even in the uplands there was much variety and a clear difference was evident between the limestone and the bleaker peaty gritstone areas, the former having richer soils and much richer plant species growing naturally. The agriculturist William Marshall (1745–1818) described the midland counties as largely a very 'fertile tract of country' characterised mainly by a 'rich middle loam' soil on which arable, animal breeding, grazing and dairying dominated to varied extents; northern and central Derbyshire, in contrast, were less productive because of the 'upland or mountainous nature' of the region.[15] The transformation of the midland countryside by agricultural improvement is evident from Young's observations on what he perceived to be good practices and criticisms of those he thought were redolent of obstinacy or conservatism, all underpinned by statistical data on the financial value of land, livestock and crops.[16]

While improvement was expected in the richer agricultural area of the midlands, Young was nevertheless 'agreeably surprised' to see how much of the Peak was generally 'enclosed and cultivated', having been 'led to expect large tracts of uncultivated country in every quarter of it'. Although there were a few unenclosed commons, 'all the southern parts' were 'rich'.[17] Looking across the county from the southern Peak hills, Young noted a 'beautiful' mixture of enclosed fields, rocks of 'shivering limestone' and trees, with the local lead-mining capital of Wirksworth in its 'fine', 'almost romantic' valley and 'a fine variety of enclosures, trees, houses, rocks, lead-mines, all in picturesque confusion', bounded by hills and 'scar'd with rocks and ruins'.[18] He praised those who introduced Norfolk husbandry and new farm buildings, removed 'scattered trees', rocks and stones from fields and planted sainfoin instead of – or combined with – grasses upon hillier wastes to make the country more productive; such paragons included the 'spirited' Wenman (Roberts) Coke MP (c. 1729–1776) of Longford Hall (and Holkham Hall, Norfolk), who promoted 'Norfolk husbandry' in Derbyshire, and Sir Robert Burdett (1716–1797), of Foremark, whose enormous North American cabbages were 'well manured with lime and dung' and kept weed-free by hand-hoeing to fatten oxen and feed sheep. Indeed, it was at the Longford family estate that 'Coke of Norfolk' 'first learned the pleasures of country life' and took 'an interest in agriculture and estate management' before inheriting the Holkham estate in Norfolk.[19]

According to Young, 'great improvements' had been obtained in the Peak highlands around Matlock, Tideswell and other places using enclosure, paring, burning and lime, and sowing turnips, oats and grasses in succession, as well as through planting legumes such as clover, trefoil, lucerne and sainfoin. The last of these was widely promoted by Young, along with Thomas Coke and the Board of Agriculture, as a long-lasting, robust, deep-rooted, reddish-flowered, weed-fighting wonder plant – originally from France – which thrived in the British climate, could enrich even fairly poor soils and produced nutritious fodder and hay, although it had been adopted along with other legumes in some places since at least the seventeenth century. According to Young, these improvement methods reduced the amount of common and fallow land, enabled more livestock to be fattened through grazing and increased productivity and land values so that landlords could increase rent to encourage the 'spirit of industry' rather than 'sloven's conduct'.[20]

The Derbyshire Tory landowners Colonel Edward Sacheverell Pole, of Radburn Hall, and Nathaniel Curzon, first Baron Scarsdale of Kedleston (1726–1804), were praised for their quality crops, their industry in tackling pests (for themselves and tenants) and their introductions of new livestock for stock and breeding, such as the former's 'fine Lancashire long horned' cattle.[21] At Kedleston the house was remodelled to a plan by Robert Adam (1728–1792), who also redesigned the gardens with the landscape gardener William Emes. With its 'magnificent mansion', 'well built' stables and other buildings, 'beautiful park and grounds, and well-drained productive estate 'clothed in verdure', Kedleston exemplified for Young how the 'nobility and gentry' could transform the countryside through better management and taste to 'great national advantage'. This advantage also involved the 'transplanting' of the village and a corn mill away from the house to a 'more distant part' and the removal of the turnpike road likewise – the more unfortunate impact of 'improvements' upon local inhabitants.[22] Similarly, at Chatsworth, the grounds and park were extensively remodelled by Lancelot Brown and others and, according to Young, the duke of Devonshire had 'advanced his estates to a much higher value' with larger rental yields. Meanwhile, the gentry around Derby could command the highest land prices of up to £3 per acre, which Young believed would get the best from farmers and challenge agricultural backwardness and conservatism (Figure 24).[23]

ure 24. Chatsworth, from J. Britton and E. W. Brayley, *The Beauties of England and Wales*, vol. 3, *mberland*, Isle of Man and Derbyshire (1802).

AGRICULTURAL IMPROVEMENT AND DARWIN'S LANDOWNING CIRCLES

As a member of the landed gentry and a much-travelled doctor, Darwin came to know many midlands landowners and farmers as friends, clients or both and was able to observe at close quarters and discuss the latest developments in agriculture, horticulture and estate improvement. During the 1780s, for example, he was often at Calke Abbey in south Derbyshire with his friend and patient Sir Harry Harpur (1738/9–89) discussing farming and mechanical as much as medical topics.[24] As Peter Worsley has shown, the Darwins were landowners and farmers of long standing in Nottinghamshire and Lincolnshire, with their own tenants, and therefore had some agricultural interests. They obtained Elston Hall and an estate of 42 acres of farmland in east Nottinghamshire near East Stoke from the Lascelles family in 1708 partly by inheritance and partly through purchase, the living of Elston Church also being part of the lordship of the manor. When he married Mary Howard in 1757 Darwin was provided with over 113 acres of farms and buildings in the Lincoln area under the terms of his deceased father Robert's will, which by 1776 were worth £1600 and were paying a rental of £53 per year, while other land included an estate at Claythorpe, in east Lincolnshire (recorded as 272 acres in 1824), which passed to him from his elder sister Susannah on her death in 1789.[25] John Bromhead (1742–1818), one of Darwin's cousins by marriage and one of his patients, was manager of the Curzons' Kedleston estate – as we have seen, a showpiece for eighteenth-century Derbyshire architecture and building, landscape gardening and agricultural improvement.[26] Darwin's second wife Elizabeth was a daughter of Charles Colyear, second earl of Portmore (1700–1785). Her half sister was Lady Caroline Curzon (1733–1812),[27] and her first husband, Colonel Pole, of Radburn, was one of those singled out by Young for embracing modern farming methods.

Darwin's development of the gardens and other lands at Lichfield and Derby – including his transformation of the boggy Abnalls site into a temple of flora – could also be regarded as smaller-scale acts of agricultural and horticultural improvement. In fact, he came to see himself as something of a farmer, describing, as we have already noted, the 'little piece of ground appropriated to agricultural purposes' on the opposite bank of the Derwent to the artist Samuel James Arnold, who visited to paint his portrait in 1799, as his 'farm'. According to Arnold, there was here 'an air, if not of novelty at least of improvement in everything about it'. Darwin had just completed *Phytologia* when he met Arnold and the conversation covered farming topics such as the creation of more effective ploughs, probably stimulated by the efforts he was making with Thomas Swanwick to improve Jethro Tull's drill plough, which were described and depicted in the book. Arnold also visited Joseph Banks on this 'professional tour' of 1799, probably at his Overton estate in Derbyshire, where the conversation also turned towards agriculture, although Darwin's reported comments – chuckling that Banks was not 'a great *mechanic*', that 'humour was his forte' and that he always gave 'very good breakfasts' – were presumably intended not to be taken too seriously.[28]

Much of Darwin's income as a medical practitioner must have come from the landed classes and he shared an improving outlook with them. Agricultural productivity and landed wealth in many ways united Whig and Tory landowners, but there were disagreements about the nature and efficacy of particular agricultural improvements supported by figures such as Francis Russell, fifth duke of Bedford (1765–1802), Arthur Young, John Sinclair and the Board of Agriculture, which did not necessarily divide along party lines.[29] Many Derbyshire landed families were sympathetic to the American revolutionaries and the limited parliamentary reform campaigns of the 1770s and early 1780s, although these affiliations came under strain during the revolutionary 1790s. Darwin had a natural Whig leaning and was politically closest to the powerful Whig family the Cavendishes, but had friends and patients of all political persuasions, including the Tory Curzons of Kedleston. As a trusted medical practitioner, treating everyone from the richest aristocrats to the poor, Darwin had close – even intimate – contact with individuals from all social classes, and was able to discuss ideas and acquire information from many different sources. This was one reason why he pursued studies of vegetable physiology and anatomy not simply for their use in medicine but also because he believed they would facilitate the practical utility of agriculture.

Some of Darwin's closest friends, such as Brooke Boothby and Francis Noel Clarke Mundy, were enthusiastic Derbyshire landed improvers. Others were manufacturers, entrepreneurs and industrialists, such as Matthew Boulton, James Watt, Jedediah Strutt (1726–1797), Wedgwood and Richard Arkwright (1732–1792) and their families, who to some degree converged with the landed elite, working closely with them on agricultural, business and enclosure schemes, relying upon them for patronage and aspiring to join them through estate acquisition and cultural emulation.[30] Arkwright and his even wealthier son Richard junior, for instance, undertook major improvements on their Derbyshire estates around Cromford, constructing Willersley Castle, the grounds of which were landscaped and planted by John Webb, a pupil of William Emes (Figure 25). From about 1795, 50,000 trees per year were planted on the estate and Richard Arkwright junior constructed extensive hothouses at Willersley filled with pineapples, melons, peaches and grapes, for which he won the London Horticultural Society's gold medal in 1806.[31] Similarly, the Strutts used the wealth gained from their textile businesses to acquire estates in Derbyshire and Nottinghamshire, including Strutt's Park adjoining Derby, the manor and estate at Kingston-on-Soar and land around Belper, developing settlements akin to large estate villages and undertaking major agricultural improvement schemes across their properties.[32] In addition, influential figures such as Banks, whose Derbyshire and Lincolnshire properties provided an income from agriculture and mining, carried out similar improvements.[33] Aspects

Figure 25.   Willersley Castle, from J. Britton and E. W. Brayley, *The Beauties of England and Wales*, vol. 3, Cumberland, Isle of Man and Derbyshire (1802).

of public culture, such as charitable ventures like hospitals, also brought the newer wealth of industrialists and manufacturers together with the more established gentry and aristocracy, with both serving together as patrons, subscribers and committee members.[34]

The Derby Philosophical Society, founded in 1783 largely at Darwin's instigation, aimed, in his words, to utilise the 'daring hand of experimental philosophy' to 'enrich the terraqueous globe' and the 'common heap of knowledge' by contributing to the 'useful arts' and sciences. About half the members in the early years were medical practitioners, but there were also manufacturers, industrialists, landowners and others with interests in furthering these objectives, and one of their primary objectives initially was to acquire a substantial medical and scientific library. The interest of Darwin and the members in agriculture and horticulture, and especially their interface with the sciences and medicine, are clearly demonstrated by the number of works on these subjects acquired between 1784 and 1815. Many of these were by Scottish authors, reflecting the significant contribution of the Scottish Enlightenment to agricultural improvement but also the medical education some members of the Philosophical Society had received north of the border. Agricultural books ordered on behalf of the society by Darwin and William Strutt, as successive presidents, included Arthur Young's *Farmer's Calendar* (1771), Jethro Tull's *New Horse-Hoeing Husbandry* (1731), James Anderson's *Essays Relating to Agriculture and Rural Affairs* (1797) and Archibald Cochrane, earl of Dundonald's *A Treatise [on] Agriculture and Chemistry* (1795).[35] The Derby Society took a close interest in the Board of Agriculture's activities, taking their *Communications* (1797–1808) and many of the county reports, as well as a series of works by the Yorkshire agriculturist William Marshall (1745–1818), such as his treatise *Planting and Rural Ornament* (1795) and studies of Norfolk (1783), Yorkshire (1788), the midlands (1790) and the West of England (1796). Horticultural works such as James Anderson's *Description of a Hot House* (1803), John Claudius Loudon's *Improvement in Hot-Houses* (1805), James MacPhail's *Treatise on the Culture of the Cucumber* (1795) and William Speechly's *A Treatise on the Culture of the Vine* (1796), and farming periodicals such as James Anderson's *Recreations in Agriculture* (1799–1802) and *Dickson's Agricultural Magazine*, were also acquired.[36]

THE WRITING AND PUBLICATION OF *PHYTOLOGIA*

Phytologia, a study of the 'philosophy of agriculture and gardening', was the last new work published by Darwin in his lifetime, and has been interpreted as a precursor of modern farming and botanical sciences, a summation of Enlightenment agriculture and horticulture and a response to wartime starvation fears and Continental exclusion.[37] It was primarily intended to be an application of natural philosophy, botany and medicine to horticulture and agriculture, informed by Darwin's contacts with farmers and omnivorous reading, which was partly obtained, as we have seen, through the Derby Philosophical Society's library. He also engaged closely with works such as Philip Miller's *Gardener's Dictionary* (1731) and various philosophical studies of vegetables, notably John Evelyn's *Sylva* (1664) and *Terra: a Philosophical Discourse of*

Earth (1674), Nehemiah Grew's *Anatomy of Plants* (1682) and Stephen Hales' *Vegetable Staticks* (1727), as well as with a body of literature and practical knowledge on farming improvements such as land drainage, enclosure, tree planting, soils and manures, animal and plant breeding and fruit cultivation, and British and international botanical works stimulated by imperial trade and rivalry. His experience of translating the works of Linnaeus into English with Boothby and Jackson provided a solid platform for this new work. Maureen McNeil has shown convincingly how the French wars, high food prices, riots, distress, dispossession and fears over food scarcity during the 1790s made concerns about farming improvement more urgent, stimulating the Board of Agriculture to act and providing the context for researching *Phytologia*, although she underplays the extent of Darwin's practical knowledge and experience of horticulture and agriculture.[38]

Darwin's approach was partly informed by Scottish Enlightenment philosophers such as Lord Kames and medical men with interests in agricultural chemistry such as James Hutton, the surgeon George Fordyce (1736–1802) and physicians Alexander Hunter (1729–1809), William Cullen and James Anderson (1738–1809).[39] Hutton, for example, whose work was partly inspired by his experiences as a farmer and his observations on British, Norfolk and European agriculture and included an unpublished two-volume treatise on 'Elements of Agriculture', shared Darwin's belief that application of the earth sciences, chemistry, natural history, botany and soil analysis would facilitate farming improvements, while the 'theory and practice of agriculture' ought to be taught in schools to young women as well as young men.[40]

While undertaking the Linnaean translations with the help of books and advice from Joseph Banks and others during the 1780s, Darwin also conducted studies of vegetable physiology when residing at Radburn Hall, Derbyshire, where he moved with Elizabeth in 1781. In March 1782 he emphasised to Banks, after reading Grew, Hales and Marcello Malpighi's studies of plant anatomy, that the 'physiology of plants' seemed to have received little attention in recent years from those sufficiently 'acquainted with the animal oeconomy'. The previous summer he observed the 'absorbent system' of the yellow-flowered Picris (*Helminthotheca echioides*, or bristly oxtongue) using a coloured dye, experiments that were later fully reported with other investigations undertaken at the time in the additional notes of *The Economy of Vegetation* and *Phytologia*.[41] Although *Phytologia* seems to have been completed in 1797, when Darwin had largely retired from medical practice, the longer gestation of his ideas is evident in his comments to Richard Lovell Edgeworth as early as April 1787 about a 'wonderful and useful and delightful' 'theory of gardening' that he was writing, and the numerous references to horticulture and agriculture in prior works such as *The Economy of Vegetation* (1791) and *Zoonomia*.[42]

As the period darkened internationally by descent into war and increased repression at home including the suspension of Habeas Corpus (1794) and passing of the Seditious Meetings and Treasonable and Seditious Practices acts (1795), Darwin lost various individuals he had known including Sir Robert Burdett, James Hutton, Lord George Cavendish and his mother at Sleaford.[43] The political unrest and international wars heightened concerns about feeding and supplying the nation. The immediate impetus for *Phytologia*, Darwin said, came from correspondence with John Sinclair, then president of the Board of Agriculture, between 1793 and 1798, to whom it was dedicated, and he praised the Board for their 'unremitted exertions', which had brought about 'many important improvements in the cultivation of the earth'. Writing to Sinclair in November 1797, Darwin described his work as a 'theory of vegetation, applied to agriculture and gardening' which concerned the 'cultivation of the earth, that great source of life and felicity', and stated that without Sinclair's 'instigation' he would 'not have attempted' the book.[44]

It was probably the problems that the Board of Agriculture experienced covering Derbyshire in their county reports that led to Darwin's correspondence with Sinclair and therefore the production of *Phytologia*. Thomas Brown's slim and vague *General View of the Agriculture of the County of Derby* (1794), which the Derby Philosophical Society did not even bother ordering for their library, was one of the worst county reports. Brown, who had almost no prior knowledge of – or connections with – the area, which he spent 'little time' investigating, admitted that he had been unable to discover much information and urged readers to send the Board additional observations in the margins of their copies so that these could be collected together for the county's 'improvement'.[45] One result of his report appears to have been the establishment of a more formal county agricultural association after Brown somewhat dismissively described an occasional livestock meeting among the local gentry as 'laudable' but little more than a 'family visit' which hardly constituted a society. The 'respectable and numerous' Derbyshire Agricultural Society, founded in 1794 to promote improvements in farming, such as sheep breeding, ran annual livestock shows during the Easter fair and in July, held competitions and awarded prizes from its inception.[46] Some individuals did respond to Brown's request for information and sent copies with comments to the author and Board. However, when John Farey's very detailed three-volume survey of Derbyshire agriculture appeared between 1811 and 1817 (which the Philosophical Society strongly supported and purchased for their library), the author felt the need to apologise on behalf of the Board of Agriculture to 'those gentlemen' who had taken 'the trouble' to do so, as none of the 'corrected copies' passed to Brown 'soon after their transmission' had ever been returned, despite 'repeated applications', which must have caused much irritation. Although not exclusively focused on Derbyshire, of course, in drawing many of its examples from the county, *Phytologia* – which Sinclair praised for its 'valuable' practical observations – therefore partly served as a study of its agriculture and horticulture.[47]

ENCLOSURE

Often regarded as a pre-requisite for agricultural improvement, the process of enclosure was believed to help increase the size, efficiency, productivity and therefore income of farms and estates, change land functions, define boundaries more clearly and turn ostensibly unproductive ground such as 'waste', forest or bog into useful land. Landowners, the Anglican Church, corporations, educational institutions, charities and other bodies generally saw enclosures as a means of making farming more efficient.

Leicester's Georgian corporation, for instance, owned 'considerable properties, especially in land' across the county, while the Common Hall of borough officials and burgesses included various 'fairly substantial' farmers, meaning that it had 'interests in common' with the county's farming gentry, perhaps impacting upon its policies on markets and support for 'improved communications'.[48] Enclosure schemes through parliamentary acts, which became the eighteenth-century norm, could be expensive and difficult, needing political support and co-operation between landowners who might disagree about individual schemes if their interests conflicted. Some enclosures were more difficult to undertake than others, including those that involved forest, which required the support of the crown, those involving many smaller landowners, and those centred upon commons, or land in which individuals or communities held certain rights.[49] But the many enclosure acts passed by parliament converted 'waste grounds' 'from barren heaths into fruitful fields', providing 'riches and support to industrious farmers' and their 'useful dependents'.[50]

Given the jealously guarded rights over common land for grazing livestock, water, firewood and other materials, as well as concerns about population dispossession, movement of villages and the large increases in rents favoured by Young and other agricultural improvers, it is not surprising that enclosures could generate serious hostility, although if any landowners objected to a parliamentary enclosure they had to present a counter-petition to amend or reject the bill.[51] In Derby and Nottingham, for example, there were major controversies concerning measures to enclose common lands surrounding the towns to pay for urban improvement measures during the 1780s and 1790s, while, as Andy Wood has shown, lead miners and other Peak inhabitants (men and women) fought against efforts by major landowners to restrict their rights, control activities and enclose lands during the seventeenth and eighteenth centuries.[52] Here, as Edward Thompson and others have demonstrated, the improving capitalist ideology clashed with an alternative moral economy with advocates in all social classes, which recognised the value of rights and long-established customs in relation to agriculture and commons, with magistrates intervening in times of scarcity to fix bread prices, government intervention through the corn laws to maintain corn prices and protests against the capitalist political economy and market values.[53]

Like Sinclair, Young and the Board of Agriculture, Darwin and his friends were so enthusiastic about large-scale agricultural improvement measures and enclosures that they believed these should be supported by government intervention if necessary. Like Jeremy Bentham (1748–1832) and the Utilitarians, Darwin believed that the right kinds of agricultural improvements would increase the health and happiness of the population and would thus be in the national interest. Unlike Thomas Malthus (1766–1834), in his *Essay on the Principle of Population* (1798), Darwin maintained that, just as the sick ailed rapidly if not offered appropriate treatments, societies might decline without effective government intervention to encourage individual farmers and landowners to increase productivity by improvement, which would lead to larger and ultimately happier populations.[54] He was a strong supporter of the campaign led by his friend William Strutt to enclose the Nun's Green common in Derby to pay for urban improvements, which resulted in an act of parliament to enable sales of land in parcels and the formation of an Improvement Commission (Figure 26). The plan provoked much hostility among those set to lose rights and privileges on the Green and others who wanted to retain it as a resource, and Darwin composed a broadside in favour of the measures, arguing that it would improve the local economy and have health benefits; he also briefly attended Improvement Commission meetings. However, his claim that tree-lined walks would be created proved to be false and the Brookside area became a run-down part of Derby.

Matthew Boulton's efforts to promote enclosures were also supported by Darwin. Like the Wedgwoods, Strutts and Arkwrights, Boulton was 'intimately connected with land ownership', 'clearly' aspired to landed gentry status and cultivated relationships with other landowners to achieve his business objectives. He used land for equity and supported enclosure acts in Staffordshire at Handsworth (1791), Cheslyn Hay

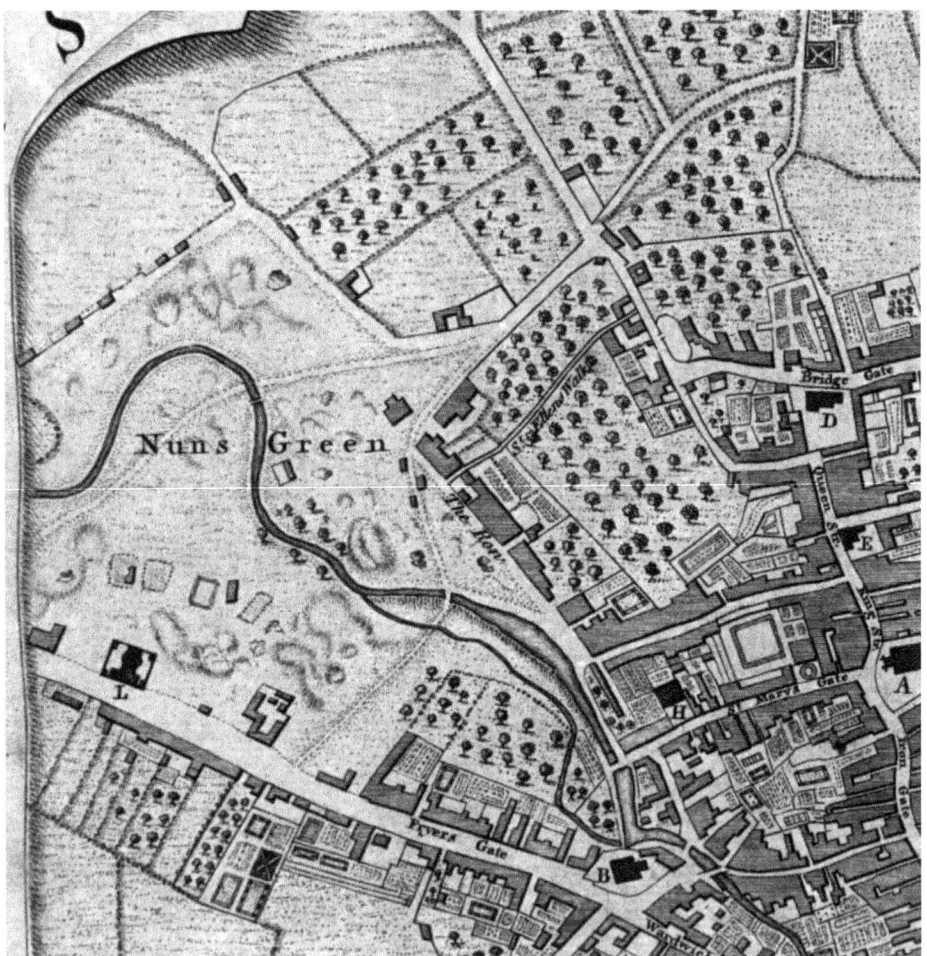

Figure 26.   Nun's Green, Derby, from P. P. Burdett, *Map of Derbyshire* (1767), inset.

(1792), Needwood Forest (1801) and Barton under Needwood (1812), although in the first and third cases he believed that these would promote greater self-sufficiency in British food production and good moral order too.[55] When the Needwood Forest enclosure was opposed by estate managers of three of the largest and most influential Staffordshire landowners, Charles Talbot, fifteenth earl of Shrewsbury (1753–1827), George Venables-Vernon, second Baron Vernon (1735–1813) and William Bagot, second Baron Bagot (1773–1856), Darwin supported his Lunar friend, who chaired the first enclosure campaign committee meeting in 1800 and promoted the measure in local newspapers. Boulton believed the Needwood enclosure and the appropriation of productive wastes (the haunts of 'poachers and deer-stealers') would help feed the country without reliance upon grain imports and reduce the national balance of payments, and lobbied ministers for crown support, which was required for forest enclosure. Here, as Brown has argued, although a 'progressive in many ways', Boulton was closely allied to 'sympathetic members of the Aristocracy'.[56] Still grieving after the death of his son Erasmus, Darwin wrote to Boulton in February 1800 noting that a meeting had taken place in Barton to consider the Needwood Forest enclosure, which he knew his friend was 'interested' in, and asking him to encourage the proponents of the enclosure to consider taking on his son's former partner Nathaniel Edwards (1768–1814) as solicitor, who employed another of his sons, Edward Darwin (1782–1829) as a clerk.[57] In November 1800 Darwin wrote again to Boulton from their mutual friend Samuel Galton's Birmingham house, asking on behalf of his friend John Gisborne of Holly-Bush in Needwood Forest that Boulton give up his allotment in the Forest 'if that forest comes to be enclosed (which I hope it will) if you mean to dispose of it'.[58]

The enclosure of land was not the only aspect of improvement that required close engagement among landowners with differing interests: careful negotiation was necessary over water rights along watercourses and other resources in relation to power for water wheels, and about turnpike and canal improvements, all of which could require enclosures and parliamentary acts. Indeed, such enclosures might also benefit agriculture; when Darwin, Thomas Bentley and Josiah Wedgwood composed a pamphlet in support of a new 'Grand Trunk' canal connecting the Mersey with the Trent in 1765, one of their arguments was that the new waterway network would help increase agricultural productivity and thereby the health, wealth and happiness of the population, and would provide new employment in distributing goods and raw materials, throwing up and distributing marle for manure, spreading foodstuffs, and boosting national and international trade by opening up new markets.[59]

HEALTH BENEFITS OF ENCLOSURE AND IMPROVEMENT

Darwin argued in *Phytologia*, as a doctor as well as a farmer, that major agricultural improvements such as large-scale land drainage in coastal areas were 'noble' rather than 'individual' public works which justified government intervention to 'increase the quantity of nutriment', the population and the 'happiness of the country'. He called for engineers and labourers to be employed by the government to dig ditches for land drainage in boggy districts below sea level such as the Lincolnshire and Cambridgeshire fens. However, Darwin believed that the wartime situation made major enclosures such as those of 'forests and commons' more difficult because of clashing private interests (such as those around Needwood) and the difficulties of obtaining 'expensive' parliamentary acts 'for every minute district', arguing that a combined 'general act', as 'so meritoriously' championed by Sinclair and the Board of Agriculture, would be more effective.[60] The enclosure and division of commons into 'private property' was nearly always justified because it increased the 'quantity of sustenance to mankind' by 'tenfold' if used to grow crops or 'even' as pasturage rather than being left in an 'uncultivated state' with only ferns, heath and gorse to nourish 'a few geese, sheep, or deer'. Enclosing pastures or meadows made them better suited for grazing and fattening cattle, dairy herds, sheep or horses and for growing their foodstuffs, as well as 'garden vegetables and fruit', hemp, flax, madder, woad, rhubarb, turnips, potatoes, carrots, cabbages and other perennials, 'esculent roots or herbage'. Darwin thought that a dispassionate and carefully considered 'political' decision needed to be taken on whether the general enclosure of arable lands was better for the nation's 'prosperity', based upon which types of cultivation produced the most 'nutritive provision'. He believed that pasturage would tend to prevail over general agriculture in all 'enclosed provinces' because meat and dairy products were 'articles of great luxury' compared with corn. However, arable farming could support ten times the population that 'animal food' production did and therefore make family life more sustainable. Enclosing arable land for animals reduced the population, resulted in fewer marriages and caused 'numbers of labouring people' to be 'much diminished' by food scarcity and unemployment, a circumstance that Darwin chose to reinforce with a passage from Oliver Goldsmith's poem *The Deserted Village* (1770), the most famous Georgian poetical condemnation of enclosure evils, which would have been very familiar to his readers:

> Worse fares the land, to hastening ills a prey,
> Where wealth accumulates, but men decay;
> Princes or lords may flourish, or may fade,
> A breath can make them, as a breath has made;
> But a bold peasantry, their country's sword,
> When once destroy'd, can never be restor'd.

Darwin therefore argued that enclosures of arable lands, and even the enclosure of 'those parts of commons' that were 'best adapted to the growth of corn', should be prevented, although division of territory generally into 'private property' remained advantageous.

Darwin was concerned that for about half a century the country had required large annual corn imports to feed its growing population despite the widespread introduction of potatoes and a decline in the 'ungraceful fashion' of powdering wigs with wheat flower. He warned against the domination of meat-producing pasturage instead of arable agriculture, which would lead to ever more 'flesh' rather than grain consumption and therefore population diminution. As a doctor – and unfashionably in a time of war and nationalism – he believed that arable farming would improve the health and vitality of both the home population and 'foreign nations', who would benefit from the renewal of British corn exports and the consequent reduction in the 'dreadful' calamity of famine. Nations dependent upon agriculture were more 'active', 'robust' and 'ingenious' than those reliant upon pasturage, such as the Egyptians, who, he claimed, had declined historically because they were less effective at inventing and using machines or cultivating soil.[61]

In Darwin's mind, the benefits of increased vegetable consumption resulting from the right kind of agricultural improvement were reinforced by the deleterious consequences of over-eating red meat, which he associated with overindulgence in alcohol, an argument that – like his concerns for the health of foreign peoples – was hardly likely to endear him to the average ale-quaffing and roast-beef-devouring Georgian John Bull figure. Despite or perhaps because of the drinking he had engaged in when young, and with the moral seriousness of age and experience, he exhorted people of all classes to abstain with the fervour of a firebrand preacher. Discouraging the 'luxurious intemperance' of 'flesh-meat' consumption and the drinking of 'intoxicating liquors' would, Darwin considered, help preserve arable farming and render the British kingdoms 'more populous, robust, prosperous, and happy than any other nation in the world'. Favouring arable agriculture over pasturage by enclosure for corn production discouraged the wastage of good produce for fermented drinks, and the nation might be prevented from 'becoming too carnivorous' by having meat abstention periods like the 'religious fast days' of the past or by levying government 'bounties' or taxation on corn exports. Darwin's strategy was also underpinned by important medical and physiological factors. Like pigs, humans were 'designed to subsist' on 'due portions' of 'both vegetable and animal nutriment' by 'nature' as the length of their intestines – longer than those of carnivorous animals like cats yet considerably shorter than those of herbivores – and the structure of their teeth, with similarities to both 'carnivorous and phytivorous' creatures, demonstrated. Those who lived 'solely on vegetables', like the 'Gentoo tribes' (Hindus), and those whose diet was dominated by flesh, like the 'fish-eaters of the northern latitudes', were 'feebler' than omnivorous British counterparts.[62]

Decades of medical practice had provided Darwin with ample opportunity to observe the deleterious effects of 'spirituous potations' upon all social classes, from gout-ridden squires to beer-swilling labourers and their families, and, in an age when almost everyone partook, many were paid in drink and small beer often substituted for water, the growing vehemence of his hostility stood out. Much of the commercial, social, political and cultural life in the midland towns that Darwin knew best was of course focused upon ale houses, taverns and inns, which provided accommodation for travellers and venues for business transactions, entertainment, charities, clubs and societies; the main annual meeting of the Derby Philosophical Society, for example, was conducted in a public house (Figure 27).[63] For Darwin, wine and spirits were products 'of art', of a 'nice chemical process' and not a 'natural food', but a 'medicine' that should be found only in apothecaries' shops as formerly. This was because the process of converting the sugars by fermentation turned them from 'wholesome nourishment' into 'poison' and the 'greatest curse of the Christian world', which was 'wisely interdicted by the disciples of Mahomet and Confucius'.[64] Certainly, beside the use of drink to dull pain in surgery, until at least the eighteenth century, there was a strong belief in the medical efficacy of beer and spirits as supports for natural healing processes and they were prescribed by healers, medical practitioners and in hospitals. The medical uses of *aqua vitae*, or spirit of wine 'rectified to the utmost degree', were long recognised. Irish whiskey was added to the *Pharmacopia Londonensis* in 1677, although spirit of wine, or what Andreas Libavius (1555–1616) called 'alkohol', was added only in 1788, being altered to 'alchohol' in 1809. Despite William Hogarth's depiction of the horrors of 'Gin Lane', compared with the delights of 'Beer Street', in 1751, distilled 'juniper water', or gin, was thought to be efficacious for stomach problems, gallstones, gout and various bodily organs.[65]

Darwin argued that a total prohibition on spirits or 'strong ale' was needed to prevent it from thinning 'the inferior orders ... by scarcity of food' and the higher classes by 'disease both of mind and body'. This would therefore doubly benefit health and agriculture by preventing the waste of productive land and increasing the health and longevity of the population.[66] Darwin himself never seems to have given up drinking entirely, but he did reduce his intake as he got older, probably because of his own experiences as well as those of patients. Anna Seward described how, on first coming to Lichfield, her friend already recognised the 'pernicious effects' of 'vinous fluid' upon even young and healthy constitutions, and had an 'absolute horror' of all spirits even when diluted, abstaining from 'strong malt liquor' and drinking wine only if diluted. He also criticised those who believed foreign wines to be 'more wholesome' that their English equivalents, saying that 'if you must drink wine ... let it be homemade'.[67]

On one occasion as a young man, according to Seward, Darwin drank more than usual when travelling along the Trent from Burton (already famed for its ales) to Newark on a warm summer's day with friends, which may have helped convince him of the evils of strong beverages. When the party neared Nottingham 'in not [an] absolutely intoxicated' state then certainly in a 'high state of vinous exhilaration', Darwin suddenly got into the water, swam across to the other bank and walked into town through the meadows, and was later to be heard, despite entreaties from a local apothecary, giving a speech on a tub 'encircled by a crowd' in which he extolled the benefits of fresh air while admitting that he'd drunk too much wine. Darwin claimed that his audience well knew his normal sober state as a 'professional man', but that an unusual degree of 'internal stimulus' had, through 'its effects upon [his] system', counteracted the air's comparative coldness on wet skin, enabling him to ignore it in his sodden state.[68]

Darwin's experiences suffering from gout, which he described in *Zoonomia*, reinforced his hostility to drink. He classified it as a disease of association centred upon the liver but with outward sympathetic manifestations in joints which moved and were subject to wide temperature changes.[69] Gout had first struck him around 1770, when he was approaching forty. His large toe had become 'much swelled and inflamed' despite bleeding, cold water and other treatments, and it was only when following the recommendation of Thomas Sydenham (1624–1689) and other physicians

Figure 27. G. Woodward, 'A convivial meeting at Nottingham', engraved by G. Cruikshank (1797) from G. Woodward, *Eccentric Excursions or Literary and Pictorial Sketches* (1817).

to stop drinking alcohol, with the exception of small beer and watered-down wine, and later abstaining from 'all fermented liquors' that he had relief. After a recurrence of the problem some fifteen years later when drawn again to spirituous liquors and beer, he returned to drinking water, which ended his trio of tribulations – the gout, piles and the gravel.[70] Seward claimed that his influence and example had 'sobered the county of Derby', while the writer Maria Edgeworth claimed he had 'almost banished wine from the tables of the rich' and 'persuaded most of the gentry' in Derbyshire and 'surrounding counties' to become 'water drinkers'. According to Edgeworth, he blamed 'almost all the distempers of the higher classes' and associated ailments of gout, jaundice and 'all bilious and liver complaints' upon drinking, views that were supported by all the 'force of his eloquence and reason' in print and his verbal 'powers of wit, satire, and peculiar humour', which gave him a 'strong ascendancy in private society'.[71]

By the 1780s even eminent aristocratic patients received a Darwinian scolding for drinking and were ordered to cut down with the zealousness of a convert, and, according to Seward and Maria Edgeworth, Darwin became renowned – or perhaps notorious – for his hostility to drink. In 1783 he told William Cavendish, the fifth duke of Devonshire (1748–1811), that over-indulgence in fermented drink 'by all ranks' exerted a 'specific influence' upon the liver, leading to 'inflammations of the internal and more vital parts' being removed to 'eternal parts' by a 'wonderful process of the animal machine' as yet not fully understood. Sometimes these inflammations were 'removed upon the joints', causing gout, or emerged as red 'pustules', as the duke had experienced, which he thought helped relieve the liver as more 'essential to life'. While there were various treatments, including lead solution applied externally, the liver would remain damaged, the risk of gout persisted and the only way to 'prolong' health was to 'obstinately' persevere in reducing consumption of 'spirituous or fermented liquor' by half on a daily basis while taking laxatives and exercise such as garden walks (presumably not together). It was 'prudent' to begin by a measured reduction, although the 'greatest good' arose only from 'complete omission'. Overeating was bad enough, as it induced 'corpulency' and 'inactivity', but inebriation was a 'deleterious' constitutional hazard which he and his fellow medical practitioners witnessed 'daily'. Darwin concluded his surprisingly forthright missive to one of the most powerful aristocrats in the country by accentuating how intoxicated individuals lost 'their understanding', becoming 'an idiot', collapsing on the floor and experiencing 'temporary paralysis' and 'stupidity in mind'; he emphasised that he expressed himself so 'strongly' because health in this was not to be 'bought at any other price'.[72]

CONCLUSION

This chapter has argued that Erasmus Darwin sought to help forge a new science of farming that would support the kind of improvements advocated by Young, Sinclair and the Board of Agriculture. A landowner and small-scale farmer himself who hailed from a landed family,, he was concerned to naturalise established power relations as well as employ his medical and scientific knowledge to increase agricultural productivity. His gardens at Lichfield and Derby and the smallholding and orchard at the latter served as laboratories and microcosms of the natural economy in action, nurturing ideas about agricultural improvements that germinated in his publications. He had an excellent practical knowledge of agriculture and gardening from his own experiences and those of family, friends and patients of all ranks, and emphasised the wild, recalcitrant resistance of a nature in which humans were just a comparatively recent part. Like his farming friend the geologist James Hutton, he wanted to 'make philosophers of husbandmen and husbandmen of philosophers'.[73]

Given his landed gentry family background and close friendship with professionals, industrialists and manufacturers who acquired land and estates to cement their new status, Darwin recognised that there was considerable disagreement among those in his circle about how the latest ideas and techniques might be applied. He came to know a great deal about agriculture and horticulture primarily through his medical practice but also though his travels around the midland countryside, talking to agricultural labourers, mole-catchers, farmers, mechanics, innkeepers, gentry, aristocrats and others with knowledge and experience of the subject, and he recognised that, like sometimes resistant patients, working with, rather than against them was crucial to agricultural improvement. He witnessed numerous changes taking place across rural England between the 1750s and 1790s, including enclosures, estate improvements, farm size increases, applications of new crop rotation systems, land reclamation and the applications of new techniques and technologies, and saw the resistance these sometimes generated.

NOTES

1   R. E. Prothero (Lord Ernle), *English Farming, Past and Present*, 6th edn with introductions by G. E. Fussell and O. R. McGregor (London, 1961), 216–17; E. J. Russell, *A History of Agricultural Science in Great Britain, 1620–1954* (London, 1966), 67–76, 107–9.

2   R. Schofield, *The Lunar Society of Birmingham: A Social History of Provincial Science and Industry in Eighteenth-century England* (London, 1963), 397–40; M. McNeil, *Under the Banner of Science: Erasmus Darwin and his Age* (Manchester, 1987), 168–203.

3   R. Williams, *The Country and the City* (Oxford, 1975), 115–16.

4   McNeil, *Under the Banner of Science*, 193–4.

5   R. Weston, *Tracts on Practical Agriculture and Gardening*, 2nd edn (London, 1773), ii–xiii, 289.

6   A. M. Duckworth, *The Improvement of the Estate: A Study of Jane Austin's Novels* (Baltimore, 1971), 47, ix, quoted in S. Wilmot, 'The Business of Improvement': Agriculture and Scientific Culture in Britain, c1700–1870* (London, 1990), 39; J. C. D. Clark, *English Society, 1660–1832: Religion, Ideology and Politics during the Ancien Regime*, 2nd edn (Cambridge, 2000).

7   F. Bacon, *Sylva Sylvarum* and *New Atlantis* in F. Bacon, *The Works of Francis Bacon*, 10 vols (London, 1824); vol. 1, 391–448 and vol. 2, 113–14; J. Steven Watson, *The Reign of George III* (Oxford, 1987), 36–66, 256–84, 360–76; H. T. Dickinson, *Liberty and Property: Political Ideology in Eighteenth-Century Britain* (London, 1979), 195–318.

8   A. and N. Clow, *The Chemical Revolution* (London, 1952), 456–502; Russell, *History of Agricultural Science*; J. D. Chambers and G. E. Mingay, *The Agricultural Revolution* (London, 1966), 34–53; Wilmot, *'The Business of Improvement'*, 8–21; J. V. Beckett, *The Agricultural Revolution* (Oxford, 1990); S. Schaffer, 'The Earth's fertility as a social fact in early modern Britain', in M. Teich, R. Porter and B. Gustafsson eds, *Nature and Society in Historical Context* (Cambridge, 1997), 124–47; T. Williamson, *The Transformation of Rural England: Farming and the Landscape, 1700–1870* (Exeter, 2002); S. Wade Martins, *Farmers, Landlords and Landscapes: Rural Britain, 1720 to 1870* (Macclesfield, 2004); P. M. Jones, *Agricultural Enlightenment: Knowledge, Technology and Nature, 1750–1840* (Oxford, 2016).

9   Schaffer, 'Earth's fertility', 124–47, 125, 133–6.

10 C. J. Withers, 'On georgics and geology: James Hutton's "Elements of Agriculture" and agricultural science in eighteenth-century Scotland', *Agricultural History Review*, 42 (1994), 38–48.

11 A. Young, *The Farmer's Tour through the East of England*, 4 vols (London, 1771), vol. 1, 158–9; Schaffer, 'Earth's fertility', 127.

12 H. Home, Lord Kames, *The Gentleman Farmer* (Edinburgh, 1776), 283–6; Chambers and Mingay, *Agricultural Revolution*; R. Longrigg, *The English Squire and his Sport* (London, 1977), 99–166; Beckett, *Agricultural Revolution*; Williamson, *Transformation of Rural England*; Wade Martins, *Farmers, Landlords and Landscapes*.

[13] Clow and Clow, *Chemical Revolution*, 471, 501–2, 582–615; S. Bending, 'The improvement of Arthur Young: agricultural technology and the production of landscape in eighteenth-century England', in D. E. Nye ed., *Technologies of Landscape: From Reaping to Re-Cycling* (Boston MA, 1999), 241–53.

[14] R. Bakewell, *Observations on the Influence of Soil and Climate upon Wool* (London, 1808); Clow and Clow, *Chemical Revolution*, 582–615; D. Hudson and K. W. Luckhurst, *The Royal Society of Arts, 1754–1954* (London, 1954); H. C. Pawson, *Robert Bakewell: Pioneer Livestock Breeder* (London, 1957); K. Hudson, *Patriotism with Profit: British Agricultural Societies in the Eighteenth and Nineteenth Centuries* (London, 1972), 1–36; G. Averley, 'English Scientific Societies of the Eighteenth and Early-Nineteenth Centuries', unpublished PhD thesis, University of Teeside Polytechnic and Durham University (Durham, 1989), 329–81; Wilmot, *'The Business of Improvement'*; S. Wade Martins, *Coke of Norfolk (1754–1842): A Biography* (Woodbridge, 2009).

[15] W. Marshall, *The Rural Economy of the Midland Counties*, 2nd edn (London,1796), 2, 3, 7; H. Arnold Bemrose, 'Geology' and W. R. Linton, 'Botany', in W. Page ed., *The Victoria County History of Derbyshire*, vol. 1 (London, 1905), 1–34, 39–50; K. C. Edwards, *The Peak District* (Glasgow, 1973), 81–134.

[16] Prothero, *English Farming*, 190–206; G. E. Fussell, 'Four centuries of farming systems in Derbyshire: 1500–1900', *Derbyshire Archaeological Journal*, 71 (1951), 1–37.

[17] Young, *Farmer's Tour*, 194–56.

[18] Young, *Farmer's Tour*, 201–4.

[19] Young, *Farmer's Tour*, 159–60, 173–6, 177–83,199–200, 206–7, 222–3, 227–8; Wade Martins, *Coke of Norfolk*, 9–10.

[20] Young, *Farmer's Tour*, 204–15, 216–18, 228–9; on the culture of sainfoin, Anonymous, *The Improved Culture of Three Principle Grasses, Lucerne, Sainfoin and Burnet* (London, 1775), 195–262; A. Young, *The Farmer's Calendar*, 6th edn (London, 1805), 203–4; Prothero, *English Farming*, 219; Beckett, *Agricultural Revolution*, 11–19.

[21] Young, *Farmer's Tour*, 177–8, 170–2.

[22] Young, *Farmer's Tour*, 190–3, 202–3; J. Pilkington, *A View of the Present State of Derbyshire*, 2 vols (Derby, 1789), vol. 2, 119–34; J. Britton and E. W. Brayley, *The Beauties of England and Wales*, vol. 3, Cumberland, Isle of Man and Derbyshire (London, 1802), 410–18; E. Rhodes, *Peak Scenery* (London, 1824), 4–5; C. Hartwell, N. Pevsner and E. Williamson, *The Buildings of England: Derbyshire* (New Haven, 2016), 469–77; D. Barre, *Historic Gardens and Parks of Derbyshire* (Oxford, 2017), 84–90.

[23] Young, *Farmer's Tour*, 160–1, 218–20; Pilkington, *Derbyshire*, vol. 2, 431–52; Britton and Brayley, *Derbyshire*, 488–93; Rhodes, *Peak Scenery*, 151–71, 260–1; J. Barnatt and T. Williamson, *Chatsworth: A Landscape History* (Macclesfield, 2005), 91–127; J. Barnatt and N. Bannister, *Chatsworth and Beyond: The Archaeology of a Great Estate* (Oxford, 2009), 67–96; Hartwell, Pevsner and Williamson, *Derbyshire*, 232–50.

[24] E. Darwin, letter to J. Wedgwood 8 March 1788, in D. King-Hele ed., *The Collected Letters of Erasmus Darwin* (Cambridge, 2007), 311–12.

[25] J. B. Firth, *Highways and Byways in Nottinghamshire* (London, 1916), 122–3; N. Pevsner, *The Buildings of England: Nottinghamshire*, 2nd edn revised by E. Williamson (Harmondsworth, 1979), 121–2; P. Worsley, *The Darwin Farms: The Lincolnshire Estates of Charles and Erasmus Darwin and their Family* (Lichfield, 2017), 40–4, 48, 65–81.

[26] E. Darwin, letter to R. Darwin, 8 June 1798 in King-Hele ed., *Collected Letters*, 519–20.

[27] D. King-Hele, *Erasmus Darwin: A Life of Unequalled Achievement* (London, 1999), 125–8, 173–5.

[28] J. Barlow and P. Elliott, 'A brush with the doctor: Samuel James Arnold's account of painting Erasmus Darwin's portrait', *Notes and Records of the Royal Society of London*, forthcoming.

[29] N. Everett, *The Tory View of Landscape* (New Haven, 1994).

30 S. Mason, *The Hardwareman's Daughter: Matthew Boulton and his 'Dear Girl'* (Chichester, 2005); D. Brown, 'Matthew Boulton, enclosure and landed society', in M. Dick ed., *Matthew Boulton: A Revolutionary Player* (Studley, 2009), 45–62.

31 Britton and Brayley, *Derbyshire*, 521–3; Rhodes, *Peak Scenery*, 257–8; R. S. Fitton, *The Arkwrights: Spinners of Fortune* (Manchester, 1989), 273–4; Barre, *Historic Gardens and Parks of Derbyshire*, 106–8.

32 Farey, *General View*, vol. 2, 219–340, vol. 3, 655.

33 J. Gascoigne, *Joseph Banks and the English Enlightenment: Useful Knowledge and Polite Culture* (Cambridge, 1994); H. Carter, *Sir Joseph Banks, 1743–1820* (London, 1988); R. Drayton, *Nature's Government: Science, Imperial Britain and the 'Improvement' of the World* (New Haven, 2000), 94–106; P. J. Naylor, *A History of the Matlocks* (Ashbourne, 2003), 148.

34 P. Elliott, 'Medicine, scientific culture and urban improvement in early nineteenth-century England: the politics of the Derbyshire General Infirmary', in J. Reinarz ed., 'Medicine and the Midlands, 1750–1950', special issue of *Midland History* (2007), 27–46.

35 Derby Local Studies Library, Derby Philosophical Society, manuscript catalogue and charging ledger, 1785–9, BA 106, 9229; Derby Philosophical Society, *Rules and Catalogue*, with supplements of 1795 and 1798; Derby Philosophical Society, *Rules and Catalogue*; P. A. Elliott, *The Derby Philosophers: Science and Culture in British Urban Society, 1700–1850* (Manchester, 2009); the society acquired J. Tull, *Horse-Hoeing Husbandry; or an Essay on the Principles of Vegetation and Tillage* (London, 1733); G. Fordyce, *Elements of Agriculture and Vegetation* (London, 1771); A. Young, *The Farmer's Calendar* (London, 1771); W. Lamport, *Cursory Remarks on the Importance of Agriculture* (London, 1784); A. Cochrane, Earl of Dundonald, *A Treatise [on] Agriculture and Chemistry* (London, 1795); J. Anderson, *Essays Relating to Agriculture and Rural Affairs* (London, 1797); R. Parkinson, *The Experienced Farmer*, 2 vols (London, 1798); R. W. Dickson, *Practical Agriculture; or Complete System of Modern Husbandry* (London, 1805); Cochrane was a corresponding member of the Derby Literary and Philosophical Society (1808).

36 Derby Local Studies Library, Derby Philosophical Society, manuscript catalogue and charging ledger; Derby Philosophical Society, *Rules and Catalogue*, with supplements of 1795 and 1798; W. Marshall, *The Rural Economy of Norfolk*, 2 vols (London, 1783); W. Marshall, *The Rural Economy of Yorkshire*, 2 vols (London, 1788); Marshall, *The Midland Counties*; J. Billingsley, *General View of the Agriculture of the County of Somerset*, 2 vols (London, 1794); W. Marshall, *On Planting and Rural Ornament: a Practical Treatise*, 2 vols (London, 1795); J. McPhail, *Treatise on the Culture of the Cucumber*, 2nd edn (London, 1795); W. Speechly, *A Treatise on the Culture of the Vine* (London, 1796); W. Marshall, *The Rural Economy of the West of England*, 2 vols (London, 1796); The Board of Agriculture and Internal Improvement, *Communications … on Subjects Relative to the Husbandry and Internal Improvement of the Country* (London, 1797–1808); J. Anderson, *Description of a Hot House* (London, 1803); J. C. Loudon, *Improvement in Hot-Houses* (Edinburgh, 1805); Elliott, *Derby Philosophers*; on William Marshall, see Prothero, *English Farming*, 196–7.

37 E. Darwin, *Phytologia; or the Philosophy of Agriculture and Gardening* (London, 1800); Russell, *History of Agricultural Science*, 62–3; Schofield, *Lunar Society*, 397–401; McNeil, *Under the Banner of Science*, 168–203; P. Ayres, *The Aliveness of Plants: The Darwins at the Dawn of Plant Science* (London, 2008), 31–54.

38 McNeil, *Under the Banner of Science*, 169, 178–88.

39 J. E. Handley, *The Agricultural Revolution in Scotland* (Glasgow, 1963), 14, 83, 104, 173–4; McNeil, *Under the Banner of Science*, 168–78; C. W. J. Withers, 'William Cullen's agricultural lectures and writings and the development of agricultural science in eighteenth-century Scotland', *Agricultural History Review*, 37 (1989), 145–7.

[40]   E. Darwin, *Plan for the Conduct of Female Education* (Derby, 1797), 40–2, 44–5, 125; J. Jones, 'James Hutton's agricultural research and his life as a farmer', *Annals of Science*, 42 (1985), 573–601; Withers, 'On georgics and geology', 38–48.

[41]   E. Darwin, letter to J. Banks, 17 March 1782 in King-Hele ed., *Collected Letters*, 202–3; Darwin, additional notes on 'Vegetable circulation' and 'Vegetable respiration', in E. Darwin, *The Economy of Vegetation*, 4th edn (London, 1799), 441–57; Lichfield Botanical Society, *The Families of Plants ... Translated from the Last Edition of the 'Genera Plantarum'*, 2 vols (Lichfield, 1787), vol. 2, 529; Darwin, *Phytologia*, 9–17; King-Hele, *Erasmus Darwin*, 179–80.

[42]   E. Darwin, letter to R. L. Edgeworth, 22 April 1787, in King-Hele ed., *Collected Letters*, 277–80.

[43]   J. Steven Watson, *The Reign of George III, 1760–1815* (London, 1960), 360–1; King-Hele, *Erasmus Darwin*, 309.

[44]   Darwin, *Phytologia*, dedication; E. Darwin, letter to Sir J. Sinclair, 8 November 1797 in King-Hele ed., *Collected Letters*, 516–17 (with somewhat different wording from the dedication); R. Mitchison, *Agricultural Sir John: The Life of Sir John Sinclair of Ulbster, 1754–1835* (London, 1962), 137–74.

[45]   T. Brown, *General View of the Agriculture of the County of Derby* (London, 1794), preface and 48.

[46]   Brown, *General View*, 48; J. Farey, *General View of the Agriculture and Minerals of Derbyshire*, 3 vols (London, 1811–17), vol. 3, 109, 129–33, 649.

[47]   Farey, *General View*, vol. 2, preface, v–vi; E. Darwin, letter to J. Sinclair, 8 November 1797, in King-Hele ed., *Collected Letters*, 517.

[48]   R. W. Greaves, *The Corporation of Leicester, 1689–1836*, 2nd edn (Leicester, 1970), 77.

[49]   Chambers and Mingay, *Agricultural Revolution*, 77–105; Williams, *The Country and City*, 65–7, 96–107; J. A. Yelling, *Common Field and Enclosure in England, 1450–1850* (London, 1977); M. Turner, *English Parliamentary Enclosure: its Historical Geography and Economic History* (Folkestone, 1980); Williamson, *Transformation of Rural England*, 7–15, 125–36; Wade Martins, *Farmers, Landlords and Landscapes*, 1–56; Jones, *Agricultural Enlightenment*, 133–41.

[50]   [T. Bentley, E. Darwin and J. Wedgwood] *A View of the Advantages of Inland Navigations with a Plan of a Navigable Canal* (London, 1765), 25.

[51]   Turner, *English Parliamentary Enclosure*, 152–3.

[52]   Elliott, *Derby Philosophers*, 112–31; P. Elliott, 'The politics of urban improvement in Georgian Nottingham: the enclosure dispute of the 1780s', *Transactions of the Thoroton Society*, 110 (2006), 87–102; A. Wood, *The Politics of Social Conflict: The Peak Country, 1520–1770* (Cambridge, 1999), 66–71, 254–66, 318–19.

[53]   E. P. Thompson, 'Time, work discipline and industrial capitalism', in *Customs in Common* (Harmondsworth, 1991), 352–403; J. Goldstein, '*Terra economica*: waste and the production of enclosed nature', *Antipode*, 45 (2013), 357–75.

[54]   Darwin, *Phytologia*, 245; McNeil, *Under the Banner of Science*, 109–12.

[55]   Mason, *The Hardwareman's Daughter*; Brown, 'Matthew Boulton', 45–62.

[56]   Brown, 'Matthew Boulton', 57–8; W. Page and N. J. Tringham, *Victoria County History of Staffordshire: Vol. 10, Tutbury and Needwood Forest* (London, 2007).

[57]   E. Darwin, letter to M. Boulton, 23 February 1800, in King-Hele ed., *Collected Letters*, 540–1.

[58]   E. Darwin, letter to M. Boulton, 9 November 1800, in King-Hele ed., *Collected Letters*, 554.

[59]   [E. Darwin, T. Bentley and J. Wedgwood] *A View of the Advantages of Inland Navigations with a Plan of a Navigable Canal* (London, 1765), 4, 6–7.

[60]   Darwin, *Phytologia*, 245.

[61]   Darwin, *Phytologia*, 424–7; Williams, *Country and the City*, 64–9.

[62]   Darwin, *Phytologia*, 426–7.

[63]   R. Porter, *England in the Eighteenth Century* (London, 1998), 22–3, 88–9, 201–2; P. Clark, *The English Alehouse: A Social History, 1200–1830* (London, 1983); P. Clark, *British Clubs and Societies*

1580–1800: The Origins of an Associational World (Oxford, 2000); C. Ludington, *The Politics of Wine in Britain: A New Cultural History* (Basingstoke, 2013).

[64] E. Darwin, letter to W. Cavendish, fifth duke of Devonshire, 20 November 1783 in King-Hele ed., *Collected Letters*, 218–22; the brewing and distilling industries did, however, support the agricultural economy in other ways, as their spent grains served as an important winter foodstuff for pigs and cattle; see P. Mathias, 'Agriculture and the brewing and distilling industries', in E. L. Jones ed., *Agriculture and Economic Growth in England, 1650–1815* (London, 1967), 80–93.

[65] J. Grier, *A History of Pharmacy* (London, 1937), 192–3; I. Gately, *Drink: A Cultural History of Alcohol* (New York, 2008).

[66] Darwin, *Zoonomia*, vol. 4, pp. 209–10; King-Hele, *Erasmus Darwin*, 47–8, 199–200, 271–2, 289–90; C. Gardner-Thorpe, 'William Paley (1743–1805) and James Parkinson (1755–1824): Two peri-Erasmatic thinkers (and several others)', in C. U. M. Smith and R. Arnott eds, *The Genius of Erasmus Darwin* (Aldershot, 2005), 68–9.

[67] A. Seward, *Memoirs of the Life of Dr. Darwin* (London, 1804); C. Darwin, *Life of Erasmus Darwin*, edited by D. King-Hele (Cambridge, 2003), 85.

[68] Seward, *Memoir of Dr. Darwin*, 65–6.

[69] Darwin, *Zoonomia*, vol. 4, 205–21; R. Porter and G. S. Rousseau, *Gout: The Patrician Malady* (New Haven, 1998).

[70] Darwin, *Zoonomia*, vol. 4, 209–10.

[71] Seward, *Memoir of Dr. Darwin*; R. L. Edgeworth and M. Edgeworth, *Memoirs of Richard Lovell Edgeworth, Esq. begun by Himself and Concluded by his Daughter, Maria Edgeworth*, 2 vols (London, 1820), vol. 2, 82–3; Darwin, *Life of Erasmus Darwin*, 85.

[72] E. Darwin, letter to W. Cavendish, fifth duke of Devonshire, 20 November 1783, in King-Hele ed., *Collected Letters*, 218–22. Darwin was a friend of the duke's uncle Lord George Cavendish since their schooldays in Chesterfield, and so knew the family.

[73] Withers, 'On georgics and geology', 40–41.

VEGETABLE PHYSIOLOGY, TECHNOLOGY AND AGRICULTURE

As well as supporting agricultural improvement, Darwin believed that there were specific ways in which the sciences could be applied to increase productivity in farming. This belief was founded upon practical observations of midlands agriculture and industry, gardening and horticultural experiments and the reading of major studies on these and related subjects, such as vegetable physiology. *Phytologia*, Darwin's study of agriculture and gardening, was divided into three parts which dealt firstly with the physiology of vegetation, secondly with the economy of vegetation and lastly with agriculture and horticulture, while an appendix contained details for an 'improved construction of the drill plough' he had made with a new design for a seed-box by Thomas Swanwick (1755–1814). The first part, on vegetable physiology, examined what Darwin referred to as the buds, absorbent and umbilical vessels, pulmonary arteries and veins, aortal arteries and veins, glands and secretions, organs of reproduction and muscles, nerves and brain of vegetables. The language utilised underscored his repeatedly asserted belief in analogies between animal and vegetable bodies, and therefore the relevance of his medical knowledge and experience for the study of agriculture and horticulture. In the introduction Darwin made this relationship explicit, declaring that his book was a 'supplement' to *Zoonomia* (which had itself been partly modelled on the Linnaean system) because it was 'properly a continuation of the subject'. Some of the material had already appeared in the latter, while the production of *Phytologia* aided him in preparing the much expanded third edition of *Zoonomia* (1801).

The second part of *Phytologia*, on the 'economy of vegetation', developed from observations made in *The Economy of Vegetation* (1791) that hinged upon what Maureen McNeil has called Darwin's notion of 'interrelated' or 'interlocking' economies of vegetable, human and natural worlds, which was similar to the emphasis upon the 'interconnectivity of life systems' advocated by his friend Joseph Priestley (1733–1804) in his experiments on airs. He noted how vegetables replenished air and water and supported animal life; and, like animals, plants were distinctive organic entities with their own laws of motion which needed to be understood holistically in relation to their 'total operations' rather than in isolation. This section focused on the growth of seeds, buds and bulbs, 'manures or the food of plants', draining and watering lands, 'aeration and pulverization of the soil', the importance of light, heat and electricity for vegetable growth and the diseases of plants. The third part of *Phytologia* considered how farming productivity could be increased in relation to the growth of fruits, seeds, flowers, roots, barks, leaves and wood and concluded with the plan for 'disposing' part of the Linnaean system 'into more natural classes and orders', which we examined in chapter 2.[1]

In his introduction to the book Darwin claimed that, although 'agriculture and gardening' were of 'such great utility' in producing mankind's 'nutriment', their potential remained unfulfilled because they were still largely regarded as 'arts' composed of 'numerous detached facts and vague opinions' unconnected by theory – which was surprising, given that other bodies of knowledge 'of much inferior consequence' were already 'nicely arranged and digested' into systems. Lack of attention to vegetable 'physiology and economy' was partly to blame for the 'immaturity' of this science, which would be remedied by a new 'theory of vegetation' founded upon the work of plant anatomists and physiologists – the 'Linnaean school' (who had tended to concentrate on taxonomy at the expense of vegetable physiology and anatomy) and his own experiments and observations.[2]

After moving to Radburn in 1781 Darwin went to some effort to obtain continental botanical works on vegetable physiology and anatomy to supplement those of British investigators such as Nehemiah Grew (1641–1712) and Stephen Hales (1677–1761), including Henri-Louis Duhamel du Monceau's *La Physique des Arbres, où il est traité de l'anatomie des plantes* (1758) and Charles Bonnet's *Recherches sur l'usage des feuilles dans les plantes* (1754), asking Josiah Wedgwood whether he could get these in London or, if not, Paris. As he observed to Joseph Banks, as a practising doctor his knowledge of the 'animal economy' provided insights into the vegetable economy unavailable to earlier researchers, as his observations on the 'absorbent system' and white 'blood' of the bristly oxtongue using a 'colour'd liquid' had made 'beautifully apparent' in 1781.[3] Later, in 1796, Sinclair encouraged Darwin to think that the current state of knowledge on plant growth and 'modern improvements in chemistry' were now 'sufficiently numerous and accurate' that a 'true theory of vegetation' could be established, revealing the most important elements, offering points of comparison and providing a template for 'future experiments'.[4]

This chapter examines four of the main ways that Darwin believed farming would benefit from scientific and technological innovations and the development of an agricultural science. The first involved employing various techniques to increase the quantity and quality of production, such as experimenting with different kinds of crop to find the most fruitful; the second involved applications of industrial technology and methods in farming; the third sought to exploit meteorological knowledge to benefit agriculture; and the fourth focused upon how improved knowledge of vegetable anatomy, physiology, nutrition and the application of chemical science might increase crop yields and reduce plant diseases, preventing food shortages.

OPTIMISING CROP YIELDS

One of Darwin's main objectives in advocating a new system of agriculture founded upon science and technology was to increase food production and thereby the bodily health and contentment of the populace. The price of corn and basic foodstuffs rose markedly between 1760 and 1800, a problem exacerbated by warfare and a series of bad harvests during the 1790s which paternalistic systems of local philanthropy and price fixing by magistrates could only partially alleviate. As more land was devoted to dairy and grazing and less to corn production, the country became an importer rather than an exporter of corn. There was a consequent rise in social discontent, and fears were expressed by Malthus in his *Essay on the Principle of Population* (1798) that if the population exceeded the rate of food production the results could only be checked by preventive governmental measures; the alternative outcomes were famine or disaster. Enclosure and improvements in farming techniques were seen as providing one possible solution, particularly if allotments for the poor were to be set aside too.[5] Informed by his reading, his observations of growing food plants such as corn and fruit trees and his examination of vegetable circulation, placentation, respiration and glandulation, Darwin offered recommendations on increasing productivity. Following methods employed by Charles Miller (1739–1817), first curator of the Cambridge Botanic Garden, he tried to improve upon Jethro Tull's horse-hoeing husbandry methods of growing wheat by demonstrating that the most productive growth occurred from cutting or 'decapitation', transplanting and burying many plants, using examples removed from a field to his Derby garden which produced 'prodigious multiplication' of 'perfect' corn ears (Figure 28).[6] He sought to determine which forms of wheat were most productive, testing an 'Egyptian' or 'Smyrna' plant obtained from Liverpool in his garden in 1799 which was recommended by Tull and Linnaeus and seemed to be particularly vigorous despite unfavourable growing conditions.[7]

Darwin used similar methods to investigate the growth of potato stems, which produced new roots even if they were shorn of young ones and transplanted, and observed that some rapidly growing grasses, bindweeds, vines, fig trees on walls (even when covered) and coltsfoot, which he grew in his garden and employed in medical treatment, behaved in a similar fashion, producing multiple shoots even from 'very minute parts of the jointed root' when cut and planted, a quality that could be problematic for gardeners and farmers but which also might be beneficially harnessed. Fruit trees in his orchard, such as the 'burr-apple', projected 'round protuberances or excrescences' similar to burrs from bark, and on branches bent or torn off and 'set in moist earth' these struck out roots so that the branch became a separate tree. Cutting circular rings of bark from trees difficult to propagate and placing earth around the branch above and below the 'wounded part' using a garden pot placed around the branch produced roots from the 'upper lip of the wound' that could be successfully planted. Cutting 'a few inches' from the end of wall trees in spring encouraged buds to grow 'near the extremity' and drove those remaining to grow with 'greater vigour',

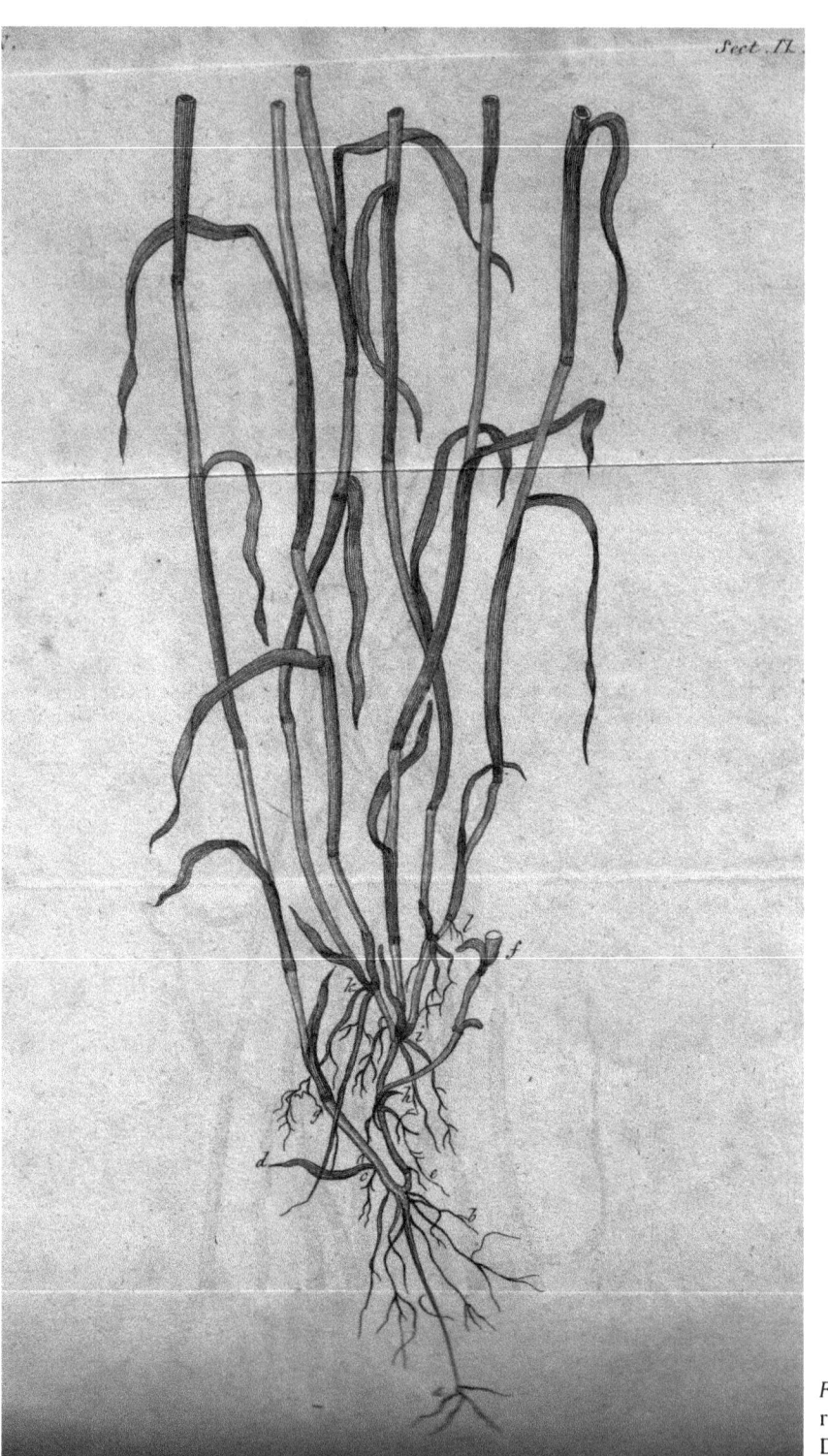

Figure 28. Transplant
root of wheat, from E
Darwin, *Phytologia* (18

as occurred when branches were cut from trees that produced 'numerous or more rigorous' 'suckers or root-scions', such as elm or apple, which Darwin attributed to the greater 'nourishment' obtained.[8]

NEW VEGETABLE FOODSTUFFS

Many of the medicinal plants and vegetable products recommended by Darwin were also consumed, as has been noted, or had useful culinary applications such as flavouring dishes, and he believed that scientific botany and agriculture could play an important role in the discovery of new food sources for the population. He was concerned at the 'present insane state of human society', as wars and their waging, especially in the period after the French Revolution sucked the 'ingenuity and labour of almost all nations', leading to destruction and enslavement with 'as little mercy as they destroy and enslave the bestial world'.[9] Despite the Royal Navy's successes, there were concerns that food shortages might occur, which, combined with poor harvests, could lead to widespread famine, and one of Darwin's major objectives for the new brand of scientific farming delineated in *Phytologia* was to not only find methods of increasing the production of familiar staples but also to drive the search for others, taking advantage of trees and plants already growing in Britain or those that might be easily introduced. Darwin's work here paralleled that of the contemporary Norwich botanist Charles Bryant (d. 1799), Beadle to the Court of Guardians, who, with assistance from Hugh Rose (1717–1792), a local apothecary, sought to arrange and identify esculent plants according to the Linnaean system rather than by the 'barbarous' local names often used by gardeners and nurserymen. Bryant and Rose also provided information concerning plants' properties, varieties and places of growth, with suggestions, based upon private experiments, for additional edible vegetables to improve 'human health and vigour' for all 'palates and pockets'. Darwin purchased Bryant's *Flora Dietica* (1783) for the Derby Philosophical Society library, which informed his work and was one of those works 'completely circulated' thro' the members' before being deposited in their rooms.[10]

It seemed to be a good bet that close relatives – or similar kinds – of plants already bred for consumption might be adapted by breeding to provide alternative staple and nutritious foodstuffs. Although ground artichoke (*Helianthus tuberosus*) 'seldom' ripened in the British climate, new varieties might be produced that fared better, while 'Pig-nut' or black cumin (*Bunium bulbocastanum*) could by cultivation from seed produce 'an agreeable and salutary' food for consumption similar to 'raw or roasted' chestnuts.[11] Other bulb-producing plants might, by 'pinching off the flowers', produce enlarged new bulbs, as was done with potatoes, which could then be eaten. Other bulbous roots were also perhaps 'worthy of cultivation', such as the green-winged orchid, *Orchis morio*, which might be prepared for the table by boiling and drying to produce a 'nourishing mucilage'. Even the roots of common snowdrops (*Galanthus nivalis*) might provide a 'nutritious' food like that of the green-winged orchid if they were dug up in winter and boiled. Darwin once cooked some of them and discovered 'on tasting them' that they had 'no disagreeable flavour'. Likewise, hyacinth roots proved to be edible if boiled, although Darwin's culinary experiments with saffron or autumn crocus (*Crocus sativus*) roots ended quickly when he found them to have a 'disagreeable taste' (Figure 29).[12]

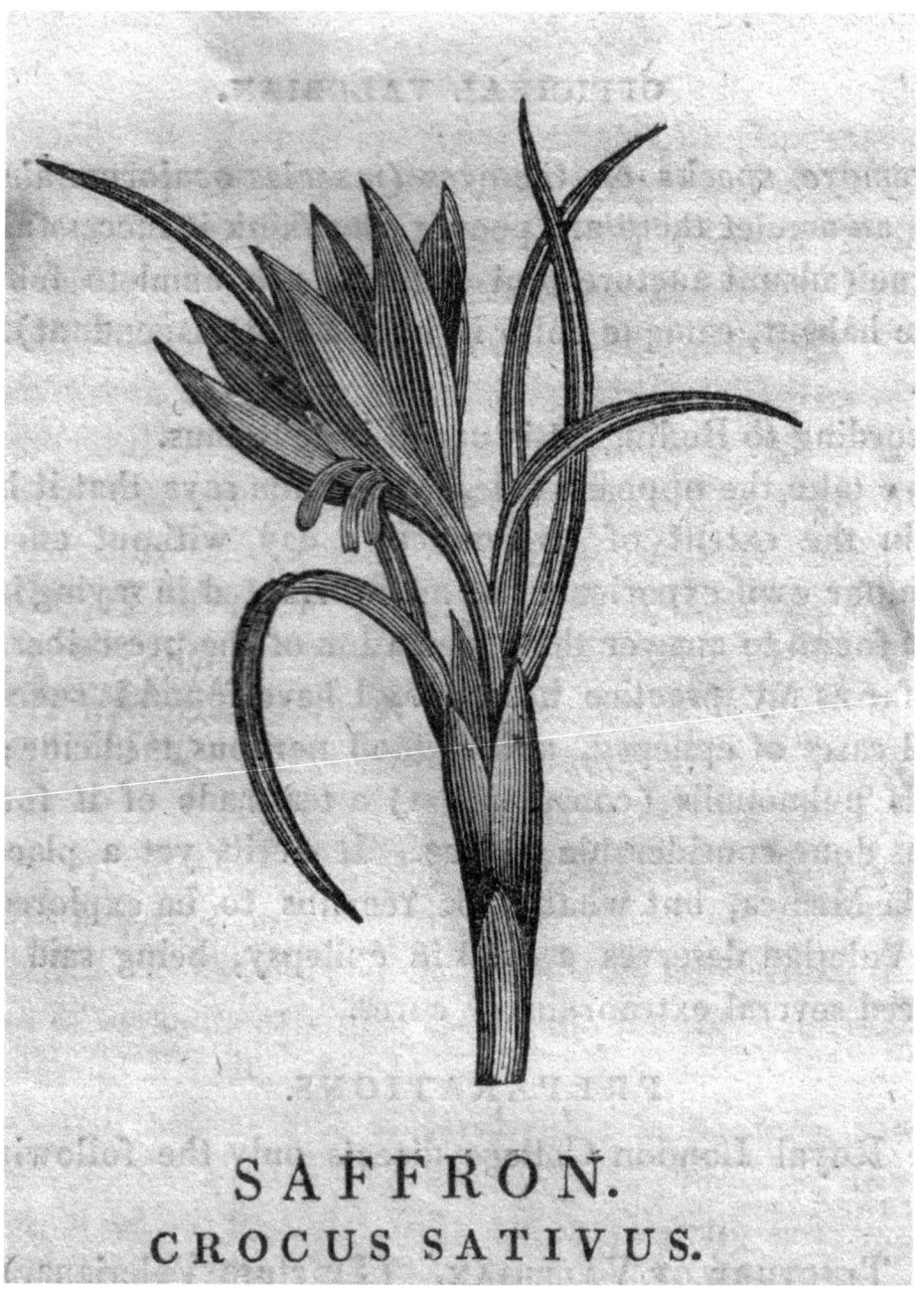

SAFFRON.

CROCUS SATIVUS.

Figure 29. Saffron, *Crocus sativus*, from R. J. Thornton, *A Family Herbal*, 2nd edn (1814).

Just as plants with close analogies to familiar foods might suggest other possibilities, so observations of consumption practices in other cultures or even among other animals looked promising to Darwin, and he noted that Johann Georg Gmelin (1709–1755), in his account of his Siberian journeys (1752), showed how the roots of the Martagon lily was consumed in that country. Though 'acrid' when raw, arum root (*Arum maculatum*) was another plant that could be made a 'palatable and salutary nutriment by cookery', as it lost its 'acrimony' on exposure to air; as Gilbert White stated in his *History of Selbourne*, it was 'scratched up and eaten by thrushes in severe snowy seasons'. Darwin suggested that the root of the perennial herb branched asphodel (*Asphodelus ramosus*), commonly fed to pigs in France, could be another tasty and nutritious human food if properly prepared, and that 'good starch' was equally available from the roots of white bryony (*Bryonia dioica*) – a climbing plant of hedges and woodland – and Peruvian lily (*Alstroemeria*). Aquatic plants known to be consumed in some cultures, such as the white- and yellow-flowered water lilies, might be encouraged to grow in all 'ditches and rivers' for eating, as these were places that were at present rather wasted because they produced 'no esculent vegetables'.[13] Some plants, such as ginger, did not lose their 'acrimony' after boiling, while others, such as those with 'alkalescent' properties (tending to alkaline), which were liable to putrefaction, such as watercress, cabbage and onions, lost their bitter taste on boiling, so Darwin tried tasting boiled leaves from the native common spotted arum (probably *Arum maculatum*) and the herbaceous perennial friar's cowl (*Arisarum vulgare*), from the Mediterranean, but found that they left his 'tongue and lips almost excoriated', probably because of some 'fixed essential oil'.[14]

The limited range of fungi commonly eaten by humans suggested that others might likewise provide delicious or fortifying sustenance, especially as Darwin believed them to be physiologically and chemically closer to animals than plants and perhaps even to be 'animals without locomotion'. Fungi had 'lacteal vessels' that were inserted into the ground with 'gills or lungs' hidden from light, like animals, but which were 'exposed to the open air' like leaves or vegetable 'lungs'. Given these similarities, which meant that some fungi were already in 'common use' at dinner tables and were known to offer 'wholesome and nutritious food', he argued that many other fungi too might therefore provide nutritious meals 'agreeable to the palate'. Of course, caution needed to be exercised because of the known 'intoxicating qualities' of some fungi, as Darwin's consumption of the 'yellow juice' of a raw saffron milkcap or red pine mushroom (*Lactarius deliciosus*) demonstrated – it blistered his tongue and turned it blue – but he felt that with boiling to destroy the 'acrimony' this and other types such as *Peziza auricula* or '[jelly] ear fungus', which had previously been used as a medicine, might form part of the human diet.[15]

Darwin's attempts to find new nutritious plants for human and animal consumption are a superb example of his efforts to create a more productive agriculture making more efficient use of the landscape, including under-valued places such as ditches and watercourses, as well as vegetables that were already often growing naturally or semi-naturally in abundance in the British climate. Though, given the cultures around Georgian eating, his suggestions might easily have led to ridicule and even hostility, they brought together his medical experience of human and animal bodies with nutrition, bio-botany and social concerns.

TECHNOLOGY, AGRICULTURE AND HORTICULTURE

One of the most important manifestations of Enlightenment horticulture was the development of hothouses, which offered winter shelter for delicate varieties. These clearly demonstrated the impact of Georgian industrialisation, changes in gardening taste, collecting and methods of display, and the importation of plants from warmer climates. Glasshouses had a major impact upon gardening, although they remained largely the preserve of middling sort and gentry because of the window tax, which was substantially increased by William Pitt's administration from the mid-1780s.[16] Darwin's hothouse dominated his Derby garden, as is clear from contemporary maps, and fortunately he provided a very full description and sketch in a letter to Richard Lovell Edgeworth in 1788. Responding to enquiries that Edgeworth had made concerning hothouse construction, Darwin explained that it was eighty-two feet long and about nine feet wide, with a glass roof of panes eight inches square and brick walls. He emphasised that the hothouse was divided into two so that 'one half may be a month forwarder than the other'. It produced an 'abundance of kidney beans, cucumbers, melons and grapes', but not pineapples, and there were pots of flowers between the vines growing up the twelve-foot rear wall. At the front, stretching from one end to the other, was a bark bed three and a half feet wide for melons, fronted by an interior brick wall two and a half feet tall with flues, behind which was a wooden walk constructed from 'old barrel staves' to protect vines' roots from injury. The hothouse was heated by two fireplaces on the back wall and slow burning 'athanor' stoves for four months only, which consumed about six tonnes of coal. The heat was circulated around the building using a series of flues entering two feet below the ground and passing through conduits in the interior wall to the east and west extremities, passing under the doorstep at one end before returning from each side to the central chimneys on the rear wall.[17]

Darwin's Derby hothouse serves as a demonstration of the impact of mechanics and industrialisation upon Georgian horticulture in various ways, including the construction of the glass roof and the design of the heating system, which provided a highly controlled environment. The attention that he gave to the subject, induced by the move to Radburn and Derby, is evident from other gardening inventions described in his commonplace book, including a 'melonometer or brazen gardener' designed to open and close windows in hotbeds or hothouses according to atmospheric conditions. The device incorporated two four-inch copper globes, one containing hydrogen gas and the other mercury, with a vacuum existing in the top half of each globe; these were joined underneath by a lengthy horizontal tube which was pivoted near the mercury sphere. In cool conditions the hydrogen sphere won the see-saw and kept the window closed, while in sunny weather the hydrogen expanded, pushing the mercury up the tube into its globe, overbalancing the see-saw and opening the window.[18] Darwin did not attempt to introduce such a device in his Derby hothouse but did incorporate oblique sashes rather than perpendicular ones.

The development of eighteenth-century air heating systems paralleled – and was reciprocally stimulated by – the evolution of air heating systems in textile manufactories, including the Derby and Macclesfield silk manufactories and the

Arkwright and Strutt cotton mills of the Derwent valley. Inspired by these, several of Darwin's friends, such as William Strutt and the clockmaker, mechanic and geologist John Whitehurst (1713–1788), produced versions for domestic buildings and public institutions such as hospitals. Whitehurst designed chimneys and hothouses and oversaw the installation of an air heating system in St Thomas's Hospital, London; likewise, Strutt designed domestic stoves and an air heating system for the Derbyshire General Infirmary. Although Darwin's hothouse cost a hundred pounds to construct because of the expense incurred by the window tax and he had to order the glass panes from Stourbridge, he declared to Edgeworth and his son Robert that he did not regret the outlay. The cost and effort expended by Darwin on obtaining the best glass from Stourbridge, the regional centre of the industry, is significant. The town had become integrated into the midlands canal network during the 1770s, which facilitated both the import of raw materials for the glass production process, such as coals, potash and lead, and the export of glass products across England. Darwin's Lunar friend James Keir (1735–1820) had moved to Stourbridge in 1770 and become a partner in a glass works, turning an old glasshouse to the rear into a laboratory and striving to both improve the industry experimentally and utilise its processes and materials for his chemical work, which included improving the production of lithage (red lead) and translating Pierre Macquer's *Dictionarie de Chimie*. It seems likely that Keir would have offered advice, or possibly negotiated the Stourbridge sale for Darwin, who refers to him in the 1788 letter to Edgeworth in which he described the hothouse. Much of the glass trade was devoted to ornamental ware and jewellery, but great attention was also given to producing larger, stronger, clearer and cheaper glass panes with immediate application to domestic, industrial and horticultural buildings where the need to maximise light and heat was paramount.[19]

Other mechanical improvements designed to increase agricultural productivity highlighted in *Phytologia* including windmills, ploughs and surveying machines, some being devised by Darwin and friends such as his Derby associates Thomas Swanwick, Major John Trowel (1744–1821), John Chatterton senior and John Chatterton junior (1771–1857).[20] The windmills included a device for 'raising water a few feet high' to drain morasses or water lands on a higher level; developed from the 1760s and first constructed full scale around 1780, the design's original purpose was to operate a flint mill for grinding colours for Wedgwood and Bentley, which was utilised at their pottery works at Etruria, Staffordshire. Darwin's windmill had a windmill sail placed horizontally 'like that of a smoak-jack' and surrounded by an octagonal tower with 'oblique horizontal boards' around at an angle of about 45 degrees which directed the wind upwards from whatever direction it came so that it would strike 'against the horizontal wind-sail' (Figure 30). The horizontal boards could be either fixed or allowed to pivot upon an axis just below their centres of gravity, thereby closing themselves on the opposite side of the 'octagon tower' from the wind. While it might be assumed that the wind would diminish before striking the wind-sail, Darwin found by experimentation, with advice and assistance from Lunar friends including Edgeworth, that the mill appeared not to 'lose power', despite the wind being 'reverted upwards' by the 'fixed plane board'. The windmill powered a 'centrifugal pump' with an 'upright

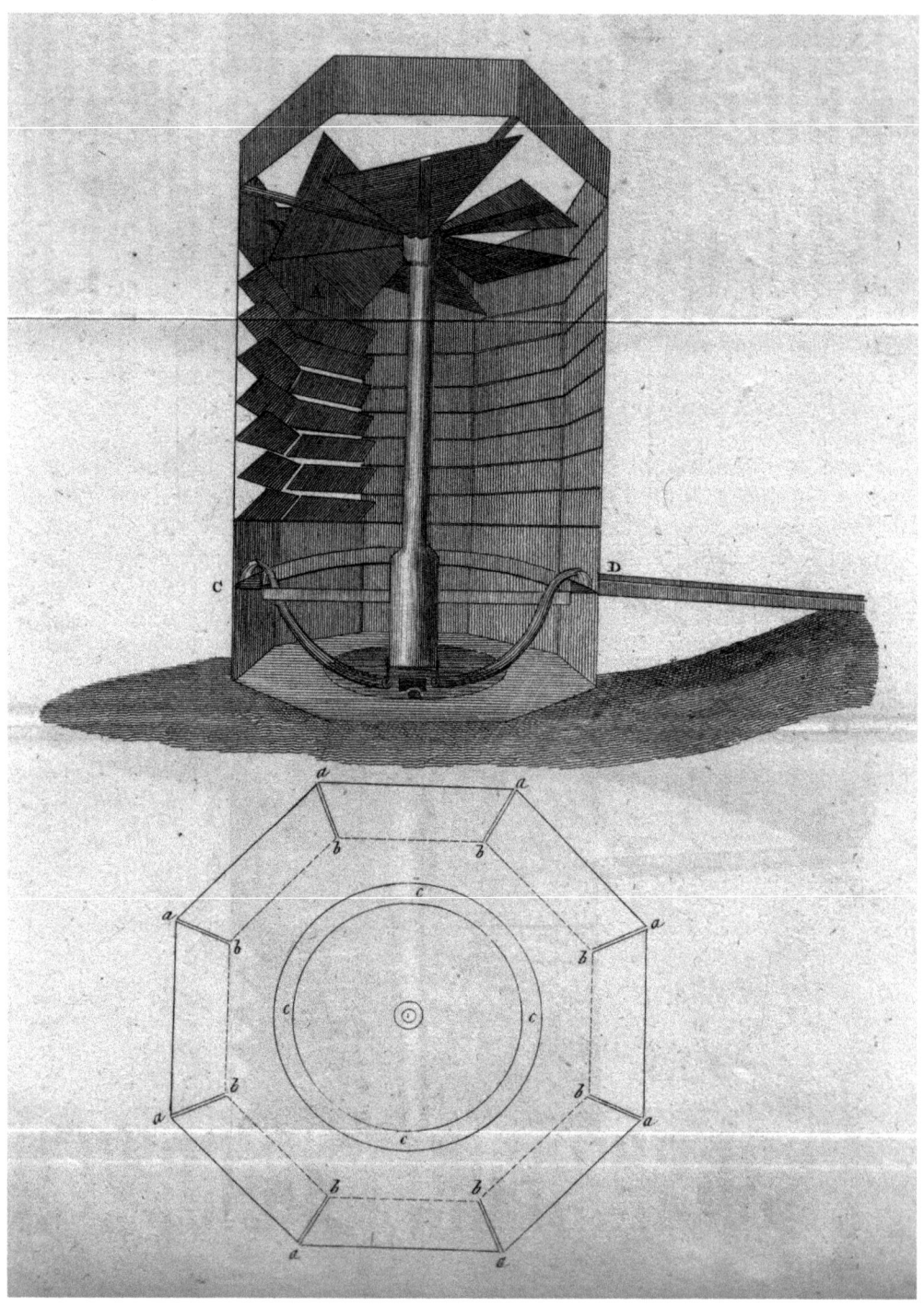

Figure 30. E. Darwin, design for a windmill for raising water from morasses, from E. Darwin, *Phytologia* (1800).

bored trunk or cylinder of lead' with 'two opposite arms' and an 'adapted valve at the bottom' to prevent returning water. Unlike the 'common windmill' used for grinding corn, this design required fewer moving parts and employed straightforward gearing, providing ample scope to significantly increase pump power by simply scaling up the size of the tower and wind-sail.[21]

The aptly named Trowel was a major in the Derbyshire militia, a member of the Derby Philosophical Society and a firm supporter of the Nun's Green enclosure campaign (mentioned above), who grew a new kind of potato praised in *Phytologia*, while the Chattertons were lead merchants, mechanics and glaziers who developed new forms of manure; John Chatterton senior promoted the use of waste ground for growing crops during the Napoleonic Wars.[22] It seems likely that Darwin also encouraged the chemist and mechanic Henry Browne's innovations and, in *Phytologia*, he praised Browne's 'ingenious paper' on composting using layers of decomposing vegetables and lime published in the *Transactions of the Society of Arts* (1798); Browne, who served as mayor of Derby in 1799 and 1808, designed a method for preserving seeds from destruction by vermin, a technique for making manure and a boiler which he described as an 'evaporator' for more effectively drying substances such as malt, and was presented with a gold medal by the Society of Arts.[23]

Darwin recognised the 'great advantages' of Jethro Tull's 'ingenious' horse-driven drill husbandry, which mechanised the processes of sowing crops such as wheat in rows at regular distances and standard depth, helping to aerate the soil and create a fine tilth (Figure 31). Darwin was personally assured by the celebrated 'Coke of Norfolk' that in all his experience working a farm of 3,000 acres, 'drill husbandry' was 'greatly superior' to any other method of sowing seeds. However, Tull's drill plough was too 'complicated' and did not deliver grain accurately enough and the comparatively slow process of hand sowing or dibbling remained popular. Another reason why Tull's system had not yet been sufficiently adopted was because the latest 'improved' and patented version of his drill plough, designed by the Rev. James Cooke (c. 1743–1817), which Coke used on his Norfolk farms, was too expensive for ordinary farmers to afford, although it cut the amount of wheat required by half compared with 'broadcast sewing'. Darwin therefore designed a drill plough on a 'cheaper' and 'simpler' plan than Cooke's machine by enlarging the section of the axle tree that delivered the grain into a specially adapted cylinder so it more effectively emitted the 'proper quantity' of seed from the hopper without bruising and loss, which he detailed in *Phytologia* with engravings. Alongside this he included an illustrated description of a seed-box designed by Swanwick, a mechanic, surveyor and 'very ingenious philosopher' who kept a Derby school for 'writing, arithmetic, and some branches of natural philosophy', compiled meteorological tables and invented a regulator or governor for mills powered by water. Swanwick's machine simplified the process of delivering grain from the seed box using a bar or slider and set of stiff brushes rather than Tull's revolving axis, with six cells to sow six rows of seeds at a time (Figures 32–34). Darwin also wrote to Samuel More of the Society of Arts encouraging them to award Swanwick a grant for his 'time and expenses', suggesting that Swanwick was prepared to travel and to provide working models and any assistance to others needed to promote the innovation.[24]

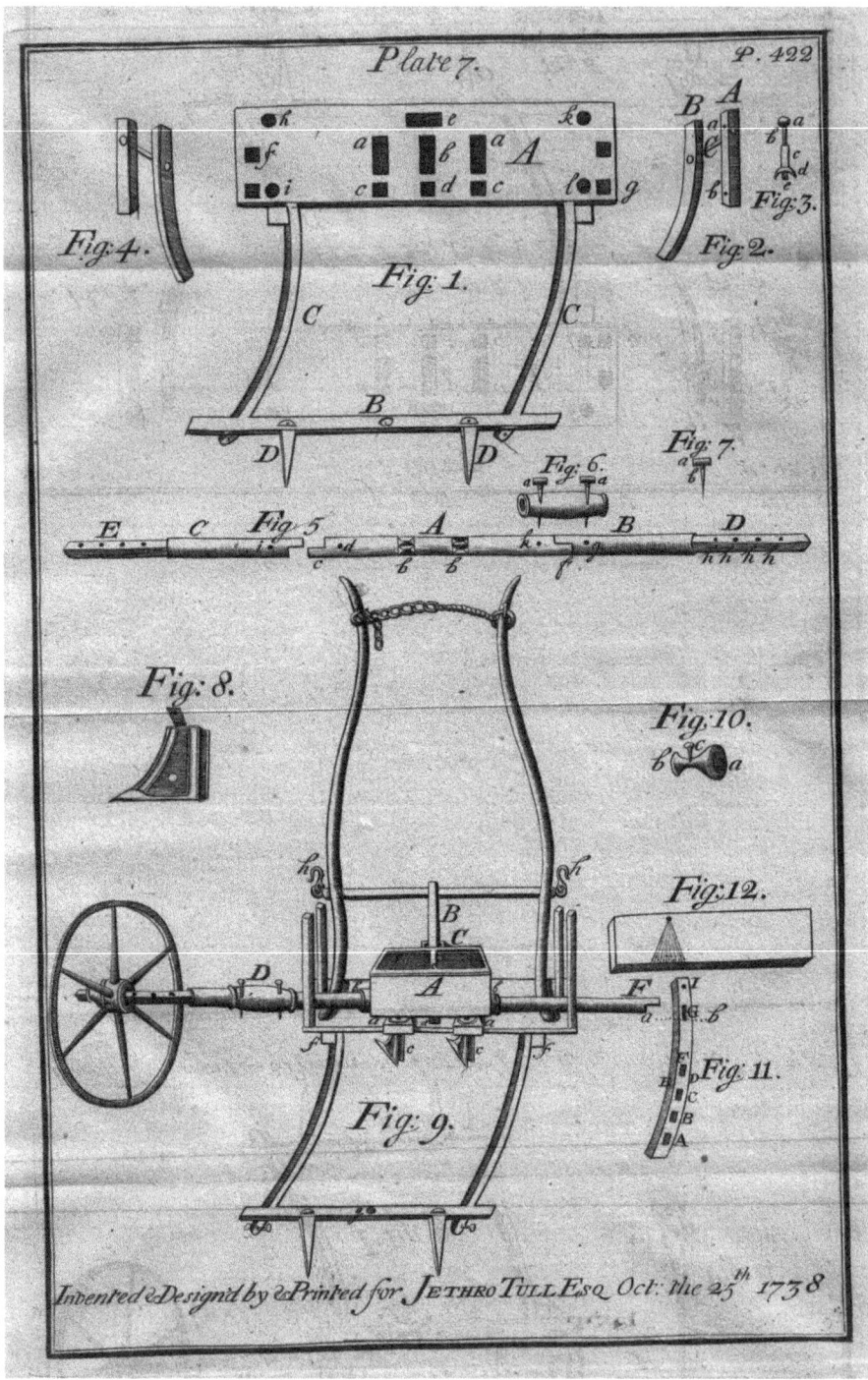

Figure 31. J. Tull, plough, plate VII, from J. Tull, *Horse-Hoeing Husbandry*, 4th edn (1762).

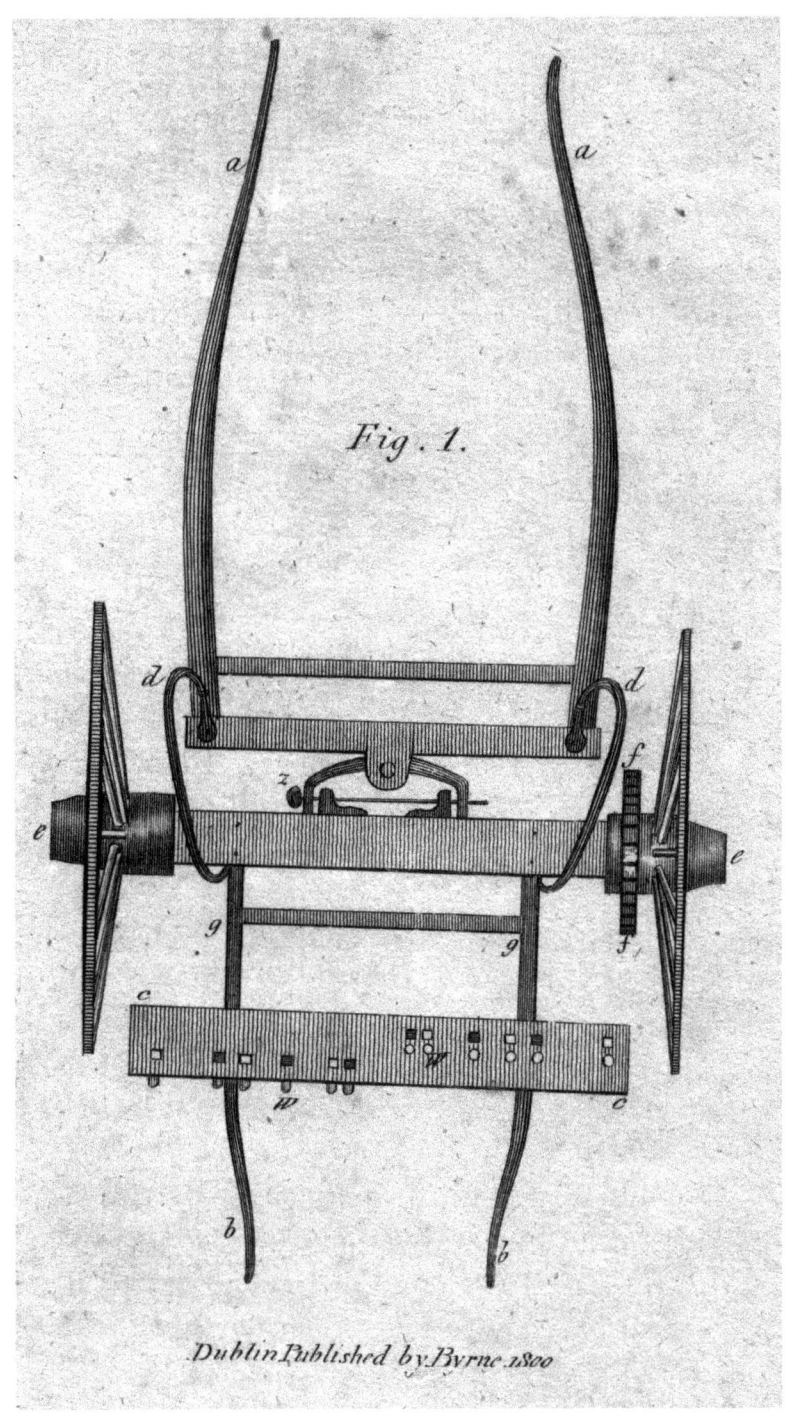

Figure 32. Drill plough designed by E. Darwin from E. Darwin, *Phytologia* (1800).

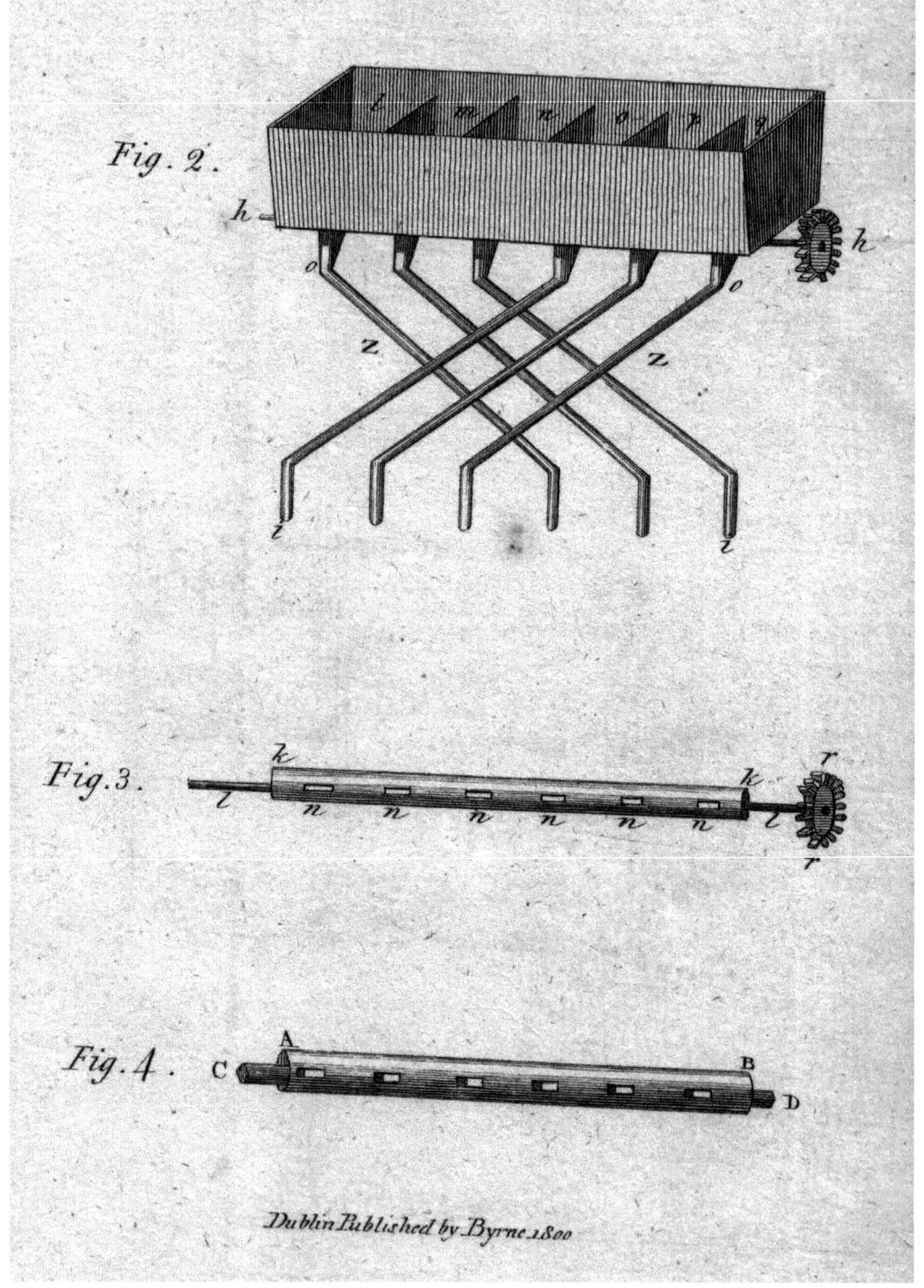

Figure 33. Drill plough designed by E. Darwin from E. Darwin, *Phytologia* (1800) 2.

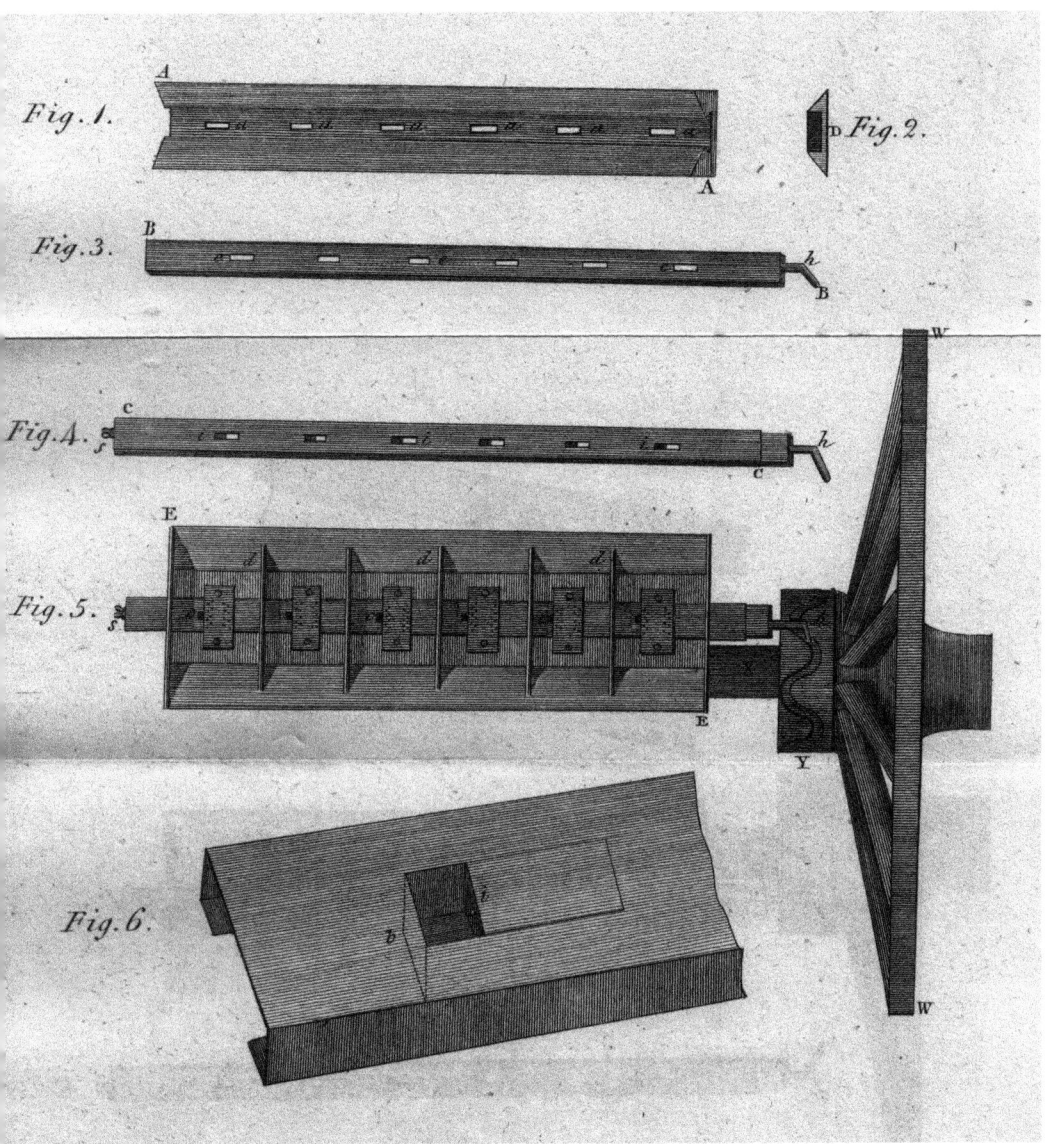

T. Swanwick, seed box, from E. Darwin, *Phytologia* (1800).

CONTROLLING THE WEATHER: AGRICULTURAL METEOROLOGY

Following the clarion call of Francis Bacon and others, Enlightenment natural philosophers became increasingly confident in their ability to understand and manipulate – as well as celebrate – agents in the natural world. As a subtle, rarified ether composed of minute particles, electricity was believed to be the most powerful active effluvia or fluid responsible for alterations in atmospheric conditions, changes in wind direction, precipitation and other weather phenomena. Understanding of meteorology was combined with locally focused studies, folk and popular beliefs and practices closely interrelated with antiquarian, chorographical and natural historical endeavours which celebrated localities and tended to emphasise the benign qualities of the British climate.[25] Encouraged by Isaac Newton's influential thirty-one queries in the second English edition of his *Optics* (1717/18), many natural and meteorological phenomena were attributed to electrical causes, including precipitation, earthquakes, meteors and, increasingly, the growth and germination of plants.[26] Scholarship on Enlightenment electricity and meteorology has emphasised how much naturalists and diarists who recorded changing weather conditions believed this information would provide the key to understanding how and why these alterations occurred, helping to make predictions possible. The prospect of being able to foretell, understand and potentially manipulate weather conditions for farming offered boundless opportunities for increases in food production and improvements in health, providing a chance of reducing or eliminating diseases caused or exacerbated by environmental factors. If 'regular journals' of atmospheric electrical 'variations' were kept, Darwin believed that numerous 'discoveries of its influence on our system' would be made.[27] For Darwin, what were perceived to be the ethereal 'fluids' of electricity and magnetism were central powers in the natural economy and acted as an 'influence on animal and vegetable bodies'. Stimulated and informed by his medical practice, he maintained that the spirits of animation and vegetation, similar ethereal fluids, were central to the vitality of organisms, interacting with the world and interfacing between body and mind.

Darwin's fascination with atmospheric electricity is fully evident in *The Economy of Vegetation* (1791) and, although some of his earliest scientific work sought to delimit the operation of the wonderful fluid in meteorology, he delighted in delineating its potent effects upon the natural world in poetry and prose and celebrating the achievements of 'electricians' such as American philosopher and statesman Benjamin Franklin (1706–1790), who 'Bade his bold arm invade the lowering sky, And seize the tiptoe lightenings, ere they fly' and was sainted with a 'crown electric round his head'.[28] Although what was regarded as the extreme subtlety of electricity made testable hypotheses difficult, the lightning conductor was a spectacular and practical invention, simple but highly effective, and in 1766 Darwin took the lead in ensuring one was fixed upon one of Lichfield Cathedral's west end spires to 'preserve' the building and 'neighbouring houses' (including his own, of course) from being struck.[29]

Darwin and other natural philosophers remained confident that, with greater understanding of the ethereal fluids involved, predictions could be made and storms, rain, or droughts altered or prevented by human endeavour. In a 1787 article on cloud formation he argued that understanding the 'very small causes' behind 'air pressure and weather' might enable devaporation of a 'great' northern 'province', which would then be filled by tempests rushing in, producing colder and then warmer south-west winds in turn. Use of the 'power of human ingenuity' to govern the winds would be a discovery of greater 'utility' than any other in human history, potentially doubling the nation's 'produce and comfort'.[30] Unfortunately, the elusiveness of electricity and difficulties detecting and measuring small charges created problems, although Franklin's demonstration – having 'snatched the raised lightning from the arm of Jove' – that laboratory and atmospheric electricity were synonymous seemed to be a major step towards understanding their operation. Franklin emphasised that both gave off light, exhibited swift motion, conducted through metals, forked in a crooked direction and set fire to flammable substances. He showed how, as a powerful atmospheric electricity, lightning was controllable through the power of points which could be utilised to direct the ethereal fluid which could be collected for experiments, resulting, of course, in the famous sentry box and key on string experiments.[31] These discoveries provided practical solutions to the problem of electrical strikes in storms, which could be averted by positioning lightning conductors to 'defend houses and ships and temples', although the question of whether these were best terminated in knobs or points proved contentious for a while.[32]

The subject of his first article of 1757 on the ascent of vapour – which he addressed to William Watson (1717–87) for his 'advancement' of electrical knowledge – meteorological electricity was one of Darwin's earliest philosophical interests. Then new to both medical practice and Lichfield, Darwin admitted to having 'no electrical friend, whose sagacity he could confide in', and the publication helped build a scientific reputation and began his association with the Royal Society.[33] A response to exaggerated claims about meteorological electricity, particularly those he believed confused effect with causation, Darwin's article displays a degree of combativeness untypical of his later works and redolent of a young experimenter and physician trying to build a reputation. The Italian philosopher Giovanni Battista Beccaria's influential two-volume *Lettre dell' electtrisimo Artificiale* (1753) argued that most weather conditions were associated with atmospheric electricity, which was therefore probably the main cause of changes. In windy conditions, little atmospheric electricity was discerned, yet rainy weather often exhibited moderate electricity. Beccaria believed that rain clouds were often produced by moderate electricity because they were uniformly spread and sometimes cast light at night. Before a storm positive electricity was demonstrated and, after the clouds had passed, his apparatus was negatively charged He thus interpreted clouds in a similar way to Franklin (who argued that there was only one electrical fluid, which could be present in a state of abundance or privation rather than two positive and negative ones), believing that electric matter accumulated in those clouds rising from the earth. Clouds bringing rain would diffuse themselves over places with 'electric fire' in abundance to locations exhausted of it by precipitation to restore the electrical equilibrium, thus reducing the air pressure.[34] Darwin, however, called for less uncritical wonder dazzled by the 'charm of novelty', where 'gazing crowds' adored what they should only 'admire', preferring subtle empiricism when dealing with electricity and other natural phenomenon in philosophy, medicine and religion. The 'vain and pompous boasts' of chemical philosophers had brought their 'art into disrespect', a situation that devotees of 'her sister electricity' needed to avoid.[35] His main target was the Irish natural philosopher Major Henry Eeles (1700–81), who argued that all vapour particles were 'endued' with surrounding 'electric fire' which was the main 'cause' of their ascent because of the 'greater space' they occupied given 'the same weight of air'.[36] Darwin argued, in contrast, that electricity penetrated the pores of material objects without increasing their bulk or moving and disturbing them. He claimed that warming an inverted electrified glass tube coated inside by gilt paper and held by a silk thread placed in oil of turpentine on wax confirmed this because the oil did not subside, which it would have done if the electric fluid had displaced the air.[37] Eeles was wrong to think all vapour was electrified and failed to recognise how long some bodies retained the electric matter despite contact with conductors. Thus the Leyden jar – an early form of capacitor or condenser made of a glass container lined with thin metal on the outside and inside, the use of which to 'store' electricity was crucial to eighteenth-century experiments – sometimes needed to be touched a few times to remove all charge; while the slower bodies were to acquire electricity, the more 'avaricious' they were in keeping it.[38] Following Franklin, Darwin argued that clouds could be electrified plus and minus or even manifest no electrical signs, and dismissed Eeles' assertion that electric ethereal accumulation supported clouds as a 'very vulnerable foundation … an air- built castle, the baseless fabric of a vision'.[39]

Much later, Darwin returned to the role of electricity in cloud formation in the third canto of *The Economy of Vegetation*, arguing that clouds consisted of condensed vapour the particles of which were 'too small separately to overcome the tenacity of the air' and which remained suspended because the 'sphereules of water' were united as a result of a 'surplus of electric fluid', only falling in violent thunder storms when this was withdrawn.[40] Darwin highlighted experiments by his friend the Wirksworth curate Rev. Abraham Bennet (1750–1799), who placed live coals in an insulated metal funnel, poured water on them and observed that ascending steam was positively electrified, while descending water had a negative charge. This demonstrated in Franklinian terms that, although clouds could sometimes be negatively charged, they usually had abundance of electricity which supported water vapour until lightning released it. Numerous metal rods pointing skywards, therefore, might induce rain, which was supported by French chemist Antoine Lavoisier's new theory of water composition, which suggested that thunderstorms might be electrically induced by combining 'oxygene (vital gas)' and 'hydrogene (inflammable air)' to produce water.[41]

When examining cloud formation in an article published in the *Transactions of the Royal Society* in 1788 and communicated by Charles Francis Greville MP FRS (1749–1809), Darwin argued that the electrical ethereal fluid was essential for 'devaporation' or condensation and that, through its own 'power of attraction' or electricity contained within, it would 'dissolve and suspend' salt, which would be 'precipitated' when colder, showing heat to be the 'immediate cause' of such solutions. Mechanically expanded 'elastic fluids', he argued, attracted or absorbed heat from nearby bodies and when 'mechanically condensed' the ethereal heat fluid was 'pressed out' and diffused around 'adjacent bodies'.[42] A few ounces of boiling water required what seemed to Darwin to be a strikingly large amount of heat fluid to be evaporated into steam, and if gases expanded into lower pressure zones without extra heat being applied, they cooled, which in his view explained cloud formation and precipitation. Mountainous regions were normally colder and wetter than valleys below because high pressure air changed to low pressure as it ascended, the cold causing rain by condensation. In order to further investigate the 'mechanical expansion of air', the causes of coldness on mountain summits and rapid condensations of 'aerial vapour' experimentally, Darwin conducted a series of experiments using air guns, thermometers and other apparatus with philosophical friends including James Hutton (1726–1797), Richard Lovell Edgeworth (1744–1817), William Strutt (1756–1830) and John Warltire (1725/6–1810), showing how rapidly expanded air from air guns was cooled and then attracted or absorbed heat from mercury. With Samuel Fox (1765–1851) in 1784, Darwin bored a hole 'about the size of a crow-quill' into a 'large … leaden air-vessel' placed at the beginning of the principal pipe in George Sorocold's century-old Derby waterworks which drew water from the Derwent by four pumps that was fed into a reservoir on St Michael's Church, about forty feet above, from where it supplied subscribing householders. The temperature of air rushing through the hole as the water passed through the lower portion of the air vessel was measured using two thermometers (to reduce the risk of one giving an anomalous reading) that had previously been placed upon the lead chamber. These registered a reduction of 'two divisions or four degrees', thus confirming, in Darwin's view, that the rapid expansion of air released into the atmosphere from compression in the vessel had been the cause of its lower temperature.[43]

Atmospheric electricity featured prominently in the poetry and notes of *The Economy of Vegetation*, partly inspired by Joseph Priestley's *History and Present State of Electricity* (1767), which included two long chapters on the subject and on 'unusual appearances in the earth and heavens'. Darwin followed Priestley in believing that electricity was a subtle but potent force responsible for 'fairy rings', meteors or 'shooting stars', 'vollied lightenings', 'rapid' fire balls and the darting 'pale electric streams' of the *aurora borealis*. Enlightenment scientific interest in the *aurora borealis*, fire balls and other meteorological and astronomical phenomena probably partly reflected their greater visibility in the night sky, when navigation by moonlight or stars was easier and before street lighting and smoke pollution obscured the heavens.[44] Other eighteenth-century meteorological events such as whirlwinds were likewise believed to be sometimes electrically generated, as air currents associated with electrified points being presented to candle flames, electrified iron bars and 'suffocating damps' and those associated with the eruption of Vesuvius seemed to demonstrate.[45] The atmosphere, Darwin argued, was formed of concentric strata, the lowest being where lightning was produced from a 'privation or an abundance of electric matter' in 'floating' vaporous fields, the next being where electrically charged shooting stars were produced and the third having air 3,000 times more rarefied than at ground level, with negligible resistance as a result. Here what were believed to be meteors or electric balls travelled a thousand times more rapidly than through a thousandth of an inch of glass, and those travelling between 'inflammable air' (hydrogen) and 'common air' caused fires as they passed through, differing in colour according to the proportion of these 'airs'. Upscaling from laboratory electricity experiments suggested that, if a mile across, such balls must be emitted from a massive surface of electric matter, which was probably partly generated by the intermixed common and inflammable air they carried along.[46] As Beccaria and John Canton (1718–1772) argued, the *aurora borealis* was probably electrical in origin, with the ethereal fluid passing from positive to negative clouds across great distances, but accumulations of 'inflammable air' over the poles uniting with 'common air' seemed another factor.[47] Darwin maintained that the *aurora* came from the same atmospheric region as meteors, where 'common air' existed in 'extreme temerity' with 'inflammable gas' ten times more rarified. The 'pale electric streams' were due either to repelling caused by a northern accumulation of the ethereal fluid, thus accounting for the light diffusion, spectacular colours and 'silence of passage', or, as Franklin suggested, to the accumulation of positive electricity in the icy poles which rose into the 'rare air' of the 'upper atmosphere' in these regions because of its imperviousness to ice, and passed in spectacular and colourful 'silent streams'.[48]

The need to investigate meteorological electricity with the hope of predicting or even controlling the weather to improve agriculture and horticulture led Darwin to ask his friend Rev. Abraham Bennet to compile an 'electro-meteorological diary' so he could investigate the impact of 'aerial currents' upon vegetable 'growth and maturity'. In order to complete the electro-meteorological record, in addition to using his gold-leaf electroscope, Bennet designed a 'doubler' to detect, augument and potentially measure small amounts of the electrical fluid present during different weather conditions, an innovation that Darwin reported to his friend Benjamin Franklin in 1787 with some excitement.[49] The results of Bennet's work – which sought to determine the relationship between electricity and various conditions, including precipitation, storms, meteors and the *aurora borealis* – were detailed in a series of *Philosophical Transactions* articles, his *New Experiments on Electricity* (1789) and Darwin's books. Bennet's atmospheric electrical theory maintained that 'transparent or clear air' always contained much water in solution, usually positively charged, the intensity of which increased with condensation, and when this occurred quickly electrical atmospheres might stretch for miles. When the 'super-incumbent' positive cloud hit the negative lower strata, with its imperfect connection with earth, the equilibrium was suddenly forcibly restored, causing ferocious thunder claps and 'vivid coruscations of lightning'.[50]

ELECTRICITY, CHEMISTRY AND THE SPIRIT OF ANIMATION

For Darwin, who made few absolute distinctions between living entities, ideas from medicine and physiology could be applied to understand humans, animals and plants, and close analogies could be drawn. The interplay between his medicine and studies of animal and plant physiology, anatomy and taxonomy likewise reflected broader Enlightenment beliefs in the interdependency of the sciences. As has been emphasised, the system of *Zoonomia* was partly modelled on Linnaean taxonomy – which was applied to zoology as well as botany – and underpinned by Darwin's intra-organic understanding of physiology. Although he acknowledged some differences, especially the lack of vegetable 'muscles of locomotion' and 'organs of digestion', the first part of *Phytologia*, on the physiology of vegetation, was primarily intended to establish that vegetables were 'inferior animals', resembling in more respects than otherwise their more mobile organic cousins.[51] In *Zoonomia* and especially *Phytologia* Darwin argued that the 'circumstances attending vegetable irritability' were 'similar to those belonging to the irritability of animals' but upon a 'less extensive scale'. Vegetable buds were 'individual beings' and constituted 'an inferior order of animals', with 'irritability', 'sensibility', 'voluntarity' and 'associations of motion', although the last three were 'possessed' to a lesser extent by 'vegetable buds' than 'more perfect animals'.[52] Likewise, a kindred 'spirit of vegetation' operated in plant bodies to translate 'external' environmental 'impressions' into 'organic motions'. This probably derived from 'uncombined oxygen' freely present in the air, which was respired by the upper surfaces of plant leaves, while that 'absorbed by their roots [was] in a more combined state'. This oxygen was 'again separated' from plant juices 'by the sensorium, or brain' of individual buds, having experienced changes in the 'circulation or secretion of it'. For Darwin, plants were therefore best seen as communities of mature and nascent interconnected and interdependent entities, which, as we will see, was essential for understanding their diseases and how these might be treated.[53]

These arguments and observations concerning plants indicated that Darwin had considerable respect for the sophistication of vegetable physiology and anatomy, which enticed him into fascinating and profound speculations on whether they could think at some level or had a degree of self-awareness which went far beyond whimsical poetical analogies or playful personification. To the 'curious query' of whether plants had 'ideas of external things' he answered in the affirmative, because, like animals, they did have organs of sense through which such ideas were received. He wrote that not only was there 'indubitable evidence of their passion of love' but they had a 'common sensorium' or brain and probably 'repeated' their 'perceptions' when dreaming, and therefore could have many 'ideas' concerning the 'external' world's 'properties' and even 'their own existence'.[54] Anna Seward was one of those who expressed admiration for this aspect of *Phytologia*, which she described as 'in parts, at least … highly ingenious', although she admitted to not having 'read it regularly' because of the subject matter; she interpreted her friend's arguments concerning plant physiology and anatomy in terms of the established 'chain of being' theory, which saw all creation as an interlinked hierarchical chain or ladder with each creature having its place below humans.[55] According to Seward, Darwin's belief that 'vegetables' were 'remote links' in such a 'chain of sentient existence', which was 'often hinted at' in the *Botanic Garden* notes, became clearly 'avowed as a regular system' by him, with evidence being presented for their 'vital organisation, sensation, and even volition'. Seward took Darwin's comments concerning the apparent 'sleep' of plants at night time – their 'low heat and cold blood', akin to that of hibernating animals – as evidence that they continued 'the descending scale of existence'. Furthermore, she believed that his theory of 'vegetable sensation' might help bring different ranks together for the good of society by reinforcing their sense of shared kinship and sympathy with other non-human beings and shared delight at nature's bounty and their ability to unleash the productivity of the land. According to Seward, landowners, farmers and improvers, if mindful of Darwin's active, animated vegetable beings, would find greater pleasure from the 'sustenance, the growth, and comfort' of their trees, crops and flowers as a 'little world of vegetation' came into being from their 'care, attention and kindness'. Likewise, gardeners and agricultural labourers might consider that they were 'nurturing' and 'cultivating' their 'fellow creatures' and become more 'worthy' of 'hire' as an 'honest heart', finding joy from contributing to the 'common stock of happiness' while growing in 'self-respect' and 'benevolence', whether these virtues sprang from 'ignorance and poverty' or refinement from 'knowledge and affluence'. Darwin's theories of plant physiology and anatomy would therefore facilitate the improvement of the mind, character, attitudes and contentment of landowners and labourers alike, just as much as they aided agricultural improvement and productivity.[56]

Darwin's portrait of vegetable passions in *The Loves of the Plants* was therefore not merely amusing metaphor but demonstrated how the fundamental similarities between animal and plant bodies underpinned the natural economy and provided opportunities for various interventions for the betterment of farming. Investigation of the powerful operation of electricity in the atmosphere offered a chance to predict and potentially manipulate weather conditions for the betterment of health and agriculture, while studies of animal and plant physiology, and particularly the chemical and electrical sciences, would have major economic and social benefits. It is striking that experiments into both vegetable electrification and medical electricity were first undertaken from the 1740s, and there are significant parallels between Darwin's belief that electricity could hasten plant growth and the role of this vital ethereal fluid in his medical treatment, underscored by his assumptions concerning the similar operations of all organic entities.[57]

In December 1746 Stephen Charles Triboudet Demainbray (1710–1782), subsequently lecturer in experimental philosophy at Darwin's *alma mater* Edinburgh University, electrified two myrtles, which he believed caused them to grow branches and produce unseasonal blossom. Demainbray's experiments were repeated and apparently confirmed in France, Germany and Italy by Abbe Jean Antoine Nollet (1700–1770), Johann Heinrich Winckler (1703–1770) and others. At Bristol, William Browning reported to the natural philosopher Henry Baker (1698–1774) his electrifying of various trees and plants, causing them to vibrate, and Baker subsequently electrified a potted myrtle tree at Ditton Park, Cheshire, in the presence of the duke of Montague, president of the Royal Society, and others. Drawing parallels between the impact of electricity upon human and vegetable bodies, the physician, antiquarian and natural philosopher William Stukeley (1687–1765) remarked in 1750 that when the earth was in an 'electric state' it hastened the spring growth of plants and that artificial electrification of vegetable bodies replicated this effect 'for the same reasons as in animals it quickens the pulse'.[58]

Benjamin Martin (1704–1782), like Demainbray a natural philosopher and itinerant lecturer, emphasised that, just as electricity had been 'successfully applied to the cure of several disorders of the rheumatic and paralytic kind', so it was 'well known greatly to promote vegetation in plants'. He believed that the main stay upon its effectiveness was that 'constant or perpetual' delivery of the ethereal fluid was required and proposed the application of a hydraulic mill (a rotating water-powered device) designed by the physician, Robert Barker, to electrical generating machines for the 'perpetual electrification of animal and vegetable bodies'. The machine provided a constant supply of water by gravity which turned two glass globes that rubbed against pads to generate (static) electricity which was then conducted through two iron arms and a long iron rod suspended by silk across an 'electrical garden' of 'any sorts' of plants and flowers in pots on a platform supported by 'small pedestals or pillars of wax and resin'. Martin noted that continuous electrification 'night and day' could be provided, alternating the use of the globes if one became too hot.[59]

Some British natural philosophers subsequently kept faith with vegetable electricity, but more work was done in France, and it became associated with French science. One exception was the horticulturist, botanist and writer Richard Weston (1733–1806), later secretary of the Leicestershire Agricultural Society, who, partly inspired by Nollet's experiments, undertook similar investigations and proposed that electricity could be used to hasten growth and destroy insects on trees and plants in stoves and greenhouses.[60] A revival of British interest later in the eighteenth century was encouraged by Darwin and others, especially after Abbe Nicolle Pierre Bertholon (1742–1800) published *De l'Electricite des Vegetaux* (Paris, 1783) which was acquired for the Derby Philosophical Society library. L'Abbe d'Ormoy's experiments during the 1780s, electrifying the seeds of mustard, lettuce, rose and other plants to accelerate their germination – which even worked with apparently old and dry seeds – led him to confidently assert that the 'influence of electricity upon vegetation' and especially the 'germination and growth' of plants was the 'best proved' property of electricity. These trials were reported by Arthur Young in the *Annals of Agriculture* in 1791.[61]

Darwin employed electricity as treatment for some conditions and had, as noted, long taken a keen interest in atmospheric electricity, believing it to be crucial to the operation of all animate bodies and arguing that the spirit of animation was a closely analogous ethereal fluid underpinning both animal and plant vitality. In the *Economy of Vegetation* (1791) he stated confidently that 'the influence of electricity in forwarding the germination of plants and their growth seems to be pretty well established'.[62] He further suggested that water was decomposed into 'oxygene and hydrogene' by the action of electricity within plants and in *Phytologia* provided a remarkable account of photosynthesis, stating that by water's decomposition within the 'vegetable system' hydrogen was combined with carbon, helping to produce all the 'oils, resins, gums, sugar' required, so the oxygen became 'superfluous', and was partly 'exhaled'.[63] After citing various experiments on plant nourishment by the Dutch natural philosopher Jan Ingenhousz (1730–1799) and other naturalists, Darwin emphasised that, because recent experiments had revealed that electricity could 'decompose water into the two airs', there was a 'powerful analogy' to support the notion that it 'accelerates or contributes' to plant growth and 'like heat' might perhaps combine with 'many bodies or form the basis of some yet unanalysed acid'.[64]

After discussions with Darwin, and perhaps partly inspired by Martin's application of the hydraulic mill to create continuous generation for his 'electrical garden', Darwin's friend Abraham Bennet at Wirksworth constructed a 'perpetual electrophorus' to keep flower pots continuously supplied with the ethereal electrical fluid (Figure 35). This used the pendulum of a Dutch wooden clock to mechanise the 'doubling' process as described by Darwin in *Phytologia*.[65] Two of the three plates (B and A) were fixed on a heavy pedestal while the third (C) swung between them on the pendulum and, by a carefully contrived system of wires 'curled into several rings to make them more elastic', the electrical fluid was supplied at each stage of the process. All three plates (ABC) were insulated with glass and two earthed wires, one on the pendulum and another connected to the base, completed the necessary doubling circuit at each stage. Bennet noted that the heavier the pendulum and the larger the plates the more electricity was accumulated by the 'small apparatus' fixed to the clock, which sometimes threw sparks between the plates, on occasion stopping the mechanism 'by

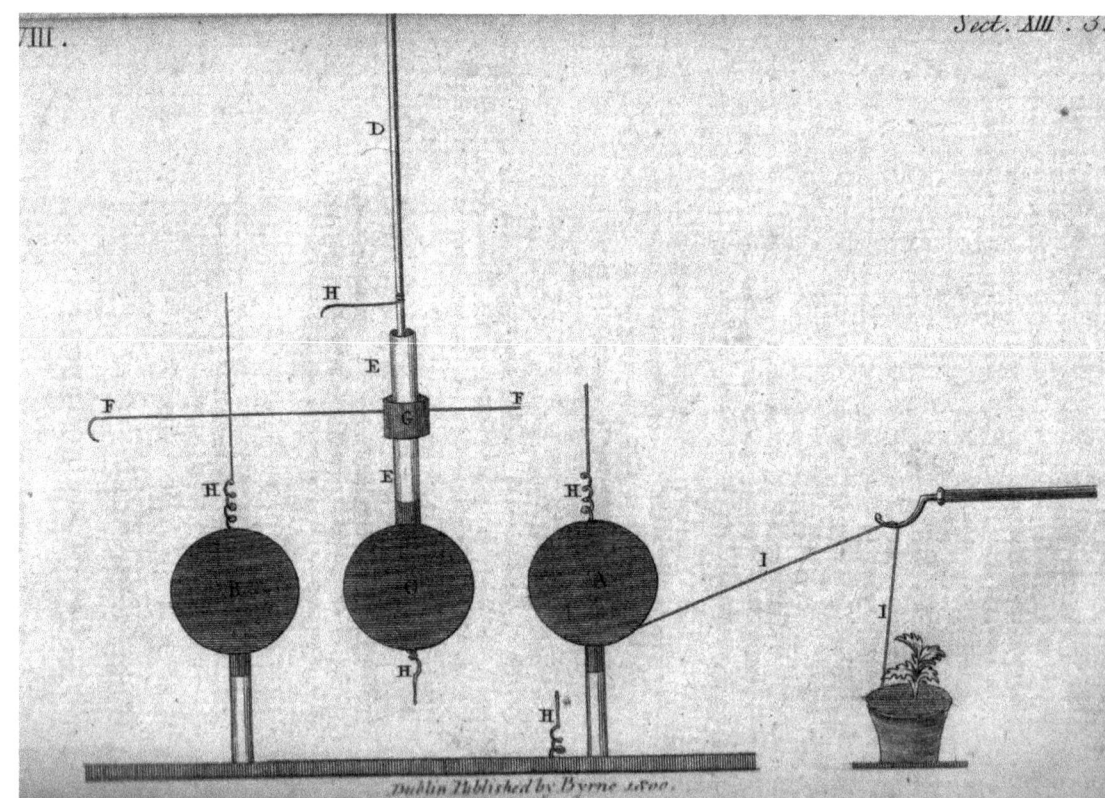

Figure 35. A. Bennet, 'perpetual electrophorus', from E. Darwin, *Phytologia* (1800).

their attraction to each other'.[66] To obviate this difficulty, he suggested that, rather than being circular in form, the plates could be shaped 'something like a lady's fan'.

Although Bennet's clockwork vegetable doubler had not been operating long enough to confirm how effective it was at hastening plant growth when Darwin described it in *Phytologia*, another protégé, the physician Dewhurst Bilsborrow, undertook a separate experiment to electrify mustard seeds with both positive and negative electric fluid, finding that they germinated 'much before' others that received none. Darwin suggested that Bennet's mechanical doubler might be contrived so that both positive and negative charges were supplied to separate plant pots if another insulated example was connected with plate B instead of the wire atop H.[67] He believed that the application of electricity might be of tremendous benefit to farming in two ways: by hastening the growth of crops directly and by inducing rain for the benefit of both animals and plants. As we have seen, he thought that electricity was vital to the operations of vegetable physiology and involved in the process of photosynthesis, and therefore believed that it could accelerate plant growth. It therefore seemed to follow that, through the erection of numerous metallic points upon the ground, the artificial production of atmospheric electricity might promote 'quicker' animal and plant growth by supplying plants 'more abundantly with the electric ether' and precipitating showers.[68] It was difficult to determine whether atmospheric electricity in its 'natural state' had a 'salutory or injurious' effect upon 'animal and vegetable bodies'. However, Darwin believed that as some creatures possessed powers to generate or accumulate the electric fluid within their bodies for defence against predators, such as the electric eel and torpedo fish, it supported the argument that it was central to animal and vegetable physiology. This was another reason why he believed that compiling more scientific records such as Bennet's electro-meteorological diary would facilitate new 'discoveries' of the 'influence' of the vital fluid upon organic and human systems.[69]

PLANT NUTRITION

The belief in fundamental similarities between animals and vegetables had a number of implications, of which one was that chemical analyses of feeding and excreting processes could be applied to all organic entities and another was that a knowledge of human physiology could inform botany and horticulture. Darwin maintained that the new science of agricultural chemistry could be used to test and develop manures and the 'food of plants', revealing optimum conditions for vegetable growth to maximise farming productivity. His analysis of plant nutrition in *Phytologia* drew upon close observations of garden plants and took inspiration from his knowledge of digestion and secretion in animal bodies, and revealed that nitrogen, carbon and phosphorus were as essential for healthy vegetables as for animal bodies.[70] While recognising the importance of the chemical experiments of Joseph Priestley and Jan Ingenhousz, Darwin rejected the concept of phlogiston, coming 'over all French in chemistry' by accepting Antoine-Laurent Lavoisier's isolation and designation of oxygen, as we have seen, employing this new information in his analysis of plant physiology. Darwin maintained that carbonic gas and water were the 'principal food[s] of vegetables' and that plant leaves, when 'exposed to the sun's light', liberated 'oxygen gas'; the main product of plant digestion was sugar. The 'wonderful effect of vegetable digestion in producing sugar' was evident from the products of sugar cane and maple trees and similar operations in animal bodies, as evidenced by a patient with diabetes at Stafford Infirmary, where Darwin was Surgeon Extraordinary, who voided copious amounts of water laden with sugar that could be extracted by evaporation.[71]

The types of soil and manure that best facilitated plant growth were examined in *Phytologia*, which recommended that manures be 'ploughed or dug into the ground' prior to sowing seeds or setting roots for best effect. This was partly because the 'atmospheric air' 'buried' with manure in the 'interstices of the earth' rendered the soil loose and easily impressible, 'gradually evolv[ing] by its union with carbon' a 'genial heat very friendly to vegetation' in the British climate. Growth was also facilitated by the production of 'fluid carbonic acid' and the 'fluid mixture of nitrogen with hydrogen', which supplied much plant nutriment. Darwin included a whole chapter on manures in *Phytologia*, with sections emphasising the importance of lime, carbon, decomposing vegetable materials and other substances such as phosphorous in plant growth and agricultural productivity; phosphorous, he believed, was universally present in all animal and vegetable materials, as heating demonstrated. He made an effort to investigate the role of putrefaction and fermentation in enriching soils for plants, consulting on these matters friends such as the young physician, Philosophical Society member and author of the *History and Present State of Animal Chemistry* (1803), Dr William Brookes Johnson, and by making phosphorous in his laboratory using the method of the Dutch chemist Wilhelm Homberg (1652–1715), which involved strongly heating a mixture of alum with sugar in a covered crucible until a 'bluish flame' appeared. Darwin's observations of midlands agriculture and the rising and falling Derwent waters between the Darwins' Full Street garden and orchard suggested that rainfall and periodic flooding improved vegetable nutrition by spreading small lumps of loose manure across the land, which was washed into the soil by 'vernal showers' and other inundations, allowing the 'essential parts' to be applied to roots before becoming 'diminished by winter rains' or 'summer exhalation' – an idea supported by contemporary Yorkshire agriculturist Richard Parkinson (1748–1815), whose *Experienced Farmer* (1798) was one of the farming texts acquired by the Philosophical Society.[72]

Experiments and observations on vegetable glandulation founded upon close analogies with animal bodies, Darwin thought, would likewise boost farming yields. Tiny glandular vessels in plants – barely visible even with magnifying glasses and microscopes – separated 'mucilage, starch, or sugar' from the vegetable 'blood' to support placentation, or the growth of seeds, bulbs and buds, and deposited 'bitter, acrid or narcotic' juices for defence against insect or animal depredations. Rather than seeing the 'odiferous essential oils of plants', balmy scents and glorious floral colours as the means of attracting insects, Darwin regarded them as defences against insect and animal attack, while flower hues resulted from the size of plants' 'blood' 'particles' or the tenuity (thinness) of their petal membranes. Honey or 'nectar' was 'one of the most curious and important' of 'vegetable secretions', and pollen or 'honey' dust was vital in the natural economy because all plants appeared to produce it using the 'complicated' floral 'apparatus' 'nature' had 'constructed' for them and, furthermore, made many efforts to expose and protect it. Darwin removed the 'nectaries' of various flowers before their petals opened or became coloured to investigate whether honey helped feed the stamens and pistils, and noted that some flowers higher up the plant produced seed while other below failed to do so, although he had insufficient opportunity to investigate whether these seeds were fertile and grew into new plants. Nevertheless, from the observations on flower structure he made, in conjunction with his assumptions about general analogies between all animate physiologies, he made what to us seems to be an extraordinary conclusion that 'honey' 'fed and nourished' the male and female floral parts while they developed their reproductive apparatus, rather than being something intended to attract insects, which was apparently confirmed by the nectary's redundancy or death following the 'birth and death' of stamens and pistils. Insects were widely seen as the most 'imperfectly formed' of animals, although Darwin had a higher opinion of them, believing that moths and butterflies that depended upon 'honey' acquired much of their 'animation' from the nutritious nectar. He noted that one of his philosophical acquaintances had suggested that the first insects might have originated as anthers or stigmas in flowers (which, as we have seen, he thought possessed higher 'sensibility or animation' than other vegetable organs because of their 'honey' or nectar nourishment) before breaking off from the parent plant and acquiring physical features such as wings and 'claws' through their 'ceaseless efforts to procure food', which was no more 'incomprehensible' than tadpoles becoming frogs or caterpillars transforming into butterflies. Hence Darwin regarded insects that fed on 'honey', including bees, as pests that plants tried to protect themselves against rather than as essential agents of vegetable reproduction. However, innumerable insects and most 'larger animals' obtained their nourishment from 'honey' and other plant parts, such as fruits, cotyledons, roots and buds, enjoying 'life and pleasure without producing pain to others' because parts such as seeds and eggs were as yet unendued with 'sensitive life'.[73]

CONCLUSION

Darwin believed that application of the agricultural science detailed in *Phytologia* would increase the range and quantity of food staples and thereby improve the health of the population. He used his gardens and smallholdings to undertake experiments and observations to facilitate this. Improvements in arable farming could be achieved by selecting and breeding the best crops and utilising technologies to optimise growing conditions by, for example, improving soils and influencing the atmosphere. His work was intended to address concerns over food scarcity and national self-sufficiency associated with the revolutionary wars and to bolster the efforts of Sinclair, Young, the duke of Bedford, the Board of Agriculture and their supporters. He approached farming improvements and horticulture in many ways as a doctor, applying methods, models and treatments honed on humans to animals and plants, underpinned by a strong belief in close analogies between animal and vegetable physiology; in turn, as a trusted and enquiring physician and philosopher, his advice was sought by farming and gardening friends and patients. Having witnessed the impacts of overindulgence, poor diet and privation upon human bodies, Darwin was keenly aware of the value of wholesome and plentiful foodstuffs for all classes and favoured arable over livestock farming as more effective and efficient. Here the wonderful sciences of electricity, chemistry and meteorology offered potentially boundless possibilities, while certain technologies, such as improved ploughs, windmills and even electrical machines, offered the exciting prospect of vastly increasing levels of agricultural production and even hastening vegetable maturity in defiance of seasonal and climatic restrictions.

NOTES

[1]   E. Darwin, *Phytologia; or the Philosophy of Agriculture and Gardening* (London, 1800), vii; M. McNeil, *Under the Banner of Science: Erasmus Darwin and his Age* (London, 1987), 194–7; S. Bowerbank, *Speaking for Nature: Women and Ecologies of Early-Modern England* (Baltimore, 2004), 137–8.

[2]   Darwin, *Phytologia*, dedication, vii–viii.

[3]   E. Darwin, letters to J. Banks, 23 February, 17 March 1782 and J. Wedgwood, 13 April 1782, in D. King-Hele ed., *The Collected Letters of Erasmus Darwin* (Cambridge, 2007), 200–4.

[4]   Darwin, *Phytologia*, viii.

[5]   R. E. Prothero (Lord Ernle), *English Farming, Past and Present*, 6th edn, edited by G. E. Fussell and O. R. McGregor (London, 1961), 195–6, 224–52, 268–70; McNeil, *Under the Banner of Science*, 185–8.

[6]   W. Watson, 'An account of some experiments by Mr. [Charles] Miller of Cambridge, on the sowing of wheat', *Philosophical Transactions*, 58 (1768), 203–6; J. Tull, *Horse-Hoeing Husbandry*, 4th edn (London, 1762); Darwin, *Phytologia*, 161–2; Ernle, *English Farming*, 169–73.

[7]   Darwin, *Phytologia*, 402.

[8]   Darwin, *Phytologia*, 162–4; Watson, 'An account of some experiments', 203.

[9]   Darwin, *Phytologia*, 479.

[10] C. Bryant, *Flora Diaetica: or History of Esculent Plants both Domestic and Foreign* (London, 1783), preface, ix–xii; Derby Philosophical Society, *Rules and Catalogue of the Library of the Derby Philosophical Society*, with supplements of 1795 and 1798 (Derby, 1793–8), 8; J. E. Smith, 'Biographical memoirs of several Norwich botanists', *Transactions of the Linnean Society*, 7 (1804), 295–301; J. Chambers, *A General History of the County of Norfolk*, 2 vols (Norwich, 1829), vol. 1, introduction, xlviii–xlix.

[11] Darwin, *Phytologia*, 435–6; Lichfield Botanical Society, *... translated from the thirteenth edition of the Systema Vegetabilium of the late Professor Linneus*, 2 vols (Lichfield, 1783), vol. 2, 637, vol. 1, 218; Lichfield Botanical Society, *The Families of Plants ... Translated from the Last Edition of the 'Genera Plantarum'*, 2 vols (Lichfield, 1787), vol. 1, 181; C. Jarvis, *Order out of Chaos: Linnaean Plants Names and their Types* (London, 2007), 365.

[12] Darwin, *Phytologia*, 435–6, 496–9; Lichfield Botanical Society, *System of Vegetables*, vol. 2, 664, vol. 1, 251, 78; Lichfield Botanical Society, *Families of Plants*, vol. 1, 214, 33; Jarvis, *Order out of Chaos*, 706–7, 529.

[13] Darwin, *Phytologia*, 436, 439; Bryant, *Flora Diaetica*, 46–8; G. White, *The Natural History and Antiquities of Selbourne*, edited by E. T. Bennett and J. E. Harting (London, 1876 [1788]), 54.

[14] Darwin, *Phytologia*, 447; Lichfield Botanical Society, *System of Vegetables*, vol. 2, 680; Jarvis, *Order out of Chaos*, 318.

[15] Darwin, *Phytologia*, 442–7. This was probably the *Agaricus deliciosus* described by Linnaeas in 1753.

[16] J. S. Watson, *The Reign of George III* (London, 1960), 288.

[17] E. Darwin, letter to R. L. Edgeworth, 1788 in King-Hele ed., *Collected Letters*, 304–7, which also reproduces the drawings.

[18] Derby Local Studies Library, E. Darwin, commonplace book, microfilm copy, 107–9, 114; D. King-Hele, *Erasmus Darwin: a Life of Unequalled Achievement* (London, 1999), 187.

[19] J. Whitehurst, *Observations on the Ventilation of Rooms, on the Construction of Chimneys and on Garden Stoves* (London, 1794); R. E. Schofield, *The Lunar Society of Birmingham: A Social History of Provincial Science and Industry in Eighteenth-century England* (London, 1963), 76–82; P. A. Elliott, *The Derby Philosophers: Science and Culture in British Urban Society, c1700–1850* (Manchester, 2009), 163–89; J. Uglow, *The Lunar Men: The Friends who Made the Future* (London, 2002), 154–65; P. Jones, *Industrial Enlightenment: Science, Technology and Culture in Birmingham and the West Midlands, 1760–1820* (Manchester, 2009), 31–2; M. Craven, *John Whitehurst: Innovator, Scientist, Geologist and Clockmaker*, 2nd edn (Croydon, 2015), 88–9, 111–12, 132–4.

[20] Darwin, *Phytologia*, 243–5, 254–5, plate VII, Appendix: Improvement of the drill plough, 541–6, plates X, XI, XII; Derby Local Studies Library, Darwin, commonplace book, 87; Schofield, *Lunar Society*, 73–4, 104–5; King-Hele, *Erasmus Darwin*, 80–1, 157–60.

[21] Darwin, *Phytologia*, 254–5, plate VII; Darwin, letters to J. Wedgwood, 18 November 1767, late February, 6 March, 9 April 1768 in King-Hele ed., *Collected Letters*, 84–7; King-Hele, *Erasmus Darwin*, 80–1, 85, 157–60.

[22] Darwin, *Phytologia*, 432–3; *Derby Mercury*, 2 October 1832; M. Craven, *Distinguished Derbeians* (Derby, 1998), 54; Elliott, *Derby Philosophers*, 241–2, 270.

[23] Darwin, *Phytologia*, 197; Elliott, *Derby Philosophers*, 241–2, 257.

[24] Darwin, *Phytologia*, 261–2, 397–401, 541–6; E. Darwin, letters to S. More, 13 October 1799 and G. Cavendish, duchess of Devonshire (November 1800), in King-Hele ed., *Collected Letters*, 533–4, 556–8; Elliott, *Derby Philosophers*, 135–6, 155.

[25] G. Pancaldi, *Volta: Science and Culture in the Age of Enlightenment* (Princeton, 2003); V. Jankovic, *Reading the Skies: A Cultural History of English Weather, 1650–1820* (Chicago, 2000); J. Golinski, *British Weather and the Climate of Enlightenment* (Chicago, 2007); P. A. Elliott, *Enlightenment, Modernity and Science: Geographies of Scientific Culture and Improvement in Georgian England* (London, 2010), 247–79.

[26] I. Newton, *Opticks: or a Treatise of the Reflections, Refractions, Inflections and Colours of Light*, 4th edn (London, 1730), vol. 3, 313–82; J. L. Heilbron, *Electricity in the 17th and 18th Centuries: A Study in Early-Modern Physics*, 2nd edn (New York, 1999), 51–5.

[27] E. Darwin, *Zoonomia; or the Laws of Organic Life*, 3rd edn, 4 vols (London, 1801), vol. 4, 239.

[28] E. Darwin, *The Economy of Vegetation*, 4th edn (London, 1799), 44.

[29] *Harris's Birmingham Gazette*, 29 September 1766, quoted in King-Hele, *Erasmus Darwin*, 72; S. Shaw, *The History and Antiquities of Staffordshire*, 2 vols (London, 1802), vol. 1, appendix of additions and corrections, 26.

[30] E. Darwin, 'Frigorific Experiments on the Mechanical Expansion of Air, Explaining the Cause of the Great Degree of Cold on the Summits of High Mountains, the Sudden Condensation of Aerial Vapour, and of the Perpetual Mutability of Atmospheric Heat', *Philosophical Transactions*, 128 (1788), 43–52; see also Darwin, additional note on winds, *Economy of Vegetation*, 403–33.

[31] Heilbron, *Electricity in the 17th and 18th Centuries*, 340–1.

[32] Darwin, *Economy of Vegetation*, 43–4; Heilbron, *Electricity in the 17th and 18th Centuries*, 380–3.

[33] E. Darwin, 'Remarks on the opinion of Henry Eeles', *Philosophical Transactions*, 50 (1757), 241.

[34] J. Priestley, *The History and Present State of Electricity*, 1st edn (London, 1767), 366, 370; Heilbron, *Electricity in the 17th and 18th Centuries*, 365.

[35] Darwin, 'Remarks on the opinion of Henry Eeles', 241.

[36] H. Eeles, 'Letter concerning the cause of the ascent of vapour and exhalation, and those of winds; and of the general phenomena of the weather and barometer' *Philosophical Transactions*, 49 (1755/6), 134; Darwin, 'Remarks on the opinion of Henry Eeles', 248–50.

[37] Darwin, 'Remarks on the opinion of Henry Eeles', 250.

[38] Darwin, 'Remarks on the opinion of Henry Eeles', 252.

[39] Darwin, 'Remarks on the opinion of Henry Eeles', 253.

[40] Darwin, *Economy of Vegetation*, 130, note.

[41] Darwin, *Economy of Vegetation*, 62–3, note; Elliott, *Enlightenment, Modernity and Science*, 272–4.

[42] Darwin, 'Frigorific experiments', 49; King-Hele, *Erasmus Darwin*, 28, 216, 226–8, 269, 271.

[43] Darwin, 'Frigorific experiments', 43–52; Schofield, *Lunar Society*, 185–6.

[44] Darwin, *Economy of Vegetation*, 11–14; H. D. Thoreau, 'Night and moonlight', *Atlantic Monthly Magazine* (November 1863), 579–83.

[45] Darwin, *Economy of Vegetation*, 187–8, note, additional note on winds, 417–21.

[46] Darwin, *Economy of Vegetation*, additional note on meteors, 249–58.

[47] Priestley, *History and Present State of Electricity*, 376; Darwin, *Economy of Vegetation*, additional note on meteors, 249–58.

[48] Darwin, *Economy of Vegetation*, additional note on meteors, 249–57; King- Hele, *Erasmus Darwin*, 259–61, 269–71.

[49] A. Bennet, 'An account of a doubler of electricity, or a machine by which the least conceivable quantity of positive or negative electricity may be continually doubled, till it becomes perceptible by common electrometer, or visible sparks … communicated by the Rev. Richard Kaye, LLD, FRS', *Philosophical Transactions*, 77 (1787), 289.

[50] A. Bennet, *New Experiments on Electricity* (Derby, 1789), 103; Elliott, *Enlightenment, Modernity and Science*, 247–79.

[51] Darwin, *Phytologia*, 1–126.

[52] Darwin, *Zoonomia*, vol. 1, 78–145; Darwin, *Phytologia*, 286–7.

[53] Darwin, *Phytologia*, 287, McNeil, *Under the Banner of Science*, 195–6.

[54] Darwin, *Zoonomia*, vol. 1, 143–5.

[55] A. Seward, *Memoirs of the Life of Dr. Darwin* (London, 1804), 410–11; A. Lovejoy, *The Great Chain of Being: A Study of the History of an Idea* (Boston MA, 1957).

[56] Seward, *Memoirs of Dr. Darwin*, 411–13.

[57] P. F. Mottelay, *Bibliographical History of Electricity and Magnetism* (London, 1922); Heilbron, *Electricity in the 17th and 18th Centuries*; P. Bertucci, 'Sparks of Life: Medical Electricity and Natural Philosophy in England, c.1746–1792', DPhil thesis, University of Oxford (Oxford, 2001); M. B. Schiffer, *Draw the Lightning Down: Benjamin Franklin and Electrical Technology in the Age of Enlightenment* (Berkeley, 2003); Pancaldi, *Volta*; S. Finger, *Doctor Franklin's Medicine* (Philadelphia, 2006).

[58] S. Demainbray, 'An application of electricity towards the improvement of vegetation', *The Scots Magazine*, 9 (January 1747), 40, also published with a sceptical editorial comment in the *Gentleman's Magazine*, 17 (February 1747), 80–1; J. Browning, 'On the effects of electricity on vegetables', *Philosophical Transactions*, 114 (1747), ii, 373–5; W. Stukeley, 'Concerning the causes of earthquakes', *Philosophical Transactions*, 46 (1749–50), 664; A. Poey, 'Report on agricultural meteorology', *Report of the Commissioner of Agriculture for the Year 1869* (Washington DC, 1870), 152–7; Mottelay, *Bibliographical History*, 179–80.

[59] B. Martin (attrib.), 'A description of a machine for a perpetual electrification', *The General Magazine of Arts and Sciences*, 1 (London, 1755–6), 116–17; J. R. Millburn, 'Martin's Magazine: the General Magazine of Arts and Sciences, 1755–65', *The Library*, 28 (1973), 221–39; J. R. Millburn, *Benjamin Martin: Author, Instrument Maker and 'Country Showman'* (Leiden, 1976).

[60] R. Weston, *Tracts on Practical Agriculture and Gardening* (London, 1773), 289–96.

61 M. L'Abbe D'Ormoy, 'Experiments on the influence of electricity on vegetation', *Annals of Agriculture*, 15 (1791), 28–60.

62 Darwin, *Economy of Vegetation*, note, 46.

63 Darwin, *Phytologia*, 283.

64 Darwin, *Phytologia*, 283–3.

65 Darwin, *Phytologia*, 284, plate VIII and note.

66 Darwin, *Phytologia*, 313, note to plate VIII.

67 Darwin, *Phytologia*, note to plate VIII.

68 Darwin, *Phytologia*, 312–14.

69 Darwin, *Zoonomia*, vol. 2, 471–2.

70 Darwin, *Phytologia*, 166–231; H. Davy, *Elements of Agricultural Chemistry in a Course of Lectures for the Board of Agriculture* (London, 1813); McNeil, *Under the Banner of Science*, 199–200.

71 Darwin, *Phytologia*, 169–79; King-Hele, *Erasmus Darwin*, 334–5; P. G. Ayres, *The Aliveness of Plants: The Darwins at the Dawn of Plant Science* (London, 2008), 43–9.

72 Darwin, *Phytologia*, 166–231, for making phosphorous, 188–9, for the flooding, 227–8; E. Darwin, letter to W. B. Johnson, late 1798 in King-Hele ed., *Collected Letters*, 526; R. Parkinson, *The Experienced Farmer*, 2 vols (London, 1798).

73 Darwin, *Economy of Vegetation*, 'additional note xxxix: vegetable glandulation', 461–71; E. Darwin, *The Temple of Nature; or the Origin of Society* (London, 1803), 63, 66–7; G. L. Leclerc, comte de Buffon, *Buffon's Natural History Abridged*, translated by W. Smellie, 2 vols (London, 1792), vol. 2, 281.

CHAPTER 5

VEGETABLE PATHOLOGY AND MEDICINE

Humans were, of course, regarded as the apex of creation in the eighteenth century, created in the image and likeness of god while the rest of nature, including animals, insects and plants, existed to glorify the divinity and supply materials for human living. Plants as material objects and their products provided everyday foodstuffs, clothing, building and domestic materials, medicines and other substances essential for human existence. Bountiful harvests and productive farms were celebrated in pastoral literature and painting and improving farming literature alike; yet, while conveying luxury in Europe and North America, plant products from tobacco and sugar to cotton were stained with the blood that had been shed growing and harvesting them.[1] Plants also had cultural significance in various ways. With their myriad changing seasonal colours, shapes and structures, they adorned gardens, pleasure gardens, parks and hothouses. Prized exotic specimens were imported from across the globe for the wealthy, educated or discerning or as part of colonial trade or horticultural experiments. Physic and botanical gardens, arboretums and pinetums provided a medical resource and specimens for taxonomic studies. Plants were also the subject of a burgeoning botanical and landscape aesthetic literature in the eighteenth century, evoking and defining particular places and changing their meanings according to context.[2]

While, as we shall see, scholars have increasingly recognised animal agency above mere passive human usage, there has also been increasing emphasis upon vegetable agency and an effort to recover the actions not only of plants themselves, their coadjutors and allies in an interdependent world, but also of their enemies. As a physician, Darwin valued plants for their medicinal value, which led him into the study of botany and natural history. Prior to the development of industrialised chemical and synthesised processes for drug production from the nineteenth century, plant and animal products and other naturally occurring substances were the key ingredients in most medicines and some botanical education was an adjunct to a university medical education as well as in the schooling of apothecaries. Darwin encountered animals and plants in multiple contexts both in wild and semi-wild habitats and nurtured on farms and estates. He observed wildflowers in the fields and hedgerows, the ancient midlands forests and extensive, richly manured and productive enclosed fields and plantations of improving landowning friends and patients and their families. The apothecaries that he worked closely with also had a keen knowledge of botany. As we will see, the prestige he acquired as a medical practitioner meant that he was consulted about diseases and other problems affecting animals and plants as well as humans. While he never had the leisure to compile a systematic nature diary in the manner of his contemporary the Rev. Gilbert White of Selbourne, Darwin's keen eye for detail shines through in his observations on agriculture and gardening.[3] Unable to devote all his time to natural history as his grandson Charles Darwin was to do, Darwin nevertheless emerges as closer to him in this respect than has previously been recognised.[4]

This chapter will argue that Darwin's medical practice and intra-organic psychophysiology inspired and shaped his philosophy of agriculture and horticulture. His medical practice provided him with an incentive to develop a philosophy of agriculture and gardening, as his emphasis upon the role of environmental factors in illness and detailed pronouncements on plant diseases demonstrate.[5] This explains why Darwin regarded *Phytologia* as a 'supplement' to *Zoonomia*, remarking to Joseph Banks that he was forging a new philosophical agriculture by applying his knowledge of the 'animal oeconomy' to the study of plants.[6] Coming from a landowning family, Darwin was concerned with the health, welfare and vitality of animal and vegetable organisms on the farms of fellow landowners. His experiences of designing, planting and managing gardens and the Derby 'farm' and orchard, discussing agricultural practices with visitors and applying medical ideas and methods to livestock keeping and horticulture led him to concern himself with vegetable diseases. Building upon our earlier analysis of his use of organic materials in medical practice, the chapter will explore Darwin's physiological and anatomical studies of plants and how these informed his understanding of vegetable ailments and the methods used to treat and control these. Plant health and pathology were important elements of *Phytologia* and demonstrated how medicine and the sciences could be applied to further improvements in farming, which were even more essential from the 1790s, as access to the continent became more difficult and British agriculture needed to be more self-sufficient to ensure the population's health and vitality.

VEGETABLE PATHOLOGY

Plant disease and depredations from insects and animals were major problems in Georgian farming, horticulture and gardening, and the remedies usually resorted to were relatively ineffective, costly and difficult to execute. The devastation wreaked upon the vegetable frame by disease was colourfully described in Darwin's poetry. In *The Economy of Vegetation* (1791) he emphasised how hard it was to:

> Shield the young Harvest from devouring blight,
> The Smut's dark poison, and the Mildew white;
> Deep-rooted Mould, and Ergot's horn uncouth,
> And break the Canker's desolating tooth.

Once disease had begun to run 'in white lines' along a field, so cankerous 'crack follows crack, to laws elastic just, and the frail fabric shivers into dust', shattering the farmer's hopes as surely as slight scratches on glass or minerals could destroy larger pieces. The 'festering wound' spread across vulnerable bodies and mined 'unperceived beneath the shrivel'd rin'd', climbing branches with 'increasing strength', spreading 'as they spread' and lengthening 'with their length'.[7] As the landed elites adorned their estates, pleasure gardens and hothouses with expensive exotics, and botanists and collectors sought beautiful or useful specimens for their collections from around the globe at great cost and difficulty, so the cost of losing plants grew and delicate and precious examples not used to the British climate or predators were vulnerable to destruction. In 1777 John Kennedy (d. 1790) highlighted the problem of pineapple destruction by a 'white insect' that eventually covered all the plants and surfaces in the greenhouse. He advocated a laborious and time-consuming method of treatment that involved removing and cleaning all pineapple plants, fumigating with smoke from brimstone powder and thoroughly cleaning, repairing and repainting entire hothouses. The problem was, of course, that the process had to be repeated if infected fruit returned or insects survived anywhere in cracked hothouse timbers, which was one reason for the replacement of wood with iron frames during the first half of the nineteenth century.[8]

But where staple crops upon which large segments of the population relied for food were destroyed, the effect was particularly devastating and went well beyond temporarily reducing the income of individual farmers. So, ergot, or 'horn-seed', for instance, believed to be caused by insects, often affected French rye crops and sometimes those in Britain, causing, as Darwin emphasised, 'great debility and mortification of the extremities' among the poor in both nations (Figure 36).[9] It seems likely that the agricultural revolution increased the prevalence, frequency and economic impact of plant pathologies as traditional practices of rotation

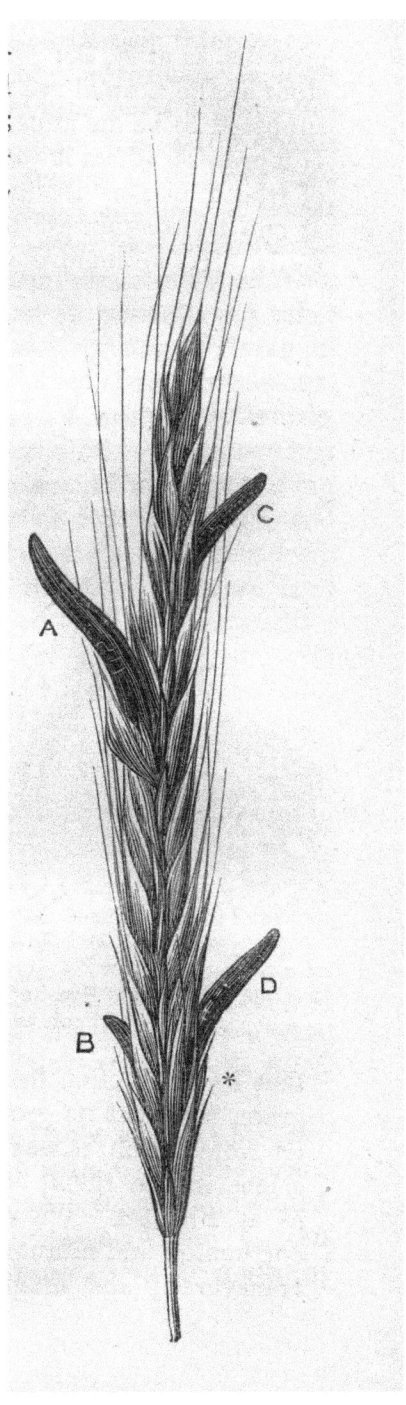

Figure 36. Ergot, from W. G. Smith, *Diseases of Field and Garden Crops* (London, 1884).

ended, farms tended to grow larger to feed a growing population, single crops were nurtured in bigger fields with fewer hedges, ditches and patches of woodland separating them and biodiversity was reduced, even prior to the development of chemical pesticides from the nineteenth century onwards. Likewise, the obsession of agriculturists such as Arthur Young and those associated with the Board of Agriculture with raising rents and the assumption that rental values were a motor – and index of – agricultural prosperity and improvement, pressurised farmers into maximising planting densities, which also furthered the spread of plant diseases. As investment in improved agriculture grew, so the costs of plant diseases became more deleterious, encouraging more systematic investigations into vegetable pathology and, indeed, the development of a specialised branch of botany and agronomy. This is one reason why, as Geoffrey Ainsworth has argued, the period between 1750 and 1850 was so crucial to the development of phytopathology, because of the extent of 'experimental evidence' that emerged through studies of disease outbreaks with the help of inducements provided by members of the elite concerned about agricultural improvement and social order.[10]

Food riots were the most numerous of all Georgian disturbances, many caused by combinations of bad weather and vegetable diseases causing poor harvests or raising expectations of food supply problems and forcing up grain prices. There was some incentive for farmers to exaggerate the likelihood and impact of poor harvests to raise food prices, which was sometimes exacerbated by hoarding and other social and political factors such as war and its attendant popular discontent, but plant diseases were unquestionably a major factor in seasonal food shortages which disproportionately affected the labouring population who were more dependent upon staple crops.[11] The agrarian troubles and food riots from 1764 were exacerbated by poor harvests in 1765 and 1766 and high grain prices. They were some of the most widespread in the eighteenth century, which the *Gentlemen's Magazine* attributed to crops being 'smutty, much blighted, and very much choked with weeds'; even the normally bountiful Norfolk and Suffolk wheat crops were meagre. Despite remedial action by some landowners, magistrates and the government to control supplies and prices, and mobilisation of the military and militia, riots spread over many southern and midland counties, including Leicestershire, Staffordshire, Nottinghamshire and Derbyshire, some of the largest being at Nottingham and Derby.[12]

The devastating Irish potato famine of the 1840s is notorious, but there were major failings of the crop for over a century before this. In 1728, for example, disease in potato crops caused riots in Cork; the destruction of the entire harvest occurred in 1739 and 1740; and further failures followed in 1770, 1800, 1807 and various years during the 1820s and 1830s, owing to diseases such as leaf curl (a virus), sometimes exacerbated by prolonged frosts or rain.[13] In England and Wales, poor harvests and attendant food shortages and high prices were regularly reported in the local and national press. In 1797, for instance, during a period of major national peril due to threat of French invasion, a 'blight or mildew' attacked wheat plants, along with worms, during the 'critical month' of June, spooking the markets; a dreadful national harvest followed, exacerbated by incessant rain and high winds, which forced up the price of flour, with poor hop, oat and barley crops and limited straw supplies for winter.[14]

Although such outbreaks provided many opportunities for naturalists to investigate, and the Society of Arts and agricultural societies regularly offered prizes for methods to prevent or alleviate vegetable diseases, making recommendations such as using salt in hay stacks to prevent over-heating and mildew (justified partly because cattle allegedly preferred its 'superior flavour'), there were few means of tackling plant pathologies once they had struck. Farmers were forced to create barriers in the manner of fire-breaks, by scything through or digging up swathes of plants, to try to stop the spread of vegetable diseases.[15] Other methods used to treat individual plants were largely ineffective or destructive, such as dressing wheat seeds with wood ash and urine, salt water, by-products of soap-making, slaked lime or even arsenic to try to prevent smut or blight. Natural philosophers such as Rev. William Kirby (1759–1850) and Joseph Banks blamed 'agriculturists' for failing to 'trouble themselves' sufficiently about the origins of different plant distempers, and the subject likewise received little attention from naturalists; where it did, the results were frequently not widely circulated, although major epidemics stimulated additional studies.[16]

Responding to serious attacks from blight or mildew in particular – the 'worst enemy' of wheat – between 1760 and 1820 a small group of British and Irish naturalists investigated plant pathologies, including the Rev. Henry Bryant (1721–1799) in Norfolk, Kirby in Suffolk (who came to it through entomology), Joseph Banks from Norfolk and Thomas Andrew Knight of Herefordshire, making observations using lenses and microscopes and undertaking experiments like those made by Bryant and Kirby in which different forms of treatment were tried and their effects compared.[17] Part of the problem was that cryptogamia, which included mosses, algae, liverworts and troublesome kinds of fungi – which were recognised as being the cause or a symptom of many vegetable pathologies – was, as we have explained, the most obscure class in Linnaean botany because, as James Edward Smith emphasised, stamens and pistils in this class could not be 'well ascertained' or numbered 'with any certainty' (Figure 37).[18]

Darwin did take a close interest in fungi, which he evocatively described as 'an isthmus between the two great continents of nature: the animal and vegetable kingdoms' – even to the extent of examining mushrooms sprouting on the path where the horse drove a grindstone in Derby's tan-yards, which left a fertile mix of equine excrement and powdered oak bark. He suggested that many more fungi might be consumed as 'wholesome and nutritive food', trying the taste of different kinds and floating the idea that the porous tupha stone of Matlock, Derbyshire, from which grottoes, walls and many houses were constructed, might be ideal for rearing mushrooms because of its similarity to the mushroom stone of Italy, which was 'indurated' (hardened) with mould and produced copious fungi in Roman and Neapolitan dwellings.[19] Yet, as the devastation caused by fungal diseases demonstrated, and the ease with which they were transmitted by airborne 'animated' spores made clear, there was 'scarcely a leaf (at least of trees and shrubs)' that fell 'to the ground that has not its peculiar *fungus*' which, 'assisted by its humidity', reduced it to 'its original earth'. While the

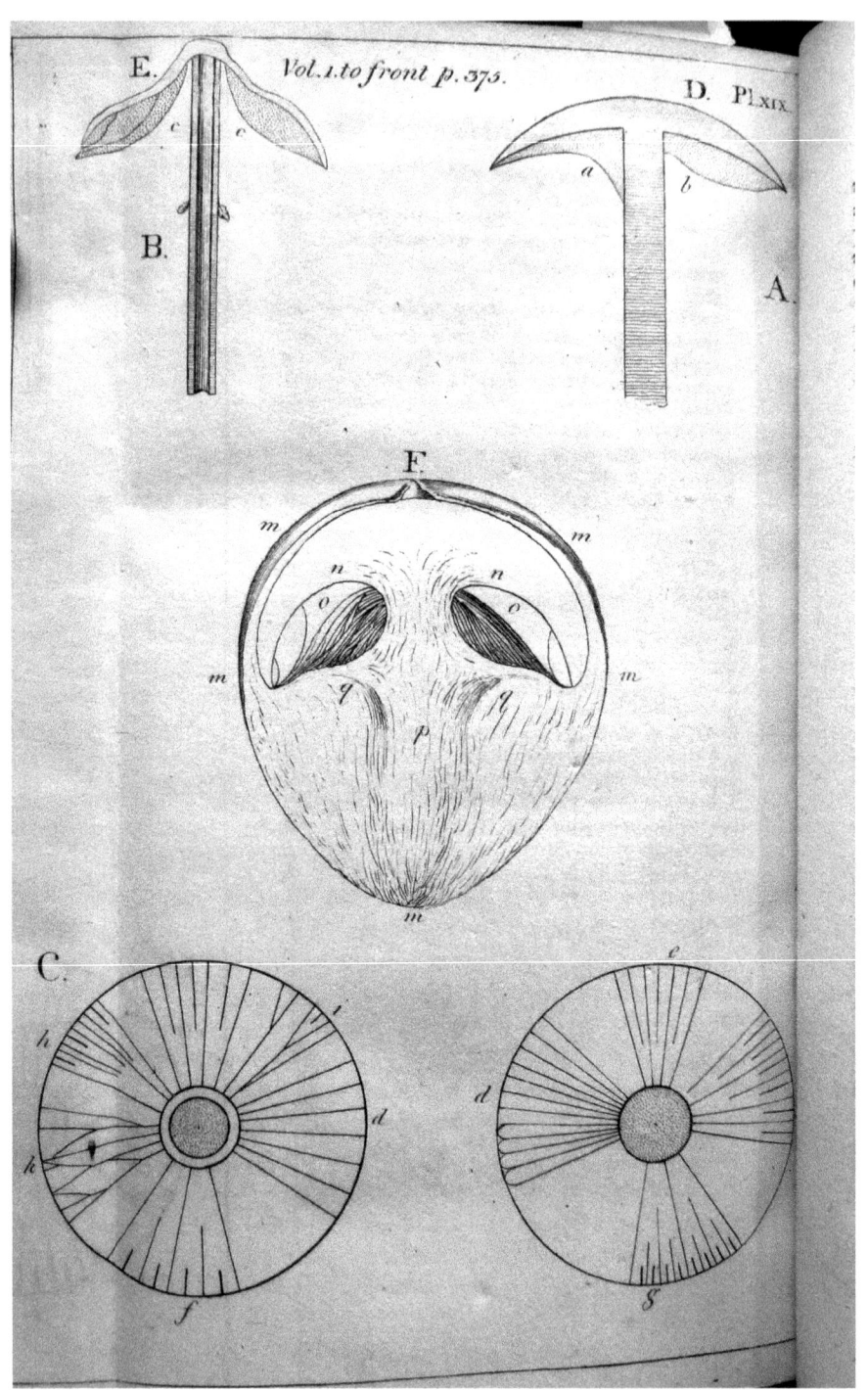

Figure 37. Fungi, from W. Withering, *Systematic Arrangement of British Plants*, 5th edn, 4 vols (1812), vol. 1, plate xix.

subject remained sketchy to naturalists, therefore, it was clear that, through their role in decomposition, fungi played a major part in the 'plan of divine providence' and were certainly not 'deficient or superfluous', despite the injurious effects of some on crops.[20]

As in medical practice, it was possible to examine the symptoms of plant diseases and consider treatment and prevention methods; and, as with animal diseases, it was easier to undertake experiments than with humans. There were, however, important differences between studies of plant pathology and medicine, including the fact that, with the important exception of especially valued specimens such as veteran or prized fruit trees, the former was normally focused upon field or crop populations rather than individual plants. Furthermore, unlike medical practice, in which the value of the individual was economically or emotionally important, or even animal pathology, the economic significance of individual field plants was relatively small, while, although Darwin repeatedly emphasised parallels between all animate physiologies, vegetable physiology and behaviour were less complex than their animal equivalents.[21]

Having briefly outlined some plant diseases in *The Economy of Vegetation* (1791), Darwin addressed the problem much more extensively in *Phytologia*, again employing favourite close analogies between animal and vegetable bodies and his medical experiences, and drawing upon his careful observations of gardening and midlands farming and horticulture.[22] The impressive range of authorities from botany, natural history, agriculture, horticulture and medicine that he cited in relation to vegetable pathology demonstrates the growing interest in the subject and the extent of his research using the Derby Philosophical Society library and other sources. These included the work of Christoph Girtanner (1760–1800), the Swiss doctor, naturalist and friend of Thomas Beddoes (1760–1808), and the German botanist and forester Julius Heinrich von Uslar (1752–1829). Equally, Darwin obtained ideas from the Swedish naturalist and explorer Pehr Kalm (1716–1779), the French botanist and mechanic Henri-Louis Duhamel du Monceau (1700–1782), the Italian naturalist Lazzaro Spallanzani (1729–1799) and British writers such as Thomas Andrew Knight.[23] Darwin also acquired the French physician and agronomist Henry-Alexandre Tessier's study of grain diseases and methods of prevention for the society library, which was based upon experiments undertaken as director of the royal farm at Rambouillet.[24]

Darwin largely confined his analysis to diseases of irritability, but, as with humans and animals in *Zoonomia*, he argued that an 'excess' of even 'the most salutory stimuli', such as 'the fluid element of heat' or electrical effluvia, was 'deleterious both to vegetable and animal bodies'. Just as the healthiest animals were those in equilibrium – with a sufficient quantity of vital spirit – and the unhealthiest those in disequilibrium – with either an abundance or privation of animal spirit – so plant health depended upon maintaining stable levels of the spirit of vegetation, which translated external environmental impressions into physiological motions through association like their bestial cousins.[25] The many human illnesses resulting from over-stimulus (such as being confined to over-heated rooms), over-eating, excessive circulation of bodily 'juices' such as blood, over-production of bodily fluids or too much growth were paralleled by vegetable ailments resulting from exposure to excesses of heat, light, water or external stimuli, which was why 'wise' gardeners placed so much 'regard' upon the 'habits of tender vegetables', accustoming or training them to moister or colder conditions by gradually cooling or watering using tepid water rather than rapidly moving them from dry to wet or warm to cold environments, and vice versa.[26] Darwin accordingly strove to categorise vegetable pathologies along similar lines to those he had adopted for animal ailments by distinguishing those internally caused from those with external origins, although it was difficult to relate the former to his emphasis upon excitability and accommodate animal and insect depredations within a model of vegetable spirit equilibrium. As in his medical practice, he tended to pragmatically favour 'traditional remedies' for plant ailments that had shown their efficacy through long-established practice and seek to accommodate these within his new system.[27]

IRRITABILITY AND PLANT DISEASES

As noted above, Darwin's analysis divided vegetable diseases into those that appeared to result from 'internal causes' or 'diseased irritability', and those with 'external elements', such as those originating from insect or animal depredation. In this he followed the French botanists Joseph Pitton de Tournefort (1656–1708) and Michel Adanson (1727–1806), who likewise distinguished between bodily and environmental factors, and also the causes of diseases identified by Linnaeus, although most botanists, including Johan Christian Fabricius (1745–1808), Johan Baptista Zallinger zum Thurn (1731–85) and Joseph Jakob Plenck (1735–1807), tended to classify plant pathologies by symptoms and the vegetable parts affected, encouraging parallels with human diseases.[28] As we have seen, for Darwin buds were 'individual beings' and 'an inferior order of animals', they therefore possessed irritability, sensibility, voluntary and 'associations of motion' in his system. Because they possessed the last three 'in a so much less degree … than by more perfect animals', he focused only upon 'diseases of their irritability'.[29] Just as animals obtained oxygen and thence their vitality and irritability from the air as well as food, so vegetables similarly obtained it 'respired by the upper surfaces of their leaves' rather than through their roots. This oxygen was 'separated' from plant juices by the 'sensorium, or brain, of each individual bud' after changes in the 'circulation or secretion of it' had occurred. There were therefore instructive parallels between diseases of irritability associated with humans and those 'attending vegetable irritability', as this was 'similar to' that belonging to animals, but 'upon a less extensive scale'.[30]

Darwin believed that a good example of such parallels was his contention that human health required 'perpetual variations' of atmospheric heat to 'preserve or restore the irritability and consequent activity' of bodily systems. This was demonstrated by the fact that human 'health and energy' was stronger and lives lengthier in the variable temperate British climate than in tropical countries, with their greater heat and climatic uniformity.[31] Using his system of psychophysiology to explain plant behaviours that were very familiar to gardeners, he argued that, as in animal bodies, recurrent stimulation of the 'vegetable fibres' by natural or artificial heat 'exhausted' the 'spirit of vegetation', making them more sensitive to smaller reductions in temperature. Hence sub-tropical plants in 'northern climates' required special protection in spring, with its tendency to sunny days and cold nights, and plants kept in warm rooms in winter ought to be 'occasionally … exposed to cooler air to increase their irritability', or spring growth would be retarded. Conditions caused by excessive stimulus, exhausting the plant, included those where debility occurred through too much sunshine or over-watering on a hot day. Irritability might also be 'accumulated' by lack of stimulus, meaning that plants carried from colder to warmer climates, for example, needed careful protection from over-stimulation (such as immediate exposure to sunlight) because they had accumulated the 'spirit of vegetation' through the 'habits' they had acquired through exposure to the previous conditions. For Darwin, 'acquired habits' were something he was particularly interested in as a medical practitioner because he believed they often caused or exacerbated illnesses. In this context, habits were powers or abilities of both animal and vegetable beings that became more 'catenated' or fixed through repetition of behaviours and physiological associations between 'fibrous contractions' induced by deliberate acts or circumstances. This meant these habituated powers, abilities or actions subsequently occurred automatically without conscious intention or act of will if, for example, similar environmental circumstances that had induced them in the first place occurred again. It also followed that the state of vegetable irritability altered seasonally and differently during the course of each day across the year, being increased after exposure to excessive cold in the morning and decreasing in the hot afternoon of a summer's day. Darwin believed that understanding of these daily and seasonal cyles of vegetable irritability meant that it was possible to draw parallels with animals to show how heat variations contributed to health (for example, in relation to hibernation) and to suggest potential treatments for some plant pathologies. Mildew, for example, which was attributed to a 'sessile fungus', could be treated by inducing greater irritability by exposing the plant to additional light and ventilation, draining the land or sowing early. Diseases that Darwin attributed to the accumulation of too much irritability included rubrigo, or rust; blight (*Uredo frumenti*); and ergot (*Claviceps purpurea*) on rye, which had been ascribed to insects by the French physician and botanist Henri-Louis Duhamel du Monceau. Likewise, smut (*Ustilago nuda*), which Linnaeus believed to be caused by insects, Darwin thought arose from 'want of impregnation' and could therefore be prevented (Figure 38).[32]

Exposure to too much cold caused the accumulation of the vegetable spirit, weakening or killing plants when sudden heat occurred, which, through 'too great increase of action' caused 'inflammation', 'mortification' and destruction, just as warming cold human limbs too quickly caused pain. Beans flooded by the cold waters of the Derwent near Darwin's house in June 1798 died because they were rendered 'much enfeebled' or 'inirritable' by summer heat. These conclusions were supported by experiments and observations such as those undertaken by Johann Julius Von Uslar (1762–1838) with light and heat on *Euphorbia*, and the fact that plants 'acquired habits' seasonally, potatoes and onions needing less heat in spring than autumn to germinate, for example, while the Scottish Enlightenment surgeon–agriculturist George Fordyce had shown in his *Elements of Agriculture* (1771) that 'grains and roots' brought to Britain from 'southern latitudes' germinated sooner than those acquired from 'more northern

Figure 38. Smut of corn, from W. G. Smith, *Diseases of Field and Garden Crops* (London, 1884).

ones'. Equally, Pehr Kalm, in his *Travels into North America*, had found that Swedish apple trees planted in New England initially blossomed too early and remained fruitless until they got used to the local climate, while hothouse vines wintered outside grew more rapidly than their cousins placed within.[33] Likewise, experiments with tree sap showed increased irritability of plants to heat after sustained exposure to cold; similarly, lucerne, olive trees and walnut died after a milder French winter when English walnuts survived after colder conditions, according to Jethro Tull.[34] Finally, Darwin believed the botanists Friedrich Kasimir Medikus (1736–1808) and Johann Julius von Uslar (1762–1838) had demonstrated the variability of vegetable irritability during the course of the day and in different weather conditions, which paralleled greater animal sensibility to heat after exposure to cold, as demonstrated by the death of insects in a warm spring after surviving underground in winter, according to the Italian philosopher Lazzaro Spallanzani (1729–1799).[35]

Primarily based upon his observations of diseases of trees and shrubs, Darwin described considerably more 'internal diseases' that the four listed by Linnaeus, which were: *'eurisiphe*, mildew; *rubigo*, rust; *clavus*, ergot, or spur; and *uslilago*, smut'. The classification of these vegetable ailments as 'internal diseases', however, presented some difficulties for his system because of the growing evidence during the seventeenth and eighteenth centuries, including the microscopic observations of Felice Fontana (1730–1805) and Giovanni Targioni-Tozzetti (1712–1783) in Italy during the 1760s, that 'plants of the fungus kind' were responsible for much of the physiological destruction. Darwin explained that *Erysiphe* was a 'white mucor', mould, or mildew that covered leaves with 'sessile tawny heads' and often attacked hops, dead-nettle, archangel and maples. While it was a fungus type 'plant', which could grow without light and penetrate host vegetable 'vessels', Darwin argued that affected plants were probably already rendered vulnerable by an 'internal disease'. Similar diseases were rubigo, or rust, which was frequently seen on lady's mantle (*Alchemilla mollis*) as a fine powder on the underside of leaves, and the black fungus labelled by the British botanist Aylmer Bourke Lambert (1761–1842) as *Uredo frumenti*, or wheat blight. Vegetable vitality and 'internal rigour' might therefore be restored to help combat fungal attacks by early seasonal sowing, thinning for light and ventilation, clearing overshadowing vegetation, draining land more effectively to prevent damp or using drier manures, such as bone or coal ash.[36]

It was believed that smut, which affected wheat, barley and oats so that they produced a 'black meal' instead of seed, might be prevented by steeping grain in brine, lime water or 'an alkaline ley' of potash and limewater before sowing. However, drawing upon work published by an anonymous member of the Bath and West Agricultural Society in their transactions in 1790, which he borrowed from the Derby Society library, Darwin maintained that as smutty ears and good ones sometimes grew from the same root so the disease must be caused by 'want of impregnation' as a result of defective male seed from the anthers, leading to putrefaction in the same way as unfertilised poultry eggs, again putting the emphasis upon 'internal' causation. The 'serious evil' of smut decimating cereal crops might therefore be prevented, Darwin thought, by sowing grain in rows widely apart and then alternating these with other kinds of wheat or the same plant a few days later so though heavy rain might damage one set of anthers, those in other rows had a better chance of surviving unscathed.[37] Finally, Darwin noted that, besides the four 'internal diseases' described by Linnaeus and Lambert's wheat blight, there were 'probably many others' that had hitherto been insufficiently studied, including 'the canker', 'gangrene', 'honey-sweat', 'miliary sweat', 'sap-flow' and 'gum secretion', which we will examine in the final chapter.[38]

Darwin's vegetable pathology, and especially his emphasis upon ailments caused by internal factors, helped to demonstrate the operation of 'acquired habits of vegetable actions' or 'associations of motions', underlining parallels between the psychophysiology of all animate beings, which was central to Darwinian evolutionary theories because complex behaviour worked through association and the propensity for animal motions to become connected by habit – so they automatically accompanied or succeeded each other. Life habits were therefore applied in turn to succeeding generations which partook of parental 'form and propensities'.[39]

EXTERNAL CAUSES OF DISEASES

Darwin also examined diseases of food crops that he believed resulted from 'external causes', which included a very wide range of inanimate phenomena that caused problems to humans and animals for similar reasons, such as excessive wind and water, lightning and the depredations of insects and vermin, suggesting various measures or treatments that might alleviate or prevent them. Again, even though the impact of extra-bodily factors was acknowledged, the role of internal physiology in determining the differential impact of vegetable pathologies was emphasised, alongside the state of the soil: the presence or otherwise of 'noxious materials' and degrees of acidity, siliceousness, sandiness and stability. Unless plants were covered over or otherwise protected, weather extremes such as excessive heat, cold, light or rain caused major problems, with heavy winters destroying early fruit or ash tree shoots, which were 'more succulent' or had less sensibility or irritability.[40] Vegetable health was also adversely affected by 'noxious exhalations diffused in the atmosphere' around 'some manufactories', which were 'said to injure' their growth or even destroy them completely; smoke from lead-smelting furnaces, potteries and lime kilns was widespread across Staffordshire and Derbyshire, while marine salt and acidity were problems in coastal areas.[41] Having seen one of his Derby apricot trees killed and an apple tree damaged by lightning while writing *Phytologia*, Darwin believed that such strikes were more common than usually realised, emphasising that the devastation was caused by a massive stimulus exhausting 'the sensorial power' and producing 'total inirritability' to the kind of 'common stimuli' normally exciting vegetable 'vital actions'. In this there were, therefore, close parallels with the action of poisons upon the 'animal system', such as arsenic, the 'contagious matter of fevers' and even the everyday, mundane emetics he prescribed as a physician to induce vomiting, which 'inverted the natural order' of 'successive irritative motions' of the stomach, duodenum and oesophagus because of 'previous exhaustion of their sensorial power' by violent stimuli.[42]

Similarly, from a medical perspective Darwin had often seen the detrimental impact upon human and animal health of what he called 'condiments', or substances that seemed to 'possess stimulus without nutriment', such as spice, salt, bitters, hops and, in all likelihood, opium and 'vinous spirit' too, which he emphasised in his correspondence and in *Zoonomia*. While some of these were derived from plants, a set of equivalent materials or substances caused parallel problems to vegetable bodies for similar reasons. Thus, though containing no nutrients and even sometimes poisonous, the use of common salt in agriculture might sometimes hasten growth in plants like fresh herbs, broccoli or artichokes by stimulating the 'absorbent vessels' of plants into 'greater action', resulting in greater food consumption or more vigorous internal circulation or secretion.[43]

Insects and animals were, of course, another major 'external' threat to plants of field, kitchen garden and orchard, and improving associations such as the Bath and West of England sought to promote original methods of dealing with these, offering prizes for those apparently most effective and publishing articles on the subject in their transactions.[44] Informed by his knowledge of the work of naturalists such as John Wagstaff of Norwich, his discussions with farmers and landowners and his knowledge of animal behaviour, Darwin recommended a mixture of interventions to control predatory insect and animal populations, including alterations to the landscape, environment and local flora and fauna, as well as the use of traps and poisons. To prevent field mice and rats from attacking wheat, peas and beans and undermining newly ploughed lands he advocated encouraging owl breeding nearby and stopping 'servants and children' from destroying their eggs, employing 'callow young' to keep numbers down, and the use of poisons and cheese-baited traps.[45] Water rats had appeared to supplant 'house' rats over the previous half-century; these lived on river and pond banks, and Darwin had seen them burrowing below and devouring water-plantain, fruits, foliage, roots and even young animals like rabbits and ducklings (Figure 39). Their 'ingenuity' enabled them to construct 'houses' similar to beavers' near river banks and pools, with two entrances, one within the grass and the other underneath the water surface, which meant that if waterside banks were kept low and free from reeds, nesting would be discouraged. Their 'lascivious nature', which Darwin likened to that of dogs, and attraction to strong scents that probably resembled their 'venereal orgasm' (again, he thought, like dogs), meant that they were attracted by smells similar to those associated with their sexual behaviours, a trait that could be employed against them by rat catchers, who mixed favoured foods with poisons, to deadly effect.[46]

VEGETABLE DEFENCES

While a few naturalists, such as Ralph Austin (c. 1612–1676), Giovanni Targioni-Tozzetti and Thomas Andrew Knight, noted differences in the characteristic responses of plants to particular diseases – such as Knight's observations that some wheat varieties escaped rust epidemics in 1795 and 1796 – Darwin placed considerably more weight upon the active agency of plants in resisting depredations.[47] His emphasis upon

Figure 39. Water rat, from T. Bewick, *A General History of Quadrupeds*, 7th edn (1820).

internal causation presented some difficulties in explaining the growing evidence for fungal pathogenicity, but the parallels he drew with animal physiology and health, combined with his extensive experience of observing patient resistance in medical practice and his confidence in the prodigality and tenacity of life, encouraged him to highlight the array of defences plants employed against their enemies. In his analysis of plant disease he argued that there was a 'power impressed on organised bodies by the great author of all things' that enabled them to 'not only increase in size and strength' from youth to maturity but to 'occasionally cure their accidental diseases', repair their wounds and produce 'armour' to defend against the most destructive attacks. Plants were therefore not merely passive victims of insect or animal attacks but agents who had armed themselves to 'prevent' 'violent injuries' that would have otherwise destroyed them and even 'repair' these, sometimes with benefits for their human cultivators that might be harnessed for agricultural or horticultural improvement. As we have seen in relation to their usage in medicine, the powers of resistance marshalled by some plants included 'poisonous juices' such as those produced by the deadly nightshade, henbane or hound's-tongue (Figure 40). Other examples included the secretion of 'viscid fluid to agglutinate the insects' unfortunate enough to fly or crawl towards 'their fructification', characteristic of plants such as catchfly (*Silene*) and sundew (*Drosera*), or the 'thorns and prickles' of holly (*Ilex*) and hawthorn (*Crataegus*), which warded off animals. Likewise, the contraction of leaves and petals to 'destroy insects', as seen in plants such as the Venus fly trap (*Dionaea muscipula*) and spreading or fly-trap dogbane (*Apocynum androsaemifolium*), fascinated Darwin almost as much as it did grandson Charles (Figure 41). Similarly, the 'very numerous bristles' on the

Figure 40. Henbane, hound's tongue and other medicinal plants, from J. Hill, *Family Herbal* (1812).

Figure 41. Venus flytrap, *Dionaea muscipula*, from E. Darwin, *The Loves of the Plants*, 4th edn (1799).

Figure 42. *Cyprepedium*, from E. Darwin, *The Economy of Vegetation*, 4th edn (1799).

tender or 'uppermost parts' of the 'young shoots' of the hazel (*Corylus avellana*) and other apparently invisible defences, such as the 'secreted juices' Darwin believed were produced by some plants, were used to trap or poison insects; this, he thought, was no 'more astonishing' than the prickles of holly warding off animals in Needwood Forest, and, where plants lacked such 'an armour purposely produced' to defend themselves from such 'destructive' attacks, such as his plum trees beside the Derwent, they were destroyed by aphids.[48]

Another example of the sophistication of vegetable defences was the brightly coloured orchid *Cyprepedium* from America, which had a large globular 'nectar about the size of a pigeon's egg'. This was of 'fleshy colour' with a 'depression' in the upper part, and the plant employed her 'successful guile' to imitate the 'bloated paunch and jointed arms' of spiders, which Darwin believed enabled her to prevent humming birds from stealing her 'honey' (nectar) (Figure 42).[49] He was convinced that bees, moths and butterflies were 'very injurious to flowers' and, therefore, fruit production because they plundered 'the nectarines of their honey' and deprived anthers and stigmas of their 'adapted nourishments'. Some aspects of floral physiology, such as 'long winding canals' in honeysuckles, trefoils, lark-spurs and delphiniums, or the eponymous hoods on monkshood, were therefore designed to defend nectar 'reservoirs' from such attacks upon their 'seminal products'. However, many flowers did produce 'more honey' than was necessary for self 'consumption' and the abundance of seeds and fruits meant that bee 'depredations' were 'not counteracted like those of other insects' but in fact 'encouraged', so that plants generally supported insect life without debilitating their own.[50]

CONCLUSION

Darwin's recommendations concerning plant and animal diseases and his understanding of the efficacy of organic substances as medicines were underpinned by his belief in the kinship of all living entities. In *Phytologia* he strove to utilise his knowledge of plant physiology, anatomy and taxonomy, which was lacking in most Georgian agricultural texts, and drew analogies between all organic entities to help place the subject upon a more scientific footing. As Maureen McNeil has emphasised, *Phytologia* was 'by far the most ambitious British project in vegetable physiology undertaken before the second half of the nineteenth century'. Darwin's 'concern for theory' differentiated it from any of his predecessors' works, in that it did not lead to mere abstraction. As in *The Loves of the Plants* (1789) – which generated some controversy – he always tried to portray animated 'vegetative processes' using images of passionate, competing plants as sexual, sensual, secreting, loving, sleeping, healthy or diseased beings that were 'graphic, abundant and unwieldy' but which 'easily related to human experiences'.[51] As the analogies he drew between the spirit of animation and the electricity produced by animals such as the electric eel (*Gymnotus electricus*) demonstrate, Darwin made few sharp physiological distinctions between humans, animals and plants, believing that all animate creatures shared common characteristics, which helped determine his approach to the role of the sciences in agricultural improvement. The spirit of animation was a property of 'animal life' which mankind possessed 'in common with brutes, and in some degree even with vegetables'.[52] Plants were 'an inferior order of animals' and the motions of species such as the Venus fly trap and *Mimosa* (or 'sensitive plant') demonstrated that there were 'not only muscles' surrounding the 'moving foot-stalks', or what Darwin called the 'claws of the leaves and petals' (which deliberately trapped the insects), but that these were 'endued with nerves of sense as well as of motion' and that there had to be a 'common sensorium, or brain, where the nerves communicate' with vegetable buds possessing irritability, sensation, volition and 'association of motion', though in 'a much inferior degree' than even 'cold blooded animals'.[53] Plants, therefore, with their moving leaves, apparent ability to sense changes in the external environment and nerve centres, had much in common with animals – a notion that inspired the younger generation of writers, such as Percy Bysshe Shelley (1792–1822), whose poem *The Sensitive Plant* (1820), as King-Hele has emphasised, depicted it in an eroticised fashion as 'full of love', delicately attuned to changes in its environment and kissing the night.[54] Other physiological similarities between organic entities, as Darwin saw it, included the significance of electricity as an ethereal fluid vital to animation for both plants and animals, which opened the door to medical electricity and potential opportunities to exploit atmospheric electricity and meteorology to improve farming.

NOTES

[1] K. Thomas, *Man and The Natural World: Changing Attitudes in England, 1500–1800* (London, 1983), 17–41.

[2] A. MacGregor, *Curiosity and Enlightenment: Collectors and Collections from the Sixteenth to the Nineteenth Century* (New Haven, 2007); P. A. Elliott, C. Watkins and S. Daniels, *The British Arboretum: Trees, Science and Culture in the Nineteenth Century* (London, 2011); P. A. Elliott, *Enlightenment, Modernity and Science* (London, 2010).

[3] G. White, *The Natural History and Antiquities of Selbourne* (1789), edited by E. T. Bennett and J. E. Harting (London, 1877); R. Mabey, *Gilbert White: A Biography of the Author of the Natural History of Selbourne* (London, 1987).

[4] See, however, P. Ayres, *The Aliveness of Plants: The Darwins at the Dawn of Plant Science* (London, 2008), which emphasises the Darwin family interest in botany through the generations. Charles Darwin's botany is examined in M. Allan, *Darwin and His Flowers: The Key to Natural Selection* (London, 1977); K. Thompson, *Darwin's Most Wonderful Plants: A Tour of His Botanical Legacy* (Chicago, 2019).

[5] R. Schofield, *The Lunar Society of Birmingham: A Social History of Provincial Science and Industry in Eighteenth-Century England* (Oxford, 1963), 398; M. McNeil, *Under the Banner of Science: Erasmus Darwin and his Age* (Manchester, 1987), 168–9.

[6] E. Darwin, *Phytologia; or the Philosophy of Agriculture and Gardening* (London, 1800), dedication, vii–viii.

[7] E. Darwin, *The Economy of Vegetation*, 4th edn (London, 1799), 229–31.

[8] J. Kennedy, *Treatise upon Planting, Gardening and the Management of the Hot House*, 2nd edn, 2 vols (London, 1777), vol. 2, 151–81.

[9] Darwin, *Economy of Vegetation*, 230; S. A. D. Tissot, 'An account of the disease called ergot, in French, from its supposed cause, viz., vitiated rye', *Philosophical Transactions*, 55 (1765), 106–25; J. Grier, *A History of Pharmacy* (London, 1937), 68–77.

[10] G. C. Ainsworth, *Introduction to the History of Plant Pathology* (Cambridge, 1981), 24.

[11] G. Rude, *The Crowd in History, 1730–1848* (London, 1981), 33–46; E. P. Thompson, 'The moral economy of the English crowd in the eighteenth century' in *Customs in Common* (London, 1991), 185–258.

[12] C. Creighton, *A History of Epidemics in Britain*, 2 vols (London, 1894), vol. 2, 130–2; Rude, *Crowd in History*, 38–43; M. Thomas, 'The rioting crowd in eighteenth-century Derbyshire, *Derbyshire Archaeological Journal*, 95 (1975), 40–5. Derbyshire lead miners rioted over food in 1757, 1796 and 1800.

[13] Creighton, *History of Epidemics*, vol. 2, 236–43, 249, 256–94; C. Woodham-Smith, *The Great Hunger: Ireland, 1845–9* (London, 1962), 38–9; H. M. Ward, *Disease in Plants* (London, 1901), 150–2.

[14] *Derby Mercury*, 15 June, 12 October 1797, quoting agricultural reports from *Morning Herald*; W. Kirby, 'Observations upon certain fungi, which are parasites of the wheat', *Linnean Transactions*, 5 (1800), 112–25; J. Banks, *A Short Account of the Causes of the Disease in Corn, Called by Farmer's the Blight, the Mildew and the Rust*, 2nd edn (London, 1806), 10; Thomas, 'Rioting crowd in eighteenth-century Derbyshire', 40–5. Similar problems were reported in 1804: *Derby Mercury*, 9 August 1804; Creighton, *History of Epidemics*, vol. 2, 156–9.

[15] *Derby Mercury*, 3 August 1786; 19 July 1792.

[16] J. Tull, *Horse-Hoeing Husbandry, or an Essay on the Principles of Vegetation and Tillage*, 3rd edn (London, 1751), 139–57; Kirby, 'Observations upon certain fungi', 114; Banks, *Short Account*, 1–2; J. C. Loudon, *An Encyclopaedia of Agriculture*, 5th edn (London, 1844), 258–63; W. G. Smith, *Diseases of Field and Garden Crops* (London, 1884); Ward, *Disease in Plants*, 159–67; E. C. Large, *The Advance of the Fungi* (London, 1940); G. L. Carefoot and E. R. Sprott, *Famine on the Wind: Plant Diseases and Human History* (Chicago, 1967); Ainsworth, *History of Plant Pathology*, 4–5.

[17] H. Bryant, *A Particular Enquiry into the Causes of that Disease in the Wheat Commonly Called Brand* (Norwich, 1783); Kirby, 'Observations upon certain fungi', 123; Banks, *Short Account*; Ward, *Disease in Plants*, 85–7.

[18] Lichfield Botanical Society, *A System of Vegetables ... translated from the thirteenth edition of the Systema Vegetabilium of the late Professor Linneus*, 2 vols (Lichfield, 1783), vol. 2, 789–838; Lichfield Botanical Society, *The Families of Plants ... Translated from the Last Edition of the 'Genera Plantarum'*, 2 vols (Lichfield, 1787), vol. 2, 737–51; J. Sowerby, *Coloured Figures of English Fungi or Mushrooms*, 4 vols (London, 1791); W. Withering, *A Systematic Arrangement of British Plants*, 5th edn, 4 vols (Birmingham,1812), vol. 3, 920–1037, vol. 4, *passim*; J. E. Smith, *An Introduction to Botany*, 5th edn (London, 1825), 394–407, quotation, p. 394.

[19] Darwin, *Phytologia*, 442–7.

[20] Kirby, 'Observations upon certain fungi', 124–5; Banks, *Short Account*, 7–8.

[21] H. H. Whetzel, *An Outline of the History of Phytopathology* (Philadelphia, 1918); Ainsworth, *Introduction to the History of Plant Pathology*, 4–6.

[22] Darwin, *Phytologia*, 285–337.

[23] N. Tsoupoulos, 'The influence of John Brown's ideas in Germany', in W. F. Bynum and R. Porter eds, 'Brunonianism in Britain and Europe', *Medical History* supplement no. 8 (1988), 63–74.

[24] H. A. Tessier, *Trait des Maladies des Grains et des Farines* (Paris, 1783); Derby Philosophical Society, *Rules and Catalogue of the Library of the Derby Philosophical Society*, with supplements of 1795 and 1798 (Derby, 1795–8), 25.

[25] Darwin, *Phytologia*, 282, 286–8; McNeil, *Under the Banner of Science*, 196–7.

[26] Darwin, *Phytologia*, 280.

[27] Darwin, *Phytologia*, 285–337; McNeil, *Under the Banner of Science*, 196.

28 Darwin, *Phytologia*, 285–5; Ward, *Disease in Plants*, 99–118; Ainsworth, *Introduction to the History of Plant Pathology*, 21–3.

29 Darwin, *Phytologia*, 287–8.

30 Darwin, *Phytologia*, 287; E. Darwin, *Zoonomia; or the Laws of Organic Life*, third edition, 4 vols, (London, 1801), vol. 1, 94–104.

31 Darwin, *Phytologia*, 290.

32 Darwin, *Phytologia*, 287–92; Darwin, *Zoonomia*, vol. 1, pp. 42–51; on the significance of Darwin's conception of habits see R. Porter, 'Erasmus Darwin: doctor of evolution?' in J. R. Moore ed., *History, Humanity and Evolution: Essays for John C. Greene* (Cambridge, 1989), 39–70, pp. 48–9.

33 Darwin, *Phytologia*, 288–90; P. Kalm, *Travels into North America*, translated into English by J. R. Forster FAS, 2 vols (London, 1771); G. Fordyce, *Elements of Agriculture and Vegetation* (London, 1771).

34 Darwin, *Phytologia*, 289; Tull, *Horse-Hoeing Husbandry*, 3rd edn, 201–2; J. Walker, 'Experiments on the motion of sap in trees', *Transactions of the Royal Society of Edinburgh*, vol. 1 (1788), no. 2, 19.

35 Darwin, *Phytologia*, 289.

36 Darwin, *Economy of Vegetation*, 229–30; Darwin, *Phytologia*, 290–1; A. B. Lambert, 'Description of the blight of wheat, *Uredo Frumenti*', *Transactions of the Linnean Society*, 4 (1798), 193–4; Ainsworth, *Introduction to the History of Plant Pathology*, 26–8.

37 Darwin, *Economy of Vegetation*, 229; Darwin, *Phytologia*, 292–3; Anonymous, 'An enquiry concerning the smut in wheat', *Letters and Papers in Agriculture, Planting, etc Addressed to the Society Instituted at Bath for the Encouragement of Agriculture, Arts, Manufactures and Commerce*, vol. 5 (Bath, 1790), 244–56, which examined its causes, the means of preventing it and remedies that might be employed to alleviate it; Smith, *Diseases of Field and Garden Crops*, 254–62.

38 Darwin, *Phytologia*, 293.

39 *Phytologia*, 123–4; *Zoonomia*, vol. 1, 61–6, vol. 2, 270.

40 Darwin, *Phytologia*, 300–4; Ward, *Disease in Plants*, 99–118.

41 Darwin, *Phytologia*, 303–4.

42 Darwin, *Phytologia*, 301–2; Darwin, *Zoonomia*, vol. 2, 519, 523–4.

43 Darwin, *Phytologia*, 304–7.

44 K. Hudson, *Patriotism with Profit: British Agricultural Societies in the Eighteenth and Nineteenth Centuries* (London, 1972), 1–23.

45 Darwin, *Phytologia*, 333; J. Wagstaff, 'On field mice and on the transplantation of wheat', *Letters and Papers on Agriculture, Planting, etc. Selected from the Correspondence of the Bath and West of England Society*, 6 (1792), 127–31.

46 Darwin, *Phytologia*, 333–5.

47 Ainsworth, *Introduction to the History of Plant Pathology*, 162–6.

48 Darwin, *Phytologia*, 315, 317–18.

49 Darwin, *Economy of Vegetation*, 228–9.

50 Darwin, *Phytologia*.

51 McNeil, *Under the Banner of Science*, 193–5.

52 Darwin, *Zoonomia*, vol. 1, 109–10; W. Henly, 'Experiments and observations in electricity', *Philosophical Transactions*, 67 (1777), 130–1, 134.

53 Darwin, *Phytologia*, 39–40, 132–6.

54 D. King-Hele, *Erasmus Darwin and the Romantic Poets* (Basingstoke, 1986), 209–10.

CHAPTER 6

AMONG THE ANIMALS

In September 1780 two pampered pussies exchanged letters which were hand delivered to their respective residences. The Persian cat Snow Grimalkin, who lived in Cathedral Close, Lichfield, explained to his fellow feline, Miss Po Felina of the Bishop's Palace, that he had spied her in her 'stately' abode washing her 'beautiful round face' and 'elegantly brinded ears' with her 'velvet paws' while swishing her 'meandering tail' with 'graceful sinuosity'. The 'treacherous' Cupid had, however, hidden himself behind her 'tabby beauties', leading Snow Grimalkin to long for more sights of his beautiful amour, watching 'day and night' from his balcony and serenading her with songs echoing through the 'winding lanes and dirty alleys' of Lichfield, hoping that the still 'starlight evenings' might induce her to 'take the air' on the palace leads. However, despite these efforts, Po Felina sat 'wrapped in fur', 'purring with contented insensibility' and sleeping 'with untroubled dreams'; nevertheless, Snow Grimalkin hoped that the offering of numerous mice 'for your food or your amusement' and 'an enormous Norway rat' that covered his 'paws with its gore' might just be the present to secure the object of his attentions. Unfortunately, while Po Felina admired the 'spotless ermine' and 'tyger strength' of Snow's 'commanding form' and 'wit and endowments', she was too wary of his 'fierceness' to return his affections.[1]

While, of course, the exchange of letters between Erasmus Darwin and Anna Seward on behalf of their cats was meant to be playful and witty, in the Augustan tradition of mock-heroic and humorous writing, with each pet assuming something of the character of their owners, it also demonstrates the importance of pets in the Georgian household and the extent to which, as we shall see in this chapter, Darwin was a close observer of animal behaviour and a strong believer in animal agency and intelligence. As portrayed by Darwin, with his blood-stained paws and smooth white fur, Snow Grimalkin combined bestial behaviours that were vestiges of a natural state with the advantages of birth, breeding and beauty. This was a miniature tiger who was 'rough and hardy, bold and free', whose 'nervous paw' could take 'My Lady's lapdog by the neck' or 'with furious hiss assault the hen and snatch a chicken from the pen', rather than a cosseted cat. With its animality, Darwin's feline friend was unlike what Bowerbank has described as the 'denatured', 'rhetorical', sentimentalised mouse that Anna Letitia Barbauld (1743–1825) created to appeal to Joseph Priestley in the 'Mouse's Petition' (1773) or the animals employed as mere 'research object[s]', 'abstracted' from their 'native environments' and even from ethical concerns by Priestley in his experiments on airs. Snow Grimalkin's rugged but refined wildness contrasted with Po Felina's delicacy – as Teresa Barnard has observed, she was 'an indulged little pet that Seward trained to live peaceably with an assortment of tame birds: a canary, a lark, a robin and a dove'. It is striking that the letter Darwin's editor Desmond King-Hele regarded as one of his most 'carefully written', addressed to one of his closest and most intimate Lichfield associates, should be the one that employed the personification of a favourite family pet as a vehicle for flirtatiousness, both an assertion of semi-tamed wild animal agency and a means of expressing admiration for a talented friend. Despite the apparent whimsy and even the ludicrousness of the subject matter, as Seward later emphasised, she and Darwin put much effort into crafting these feline missives.[2]

While humans were idealised by the Georgians as ultimately moral and rational creatures, other animals were believed to operate through irrational emotions and baser instincts, a division reinforced by what was seen as a Christian emphasis upon creation being at the service of humanity.[3] Many creatures, such as 'exotic' or prized agricultural examples, were exhibited alive in menageries, zoos, museums, estate parks, as dead, dried, pickled or skeletal preserved parts, or as artistic representations in paintings or engravings also depicting places, their meanings changing according to context and serving as 'material knowledge in transit'; in Sam Alberti's memorable phrase, such images were 'chunks of mobile landscape (or seascape) ripped asunder and transplanted' to other places.[4] As the physician and writer John Aikin (1747–1822) observed in his 1777 essay on natural history and poetry, unlike the relatively static beauties of the vegetable and mineral kingdoms, animals had much 'in common' with the humans whose lives they were intertwined with through their wonderfully 'varied' 'habitations and pursuits'. They provided close 'companionship' and mutual attachment' and were sources of 'pleasing and sublime speculation' and of amusement and instruction, while bringing picturesque 'animation to the objects around them' in rural landscapes.[5] In this chapter we will argue that in building upon the analogies between animal and vegetable bodies that underpinned his understanding of physiology and anatomy Darwin broke down the barriers between humans and other living beings. In an age of revolution, he effectively dethroned humanity from the privileged semi-divine status at the apex of creation that it had presumed to hold for many millennia.

ANIMALS IN LATER GEORGIAN SOCIETY

The main objective of agricultural improvement was to increase the quality and efficiency of farming production from a human perspective and not directly to improve the lives of animals. Improved livestock farming, for instance, though only a relatively minor part of agriculture in the period, aimed to breed animals that produced more, and sometimes better quality, wool, meat, fat and other products. Robert Bakewell improved his Leicestershire flocks and herds by 'the marriages of those' with most desirable and evident 'properties' without 'regard to consanguinity of incest', the skeletons and animal parts on display for visitors at his Dishley farm near Loughborough

Figure 43.   The Leicestershire improved breed, from T. Bewick, *A General History of Quadrupeds*, 7th edn (1820).

being measured, weighed and celebrated as disembodied material products rather than as parts of sentient, intentional, living entities (Figure 43).[6] However, the kind of livestock breeding that privileged meat over fat promoted by Arthur Young, Bakewell, Francis Russell, Thomas Coke and King George III himself, supported by the Board of Agriculture, the Smithfield Club and county agricultural societies, was beyond the means of most farmers.[7]

Animals were largely judged on how economically useful they were, more rarely on the degree of pleasure they gave as domestic pets, in sport and, increasingly, from the later eighteenth century, as sweet, sentimentalised – and often needy – denizens of divine creation and conveyors of moral messages. Fashionable, amusing dialogues on the study of nature by writers such as the novelist Charlotte Turner Smith (1749–1806) and Quaker philanthropist Priscilla Wakefield (1751–1832), author of the popular epistolary *Introduction to Botany* (1796), aimed at children and 'young persons', intended to inculcate ideas concerning the divinity and 'order and harmony of visible creation'. Horses were at the apex of the bestial tree, with their noble equine forms celebrated in art and among breeders for their role in sport. Ralph Beilby (1744–1817) and Thomas Bewick (1753–1828), for instance, began their *General History of Quadrupeds* (1790) with those animals that 'materially contribute to the strength, the wealth, and the happiness of this kingdom': they started with the 'noble' horse, which included 'the hunter', the 'most useful breed of horses in Europe' (Figure 44). Domestic animals such as dogs and cats were also admired, although horses and dogs were seen as more fitting companions for humans in matters of hunting, whereas cats were regarded as more devious and selfish.[8]

Figure 44.   The hunter, from T. Bewick, *A General History of Quadrupeds*, 7th edn (1820).

The emphasis on economic and domestic utility was combined with intensely personal, everyday, intimate encounters between humans and beasts. Agriculture was the mainstay of the national economy and the majority of people lived in the countryside, although animals and humans resided in close proximity in towns too. Many kept chickens, goats and pigs, while horses were, apart from walking, the main mode of transport and trusted beasts of burden. Despite the fashionable impact of notions of sensibility evident in the nature study dialogues for children and young people alluded to above, with their encouragement of curiosity, sympathy and empathy and their condemnation of human abuse of God's creation, most people felt no contradiction between loving and tending animals, slaughtering them for food or other purposes or placing bets so they would race to exhaustion or fight to the death in front of baying crowds. As a major part of rural culture and countryside management, shooting and hunting ensured the tolerance and nurturing of many animals, such as horses for riding, foxes to chase, and dogs – such as new, more determined foxhound breeds – to assist, but animals were also required for a range of products, from leather for boots and fur for hats and coats to grease for muskets. Indeed, animal products underpinned much of the broader economy, domestic life and culture, from foodstuffs to clothing and medicines.[9]

Although there was clearly a marked asymmetry of power and a strong rhetoric of domination, humans and animals in early modern society lived an intertwined, interdependent existence, the latter helping to define the meaning of the former and contrawise. Idealised representations of especially prized or celebrated animals or animal and human interactions, such as the paintings of horses, dogs and hunting scenes by John Wootton (1682–1764) or the equine portraits and studies of anatomy and physiology of George Stubbs (1724–1806), highlighted human and animal homologies, as did many cartoons depicting political figures, usually unflatteringly in the case of the latter. Furthermore, some animal actions and behaviours tended to be anthropomorphised, although humans remained at the apex of God's creation, endowed with a degree of rational thought and moral elevation not possessed by other animate entities.[10]

The way that animals were presented in natural history changed from the seventeenth century, moving from collections of wondrous facts associated with bestiaries towards more systematic descriptions like the binomial classification system of Linnaeus, which was extended from plants to animals, though elements of the bestiary remained, if disguised. More traditional schemes also continued to be employed to name and locate animals in creation, such as those based upon the chain of being, which, as noted previously, posited an interlinked hierarchy of life and matter from the divine through humans and animals to plants and minerals. Likewise, the designation of 'kinds' such as the 'poultry kind' and the 'serpent kind' was used at a more popular level and the term quadruped was favoured to designate four-footed beasts, which were regarded as next in rank below human beings and above other creatures. Animal taxonomies were challenged by developments in breeding, researches into physiology and anatomy, analogies drawn across animate creation and western encounters with novel or exotic animals such as the Australian kangaroo and duck-billed platypus. More 'natural' taxonomies based upon structural affinities, following John Ray (1627–1705), rather than 'artificial' ones founded upon more limited 'essential' characteristics, as in Linnaean botany, encouraged the collection of a broader range of information, including appearance, size, habitat, geographical location and behaviour, to differentiate and identify creatures. The human position within taxonomies presented problems, with some naturalists, such as Linnaeus, emphasising similarities between humankind and apes with a designation of 'man-like apes' in his *Systema Natura* (1735), while the Scottish philosopher James Burnett, Lord Moboddo (1714–1799), argued that orang-utans (and other anthropoid apes) were a 'barbarous nation' of the human species as yet uncivilised by acquisition of language.[11]

The most influential works of natural history in Britain later in the eighteenth century, such as the *Natural History* of Count George Louis-Leclerc de Buffon (1707–1788), the *History of the Earth and Animated Nature* (1774) by Oliver Goldsmith (1728–1774) and Beilby and Bewick's *General History of Quadrupeds*, contained some fabulous information about beasts from across the globe intermixed with facts concerning their geography, appearance, behaviour and so on (Figure 45). These works were also increasingly provided with illustrations that sought to capture the special characteristics of each creature, such as the copper engravings that adorned editions of William Smellie's English translation of Buffon's *Natural History* or Bewick's wonderful woodcuts for his *General History of Quadrupeds*, while animal descriptions were also included in geographical works and accounts of explorations, such as Joseph Banks's

Figure 45.   The kangaroo, T. Bewick, *A General History of Quadrupeds*, 7th edn (1820).

description of his voyage as a naturalist with Captain James Cook (1728–1779). Exotic animals or ordinary animals doing extraordinary things were also encountered in wild beast shows, zoos, as stuffed exhibits in menageries and museums and as works of art.[12] The Derbyshire animal dealer and showman Gilbert Pidcock (1743–1810), for instance, toured Britain for three decades with various exotic beasts and ran a famous menagerie at the Exeter Exchange, London, after exhibiting a collection of preserved animals and live birds with musical instruments in his home town of Buxton during the early 1780s. His 'Grand Menagerie' made various visits to Derby on tour – exhibiting, appropriately enough, at the White Lion between December 1789 and January 1790 for a shilling admittance – and included two 'lion-tygers' (mountain lions), a Bengal tiger originally intended for the Prince of Wales, a Barbary ape, 'two Satyrs or Ethiopian savages' (probably baboons) and a 'talking cockatoo'.[13]

Natural histories tended to follow a hierarchy of animals, with horses, dogs and those of greatest agricultural or domestic significance receiving fullest and most favourable description. There was some continuity with earlier bestiaries in the range of information presented, with 'exotic' wild animals viewed as strange and sometimes dangerous, although increasingly subject to human control.[14] Some creatures, because they were perceived to cause diseases or depredation of crops or livestock, were regarded as inimical to improvement and agricultural societies offered prizes for effective methods to combat or destroy these. The geologist and writer John Farey (1766–1826) followed the Board of Agriculture in describing slugs, wasps, rooks, tom-tits and sparrows, rats, mice and moles as not only 'obstacles' but 'enemies to improvement', recognising the scale of threat these animals were believed to pose but also implying a strong degree of agency opposed to – or at least independent of – human will.[15]

DARWIN ON ANIMAL PHYSIOLOGY AND BEHAVIOUR

The strength of Darwin's belief in the close analogies between all creatures is evident in his anthropomorphic attitude towards animals and their interactions with each other and humans. This was much more than abstract analogy; it informed and coloured his approach to living beings, and examples from the animal world underpinned his educational and moral philosophy. Darwin assembled much information from around the globe about animals for his analysis of instinct and learned behaviour, talking to pet owners, landowners, farmers and others who worked (or had to deal) with animals, such as mole catchers. Friends, patients and their families provided other information. He also mined printed works and the transactions of learned societies, some obtained through the Derby Philosophical Society library. These included, as might be expected, books by naturalists and travellers such as Buffon, Linnaeus, Goldsmith, Thomas Pennant, William Gilpin, Gilbert White and Michel Adanson, as well as studies of natural theology such as John Ray's *Wisdom of God Manifest in the Works of Creation* (1691) and William Derham's *Physico-Theology* (1713), classical works such as those of Aristotle and Plutarch and biblical sources.

The terms Darwin used in *Zoonomia* made little distinction between humans and other beasts and underscored the interdependency of animate beings in the natural economy, while his chapter on instinct crossed seamlessly between animal, vegetable and human forms. According to Darwin, all the 'internal [unconscious] motions of animal bodies' were based on irritation and proceeded with 'equal regularity' in both vegetable and animal systems. Other creatures were 'fellow animals' with bodies 'supported with bones' and 'covered with skins', which were 'moved by muscles', possessed 'the same senses' and 'appetites' and were 'nourished by the same aliment with ourselves', so on the 'strongest analogy' many of their 'internal faculties' were probably substantially similar to those of humans.

Darwin's medical perspective and observations of animal bodies and behaviour reinforced this emphasis upon physiological similarities, which was underpinned by the concept of the 'spirit of animation' as a vital fluid interfacing between mind, brain, nervous system and body, operating in a manner partly analogous to electricity. From the start of his career during the 1750s, Darwin drew parallels in letters to his friend Albert Reimarus (1727–1814) between the operation of electricity and the conveyance of motion through the nerves that were inspired by the Swiss physiologist Albrecht von Haller (1708–1777) and the Dutch physician Hermann Boerhaave (1668–1738).[16] These parallels were encouraged by the Enlightenment fascination with creatures that appeared able to generate and administer large charges of electrical effluvia at will, such as the electric eel and electric ray (Figure 46).[17]

Natural philosophers such as Darwin's friends Benjamin Franklin and Joseph Priestley also experimented with some creatures to observe the effects of electricity upon their bodies, including birds, rodents, dogs, cats, frogs and even larger beasts, such as sheep just as they observed the effects of vacuums and gases upon their bodies in pneumatics and chemistry. After relating how turkeys had been tested to destruction in order to observe the effects of foxglove, the Lunar physician and botanist William Withering likewise called for more medical experiments using 'insects and quadrupeds', which he thought had been previously insufficiently attended to.[18] Although such investigations demonstrate some contempt for creatures as mere experimental instruments, or at the least a belief that their suffering could be justified in the name of scientific progress, moral scruples were sometimes evident, and in 1767 Priestley expressed his discomfort at using animal experiments because it was 'paying dear for philosophical discoveries to purchase them at the expense of humanity'.[19] However, he continued to do so in his work on airs, believing the practice was justified for the wider scientific good, and in her poem 'The Mouse's Petition' (1773) Anna Letitia Barbauld (1743–1825) used the device of a personified mouse to appeal to Priestley not to sacrifice creatures in his experiments, expressing, as Sylvia Bowerbank has emphasised, 'an ecological principle of natural compassion uniting all life requiring humanity to treat animals ethically'.[20]

Darwin's opposition to animal cruelty probably helps to explain his general reluctance to conduct experiments that entailed the killing of animals (except where medically necessary). Where he does refer to the results of such experiments, such as the fact that the hearts of vipers and frogs continued pulsating after removal from their bodies, this information was obtainable from the scientific literature. In any event, his extensive use of medical electricity provided him with opportunities to observe how it might be efficacious in conditions such as limb paralysis, as pain relief or where invasive creatures such as tapeworms needed to be expelled or killed. Even here, however, his admiration for the tenacity and adaptability of such parasites is demonstrable. He believed that the simple 'structure' of roundworms enabled them to live in widely varying temperatures and their 'powers of life' prevented destruction by the process of digestion, while the tapeworms that thrived in human and many animal bodies, consisting of a 'chain of animals extending from the stomach to the anus' possessed a 'wonderful power of retaining life'. Two worms of 'several feet in length' voided by a pointer dog after 'violent purgatives' were little inconvenienced by boiling water, strong gin or whisky, while thread-worms reappeared weeks or months after apparently being destroyed and were likewise 'very tenacious of life'. Darwin observed the latter under the microscope surviving 'without apparent inconvenience' after immersion in a 'strong solution of sugar of lead' (lead acetate).[21]

The idea of an 'animal electricity' or nervous fluid secreted by the brain, promoted by the Italian natural philosopher Luigi Galvani (1737–1798) and his followers, was

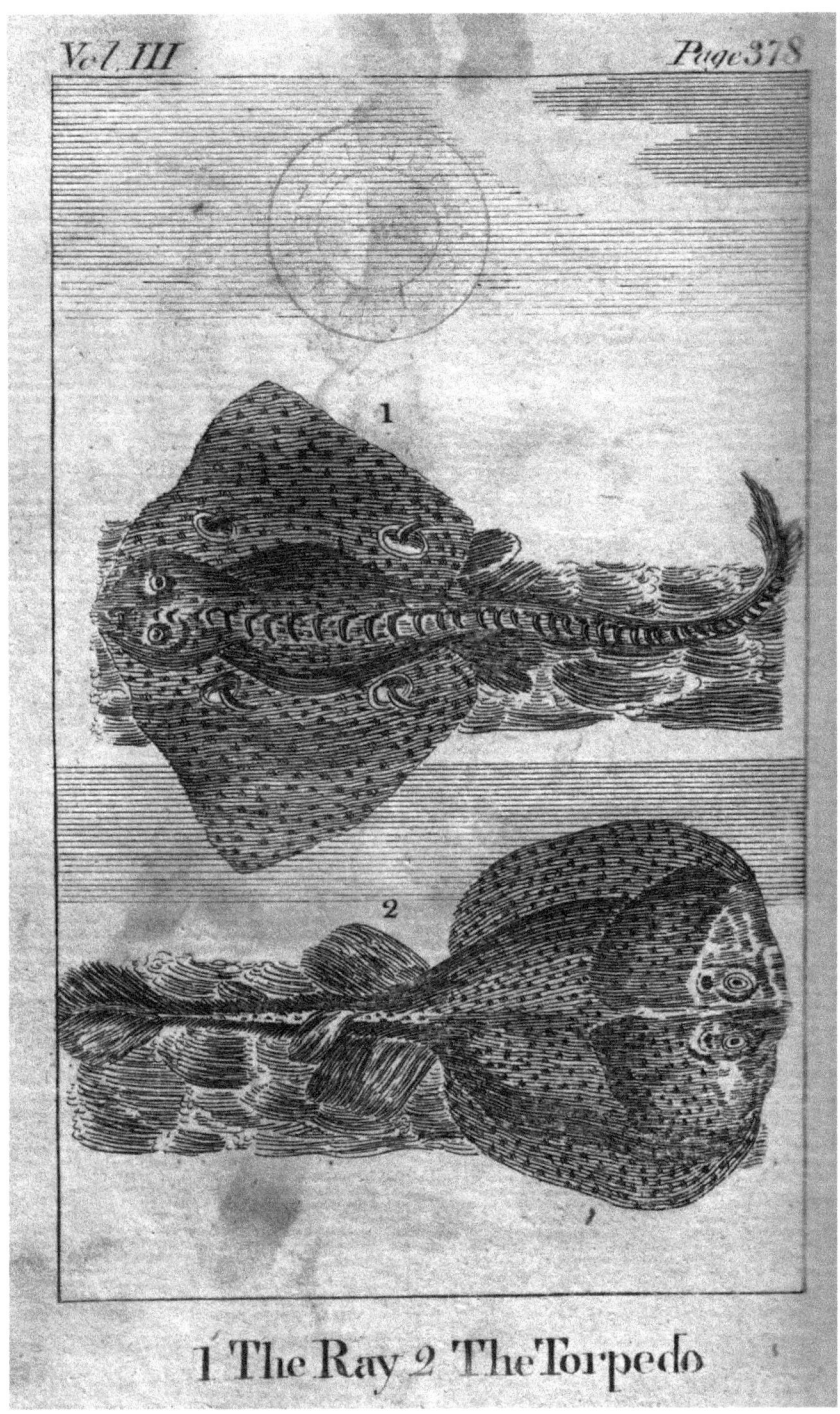

1 The Ray 2 The Torpedo

Figure 46. Ray and torpedo, from O. Goldsmith, *A History of the Earth and Animated Nature*, 4 vols (1804), vol. 3.

based originally on experiments with convulsing frogs' legs and appeared to underline similarities between human and animal physiology. However, it was opposed by Alessandro Volta (1745–1827) and his followers, who argued that no special or different form of electricity had been uncovered.[22] Darwin did not identify fully with either Volta or Galvani, although he did refer to the former's newly invented pile, or (battery, as a 'Galvanic Pillar' and there were many similarities between his 'spirit of animation' and the 'animal electricity' of the galvanists. The spirit of animation helped to differentiate between animate (including humans and all other animals and plants) and inanimate forms, helping to explain how the former interfaced dynamically with their changing environments and responded to sensory stimuli and catenations of association.[23]

In Darwin's view, John Locke's argument that animals were distinguished from humans because they 'possessed no abstract or general ideas' was refuted by Bishop George Berkeley (1685–1753) and David Hume (1711–1776), who showed that these 'abstracted ideas' had 'no existence in nature' or the 'mind of their inventor'. Language acquisition, 'labouring for money', 'praying to the deity' and tool-making abilities distinguished people from animals, but many creatures were more intelligent that often supposed, with forms of community, language and a talent for tool-fashioning, especially evident among those living wild.[24] Ultimately, it was love or 'animal attraction' that constituted the 'purest source of human felicity, the cordial drop in the otherwise vapid cup of life' that rewarded 'mankind for the care and labour' entailed by the 'pre-eminence of his situation above other animals'; yet, as Darwin's designation of love as 'animal attraction' demonstrates, it also showed how much he believed in the fundamental kinship between humans and other living beings.[25]

The long section 'Of Instinct' in *Zoonomia* might have been better titled 'against' or 'the limits of' animal instinct, as it sought to reinterpret the actions of animals usually ascribed to instinct as learned behaviours to blur or efface boundaries between humans and other organic beings. This is because Darwin opposed naturalists and philosophers such as William Paley (1743–1805) and Gilbert White, who tended to attribute much animal behaviour to instinct brought about as a result of divine creation, although the latter had a nuanced understanding of 'that secret influence' by which all species were constantly 'impelled naturally to pursue … the same way or track, without any teaching or example', whereas reason often varied and used different methods to accomplish particular ends. White's intense natural history observations led him to be critical of instinct as 'that wonderful limited faculty', which sometimes raised 'the brute creation' above reason and on other occasions left them 'far below it', noting that this generalisation required qualification because this faculty did sometimes 'vary and conform to the circumstances of place and convenience'.[26] Darwin complained that the role of instinct in animal behaviour was over-estimated, while the degree of bestial learning and intelligence was under-estimated. All animal 'actions … attended with consciousness' that seemed 'neither directed by their appetites, taught by experience', nor deduced from observation or tradition' were simply attributed to 'the power of instinct'. Yet this apparent instinctual element was in turn 'explained to be a divine something' and a 'kind of inspiration', although the 'poor animal that possessed it' was regarded as 'little better than a machine'.[27]

One 'ingenious philosopher', whom Darwin did not name, argued that animals did not enter into mutual contracts or agreements and that this was an 'essential difference' between humans and animals. He argued, however, that 'daily observation' demonstrated how the latter formed 'contracts of friendship' with each other and with mankind that did not rely upon understanding 'each other's arbitrary language'.[28] Darwin opposed the denial that animals could 'enter into contracts' because it was clear from 'daily observation' that they could form agreements 'of friendship with each other' and humanity. Domestic pets provided the best example of the latter, but elephants used to human contact would protect their master's children and could also be trained to entrap fellow wild elephants.[29] The fact that birds could 'enter into' 'marriage' contracts, supply food to their progeny and undertake 'joint labour' to fashion a nest or 'bed for their offspring' was something that had excited the 'admiration' of naturalists for centuries. Yet the attribution of this 'to the power of instinct', like the 'occult qualities of the ancient philosophers', was something that 'prevented all further enquiry' rather than genuinely illuminating the subject.[30]

Despite being perhaps unfamiliar with his work, Darwin's approach to creature intelligence, behaviour and language paralleled Bentham's argument that animal interests had been unjustifiably neglected and likewise cast doubt upon common perceptions that they were irrational and brutal beasts. Bentham maintained that while 'under the Gentoo [Hindu] and Mahometan religions' non-human animal interests 'met with some attention', in other societies the 'less rational animals', though 'sensitive' beings, were treated with cruelty, like slaves, and were 'degraded into the class of *things*', while Darwin contrasted the behaviour of 'free' animals in the wild with those 'enslaved' by humanity and complained that they were seen as mere machines. Yet the assumption that physical differences or the lack of a 'faculty of reason or ... faculty of discourse' justified this collapsed, according to Darwin and Bentham. Darwin argued that animal sagacity bore 'a near resemblance to the deliberate actions of human reason' while Bentham maintained that a 'full-grown horse or dog' was 'beyond comparison a more rational, as well as a more conversable animal, than an infant of a day, or a week, or even a month, old', but in any event the question ought not to be 'can they *reason?* nor, can they *talk?* but, can they *suffer!*'[31]

Underlining the fundamental similarities – and mutual reliance – between all animate forms emphasised by Bentham and Darwin also reinforced the latter's recommendations concerning human moral and educational development, which closely accorded with the advent of cultures of sensibility focused on the cultivation of a strong moral and spiritual self-awareness especially, but not exclusively, for women.[32] In his *Plan for the Conduct of Female Education* (1797) Darwin argued that the 'greatest part of adult mankind' learnt all 'common arts of life' by imitating others, which was also how animals acquired knowledge and which was often more effective than training by humans. The 'readiest way to instruct all brute animals' was to utilise those of the same species who had already learnt the necessary arts, so sick dogs, for instance, learnt from 'each other to eat grass as an emetic', while cats likewise moistened their paws to wash their faces.[33] He therefore had little time for those who were wantonly cruel to other creatures and maintained that young people needed to be encouraged to show sympathy and compassion for 'brute creation', at least as much as 'our necessities will admit', because mankind could not exist without some 'destruction of other animal or vegetable beings either in their mature or embryon state'. With mortality came the 'first law of nature' – 'eat or be eaten' – which underpinned survival in a competitive world, so it might be supposed that humans had a 'natural right to kill those brute creatures, which we want to eat, or which want to eat us' in order to survive. However, killing animals 'wantonly', 'even insects', showed an 'unreflecting mind, or a depraved heart'.

Latent sympathy with animate nature could be engendered or fostered by education. Darwin once observed a mother with 'apparent expression of sympathy' ask her daughter, as she cruelly pulled the legs off a fly, whether she would like her own legs pulled off similarly; 'would it not give you great pain?' He believed that this lesson would 'make an indelible impression' upon the child and 'lay the foundation of an amiable character'. When brutality was inflicted upon animals this invited revulsion and condemnation. Darwin recounted how, on another occasion, he had been told by a 'young gentleman' that, having fallen in love with a young lady, when they took a summer evening walk together she had taken 'two or three steps out of her way on the gravel walk to tread upon an insect', a 'picture of active cruelty' that he could never henceforth get out of his mind, and so she 'ceased to be agreeable' and he 'relinquished his design of courtship'.[34]

Encouraged by his healing ministrations, Darwin argued that observations of foetal and baby development, such as initial struggles to breath when released from the womb or egg, the first swallow and first suckling, demonstrated that the 'internal motions' of all 'animal bodies' were fundamentally similar. He maintained that throughout nature there was ample evidence of animals learning and adapting, forming communities to obtain food and protect themselves from danger and cooperating to migrate over long distances across water. If 'the sense of smell and taste' were considered, numerous animals 'greatly' excelled mankind: the ability of dogs to pick up scents that could not be detected by humans was one clear example and, in many cases, animals had 'almost an additional sense', with cat's whiskers operating like the antennae of moths and butterflies, helping them to move in tight spaces or dark conditions.[35] Some behaviours were more obvious among wild creatures free from human enslavement, although observations of domestic animals showed how many traits were learned rather than instinctual. Darwin assembled evidence concerning a range of creatures, including insects, to show that they experienced a similar spectrum of emotions or passions to humans, such as love, fear, anger, grief and pleasure. From birth, all animals learnt to discern signs of anger or displeasure in others and to respond physiologically and emotionally, just as children learnt to understand parental expressions and to associate these with particular behaviours: birds such as fighting cocks, for example, recognised rising hackle feathers around the neck as indications of anger. Such 'natural signs' formed the basis of the language by which all creatures sought to understand each other.[36]

Animals also adapted and changed their behaviours according to context, with dogs and cats, having been 'forced into each other's society', picking up habits from each other; one of the family terrier bitches at Full Street apparently learnt to wash her face from two feline friends with whom she shared the parlour. In contrast, domestic animals that escaped or were let loose from the 'delicacy of domestication' would revert to wild behaviours; thus a stray cat rediscovered through necessity how to catch fish in a Weaford millpond near Lichfield.[37] Physiological responses to fear and grief were apparently similar across the animal world, as suggested by Linnaeus's account of a bear, among other animals, shedding tears in sorrow. Having been told that it fainted whenever its cage was cleaned, Darwin observed George Harvey's canary begin trembling as the bottom of the cage was drawn out before turning 'quite white about the root of his bill', breathing more quickly, moving his wings, closing his eyes and then remaining in a 'cataleptic' state for nearly half an hour before returning to normality.[38] As children came to associate particular sights, smells and touch with their mother's milk during feeding, so the salivary glands were excited by the smell and taste of food. Animal's sexual organs became excited by the anticipation of sex through association, while, as the experience of dogs, lambs and other animals demonstrated, the first freeing of excrement in the bowels became linked with pleasurable sensations and tail wagging, subsequently denoted contentedness. Darwin claimed that the association of smiling with pleasure originated from the mouth relaxing after the muscular effort of closing the lips around the maternal nipple and suckling, and was encouraged by imitation of the pleased countenances of adults present; he thought this to be visible alike in kittens and puppies when 'played with and tickled', as well as humans.[39]

But it was not only that animals and humans registered their emotions through similar behaviours and expressions. Having argued that, in addition to humans, 'all other animals' communicated 'in signs or tones', and that there was a 'great resemblance in the natural language of the passions' among all creatures, Darwin sought to demonstrate that those animals who had best 'preserved themselves from human enslavement' and lived together in 'societies' also developed forms of 'artificial language' and 'traditional knowledge'.[40] Buffon and naturalists drawing upon him emphasised how different the qualities of horses were around the world and noted that, although so many lived in 'slavery and servitude' with humans, when found in a wild state these 'proud and haughty' animals remained 'proud' of their 'independence'.[41] A 'community' was evident among horses in the Tartarian and Siberian deserts, where they 'set watches' against human attack and had 'commanders' who directed their flight. Likewise, horses moving in line responded to each other and, when wanting to scratch itchy parts would bite the equivalent area needing relief on another animal, who responded by obliging them. Horses worked together in the New Forest, Hampshire, and on Staffordshire moorlands to stamp upon prickly gorse bushes so they could eat 'without injury', and when those from more fertile parts tried the same method they injured their mouths, not being used to it.[42]

There were also plenty of examples of avian skills, resourcefulness and ingenuity. Migrating birds travelled vast distances to different climates seasonally and formed 'large societies' with leaders, using their cries to keep formation and probably learning routes 'like the discoveries of mankind in navigation'. These behaviours were, therefore, not a result of 'necessary instinct' but were 'accidental improvements' 'taught' by contemporaries or 'delivered by tradition' between generations, in the manner of human achievements.[43] Wild 'tribes' of birds who knew their enemies well by observing their attacks upon their young were much more alert to threats than 'domestic birds' who had few opportunities to see this, so their 'knowledge' of 'distant enemies' was probably 'delivered by tradition' through 'many generations'. Mother turkeys alerted their young to predatory kites using 'gesture and deportment' and particular cries, which formed an 'artificial language', like that which rabbits and dogs taught their young, which explained how domestic animals learnt to come for food or avoid humans when angry according to changes in their tone of voice.[44] Wild animals in 'cultivated countries' appeared to learn 'very early in their lives' from each other or 'from experience or observation' to 'avoid mankind', meaning younger ones were easier to tame than older creatures. However, the 'resemblance' in animal language meant that humans and beasts generally knew each other's 'humour', so could judge, for instance, whether peaceful or angry. Other evidence for animal 'societies' included rooks, with their 'cities over our heads', who cooperated to warn each other when armed men were approaching their rookeries. Other birds, such as the fieldfairs who bred in northern Europe, and the uett-uett (squaller) of the Senegalese forests warned each other against predators and humans, while birds such as the lapwing appeared to distract predators, including men or dogs, from their nests by deliberate design.[45]

Some understanding of animal behaviour and their 'communities' could also help where they caused problems for agriculture or horticulture. Moles were a major concern in some areas, especially because they seldom appeared at the surface, being content to 'feed on the roots of vegetables' and on 'subterraneous insects (Figure 47). They were therefore hard to kill and the established methods of placing poison

Figure 47.   Mole, from T. Bewick, *A General History of Quadrupeds*, 7th edn (1820).

or injecting 'burning sulphur' or tobacco into their 'subterraneous mansions' were fairly ineffective because the earth frequently fell behind them as they went through or was accumulated 'behind them by their hindermost feet', creating impenetrable compartments. On one occasion when he was back in the family village of Elston, Darwin accompanied Francis Paget, the local 'very popular and successful mole catcher', around 'in his occupation' to observe his methods. Paget 'cleared' many 'neighbouring parishes' of moles while taking an 'annual tax' from farmers for future 'defence of their territories', a method pressed further by other mole catchers, who threatened to reintroduce the animals on to the land of farmers who refused to provide this 'annual stipend'. Darwin therefore described Paget's methods in order to 'enable every farmer' to become 'his own mole catcher' or 'teach the art to his servants'. Although apparently not so much of a problem among Derbyshire farmers, moles were regarded as a pest in parts of Nottinghamshire and the Vale of Trent, where the Darwin family farmed. According to Darwin, moles lived in 'cities underground' consisting of 'houses or nests, where they breed and nurse their young', joined by 'wider and more frequented streets' created by the 'perpetual journeys of the male and female parents' and less-traversed 'bye roads' with 'many diverging branches' that were 'daily' extended to 'collect food for themselves or their progeny'. Moles were 'more active' in spring, when males were 'courting', creating numerous burrows for their 'meeting with each other', and though 'commonly esteemed to be blind', the fact that they began 'their work' with morning light before the sun could warm the ground suggested they had 'some perception of light' even in their 'subterraneous habitations'. Paget rose before sunrise and went to areas he knew moles were active; where he saw 'the earth move over them, or the grass upon it' he would sometimes cut them off with spades and dig them up, while lying upon mole hills with his ear to the ground revealed their scratching, conveyed through vibrations in the 'solid earth' more effectively than through the 'light air'. He dug up mole hills, below which were deep nests with 'their streets or bye-roads', and severed interconnecting roads to block 'the retreat of the inhabitants'. The 'most frequented streets' were determined by carefully making foot marks on new mole hills and observing whether they were still present the day after, and then laying wooden traps along these highways.[46]

ANIMAL INTELLIGENCE, INGENUITY AND ADAPTABILITY

Darwin strongly believed that many creatures were more intelligent and adaptable than often supposed and collected accounts of their actions from a range of sources to support his arguments. His brother, the Rev. John Darwin (1730–1805) of Carlton Scroop, Lincolnshire, described rook behaviour to him, utilising information from his own social circle, and facts about sea creatures were likewise obtained from Darwin's 'ingenious friend' the cartographer Peter Perez Burdett (c. 1734–1793), who had surveyed parts of the Lancashire coast. The botanist and writer Maria Elizabetha Jacson (1755–1829), who had strong Derbyshire connections and was encouraged by Darwin in her plant drawings and publications, such as the *Botanical Dialogues* (1797), described how an intelligent bird had contrived to shake a poppy so the seeds could be taken from the ground and consumed.[47] Another friend explained how he saw many crows along the northern Irish coast dropping mussels on to rocks from a height in order to smash the shells and extract the flesh, while the Rev. Robert Wilmot (1750–1803) of Morley, Derbyshire, sent a detailed account of cuckoo nests on coal-slag hills near his farm.

Other acquaintances and friends happily recounted the details of animal behaviours they had witnessed on their trips abroad, which helped to make up the fact that Darwin never travelled beyond England himself and informed his understanding of animal behaviour and intelligence. His 'late friend' the Rev. James Chambers (d. 1790) of Derby, rector of Stretton-in-the-Field, Derbyshire, and Barwell, Leicestershire, explained how he had been told while on the Island of Caprea in the Bay of Naples that large numbers of quails settled there annually after becoming exhausted on their travels between Africa and Europe. Dr John Marten Butt FRS (1738–1769), who had been physician general to the Jamaican militia at Kingston and tutor to Thomas Erskine, the future lord chancellor, showed Darwin a preserved spider's nest he had brought back from the Caribbean and described its operation when the two met in Bath during the 1760s, where Butt had retired for his health. The physician Benjamin Smith Barton (1766–1815), professor of natural history at the University of Pennsylvania, Philadelphia, sent Darwin his manuscript on how rattlesnakes and the 'black serpent of America' caught birds before it was published in the US *Philosophical Transactions*.[48]

Darwin believed he once witnessed a demonstration of 'the power of reason … as it is exercised among men' by a wasp, which caught a fly 'almost as large as itself and then separated 'the tail and the head from the body part' on which the wings were attached. However, as the wasp flew away with the fly body, a breeze caused the fly wings to turn it around in the air. Once the wasp had 'settled again with its prey', Darwin then 'distinctly observed' it cut off 'first one of the wings and then the other' with its mouth and fly away 'unmolested by the wind', causing him to exclaim with wonder: 'Go, thou sluggard, learn arts and industry from the bee, and from the ant! Go, thou proud reasoner, and call the worm thy sister!'[49] Darwin also had a high opinion of beaver intelligence. Beavers were recognised by Buffon and Goldsmith as a 'monument of brutal society', although Goldsmith believed them to be 'without language or reason'. Darwin, however, thought they showed great industry and 'amazing ingenuity' in forming dwellings and dams, especially in 'thinly peopled' countries where rivers flowed uninterrupted.[50] Equally, the hands of monkeys were, as Darwin explained, the most 'well adapted for the sense of touch', giving them a 'great facility of imitation', although the fact that their thumbs were not used to counteract figure pressure limited their ability to acquire 'the figures of objects' or determine 'the distances or diameters of their parts', while, as the French philosopher Claude Adrien Helvetius emphasised, pressure from humans and geographical situation combined 'to prevent' their 'improvement'. Darwin cited the example of 'an old monkey' shown at the menagerie of 'exotic' animals at Exeter Exchange in London, which had lost its teeth through age (and perhaps diet); it took up a stone when provided with nuts 'and crack[ed] them with it one by one', so 'using tools to effect his purpose like mankind'.[51]

More evidence that animals' behaviour was misinterpreted and their intelligence underestimated came from the experience of pigs, of which, like bees and moles, Darwin was able to talk with some experience. Generally seen as stupid, docile and dirty animals who only existed at the sufferance of humans because of the functions that they were able to perform for them, pigs supplied excellent meat that could be easily stored by salting and other products such as fat while being able to eat almost anything, including many domestic waste products such as peelings and leftovers.[52] In Goldsmith's popular *History of Animated Nature* the hog was described as the 'most sordid and brutal animal in nature' with an 'awkwardness' of form that influenced 'its appetites'. Its sensations were entirely as 'gross as its shapes are unsightly' and it appeared to possess merely 'an insatiable desire of eating', choosing food that 'other animals find the most offensive'. It was so lowly it was despised not only by humans but other animals too. Goldsmith accepted that the hog's 'sordid and brutal' nature partly arose from a lack of 'proper nourishment which it finds in the forest' and that swinish 'indelicacy' was therefore 'rather in our apprehensions than in its nature', but it was 'naturally formed in a more imperfect manner' than other domestic animals, and both 'less active' in its movements and 'less furnished with instinct in knowing what to pursue and avoid'. It was 'by nature, stupid, inactive, and drowsy' and, 'if undisturbed', would 'sleep half its time' until wakened by 'the calls of appetite'. Its 'whole life' was therefore merely 'a round of sleep and gluttony' and a 'helpless instance of indulged sensuality'.[53]

Although the presentation of pigs in George Smellie's translation of Buffon's *Natural History* was more nuanced, they were still seen as ungainly and stupid creatures susceptible to disease. It was asserted that, of all beasts, 'the hog appears the most rough and unpolished', with a 'voraciousness' that apparently depended upon 'the continual want which he has to fill the vast capaciousness of his stomach'. The 'roughness of the hair, the hardness of the skins, and the thickness of the fat' rendered pigs 'so insensible to blows' that, in an account much repeated in natural histories, mice were known to lodge in their backs and 'eat their fat and their skin without their seeming sensible of it'. Swine were 'imperfect' in their 'senses of the taste and touch' and particularly susceptible to disease, although this was partly a product of the ways they were kept and fed. Clean stables 'without litter' with 'plenty of wholesome food' would keep them from disease and provide flesh that was 'excellent to the taste'.[54] Similarly, while Beilby and Bewick underlined the commercial and naval importance of pork, they described common pigs as the most 'filthy and impure' of domestic animals (Figure 48). Like Buffon, Smellie and Goldsmith, they emphasised that their form was 'clumsy and disgusting', and their appetite 'glutinous and excessive', meaning 'nature' had 'conspicuously shewn her economy' by creating a beast 'whose stomachs are fitted to receive nutriment' from many things that would otherwise be wasted, namely the 'luxuriant repast' of 'refuse of the field, the garden, the barn, or the kitchen'. 'Naturally stupid, inactive and drowsy' and 'much inclined to increase in fat', pigs were 'useless during life, and only valuable when deprived of it'.[55]

The denigration of pigs assumed a political dimension during the 1790s after Edmund Burke bewailed in his *Reflections on the Revolution in France* (1790) how, 'along

Figure 48.   Sow of the improved breed, T. Bewick, *A General History of Quadrupeds*, 7th edn (1820).

with its natural protectors and guardians', 'learning' would be 'cast into the mire, and trodden down under the hoofs of a swinish multitude', doubtless encouraged by the traditional low status of pigs in natural history as well as biblical references to demons being cast into the Gadarene Swine and running off cliffs to their doom. Burke's reference to the 'swinish multitude' was taken up by political reformers, although, ironically, it was turned against him and his reactionary supporters.[56] Nevertheless, the phrase's interpretation continued as Burke intended it: as an insulting reference to the destruction of social order by a common, ignorant rabble striving to subvert the 'natural' order. The political campaigner Thomas Spence (1750–1814), for instance, entitled his radical weekly publication *Pig's Meat; or Lessons for the Swinish Multitude*. In Derby, allusions to pigs were employed contemptuously by both proponents and opponents of the sale of Nun's Green common in the early 1790s. For example, the 'Celestial Bard', a supporter of the proposals, wrote a mock poetical petition against the sale on behalf of the jack-asses, geese and pigs – a 'truly honourable race' – who feared for the loss of their green home, while opponents of the measures described themselves ironically as the 'swinish multitude' in ballads and broadsides.[57]

Most livestock improvement in Georgian agriculture was focused upon cattle and sheep rather than pigs. However, in Derbyshire, like many similar areas, most farmers kept some pigs for pork, bacon and ham, which were fed on the skimmed milk and whey that were by-products of the dairying process and scraps or foraged food. No particular breed dominated but those called Derbyshire, Burton and Tamworth pigs were renowned from being usually free from distempers, and some landowners, such as Darwin's friend Francis Noel Clark Mundy at Markeaton, who, according to John

Farey, was 'long … famous for a fine white breed of pigs', were well known for keeping these. Farey believed that swine were kept most successfully in secure sties with strong fences so they could feed on 'small portions' of land at a time and not use their snouts to dig up all the roots and fences in the vicinity. While farmers were becoming more aware of the 'great benefits' of feeding pigs on 'green crops' such as clover, lucerne, chicory and grass, he thought that, 'except for the refuse of the garden', this was not practised as much in the county 'as it ought to be'.[58]

Darwin's pigsties on his 'farm' beside the Derwent were visited by the young artist and dramatist Samuel James Arnold, who had come to paint his portrait. He was struck by their cleanliness and commented on this to his host, who launched into a lengthy discourse on the benefits of cleanliness in pig husbandry. After listening to Darwin, Arnold argued that 'the animal had acquired a character for dirty propensities by its notorious love of wading in the mire', as Buffon, Goldsmith and Bewick asserted. With a hint of asperity, Darwin challenged these assumptions, arguing it was more a 'direct proof of the natural cleanliness of the creature' when asses were seen 'rolling on the road or in the common', which was not to 'dirty his hide'. In fact, it was his way of 'dusting his jacket – the only method he can contrive to comb his hair and rid himself of the vermin which vex and perhaps torture him'. Likewise, it was 'foul-feeding', encouraged by humans, that caused pigs to develop a skin disease that resembled the scurvy, suffered by sailors and others. Swine scrubbed themselves 'bare against every post and roll in the mud and mire to cool their surfaces and allay the irritation under which they are suffering', but this was not a 'natural propensity', as commonly imagined, but a method they had learnt of 'allaying suffering or of removing inconvenience'.[59] Darwin must therefore have been annoyed when his son Francis and a friend, George Bilsborrow, caused trouble on one occasion around 1796 by shooting at the pigs with arrows, although he may have been distracted by the subsequent appearance of a rabid dog, which attacked the swine and then a horse as the boys hastily climbed the trees to escape, before the dog was shot dead in neighbour Charles Upton's garden by a crowd in pursuit. The whole extraordinary episode was sketched decades later by Violetta Darwin based upon the recollections of her father Francis (Figure 49).

In *Phytologia*, Darwin examined pig farming and argued that 'wooden movable swine troughs' 'placed on the summit' of 'heaps of dry straw' were preferable to traditional methods of keeping them, where food was provided in 'fixed stone troughs' from which refuse was occasionally washed or swept away. Movable wooden troughs facilitated 'more perfect' decomposition of 'vegetable recrements' as 'broth, whey' and other animal and vegetable substances in the swill mixed with their 'urine and ordure' to 'generate early putrefactive processes'.[60] His respect for pigs and his belief that they were underestimated because of their usual treatment, which may also have partly been a response to Burke's notorious comment, is also evident in *Zoonomia*, where Darwin argued that they acquired knowledge as a result of changes in their environment. Swine had a 'sense of touch as of smell at the end of their nose', which they employed 'as a hand' to 'root up the soil' and to 'turn over and examine objects of food', in the manner of smaller elephant's trunks. He thought that the cold British

Figure 49. Pen and ink drawing by V. H. Darwin, 'Francis S. Darwin and George Bilsborrow while engaged shooting pigs with arrows are disturbed by a mad dog, which communicates hydrophobia to the pigs and a horse, it is eventually killed by the mob'; 1796, from K. Pearson, *The Life, Labours and Letters of Francis Galton*, vol. 1 (1914).

winter climate had induced them to 'collect straw in their mouths to make their nest', assisted by 'their companions', who were summoned by 'repeated cries' and shared the snug bed with 'numerous bed fellows'. Although pigs were 'esteemed so unclean', they 'learned never to befoul their dens' with 'their own excrement' where they had the 'liberty' of not doing so, an 'art' that cows and horses had failed to acquire. He claimed to have 'observed great sagacity in swine', who were underestimated because of their 'confinement' and the 'short lives we allow them'; otherwise they might prove 'greater' than dogs.[61]

Darwin's observations on porcine perspicacity may have been partly prompted by performances from the 'Learned Pig', trained by Samuel Bisset and John Nicholson, which toured Britain and exhibited at the Old Assembly Rooms in Full Street, Derby, for six days in October 1784. A notice in the *Derby Mercury* proclaimed that Nicholson, who trained other animals, such as a turkey-cock that performed a country dance, had conquered 'the natural obstinacy and stupidity of a pig' and claimed that Derbyshire ladies and gentlemen would witness the 'greatest curiosity in the Kingdom', as it united 'the letters of any person's name', counted the number of people present and 'the hour and minutes by any watch'.[62] The 'wonderful learned pig' was seen at

Nottingham by Darwin's Lichfield friend Anna Seward, who claimed it had done 'all that we have observed exhibited by dogs and horses'. Her account of the pig amused the lexicographer and literary celebrity Samuel Johnson, who, like Darwin, observed that pigs were 'a race unjustly calumniated' and would display their intelligence more readily if they were saved from the pot for longer and 'educated'.[63]

INSECTIVOROUS INDUSTRY

The Linnaean system encouraged, in many ways, a static, passive view of life. However, many years attending to human and sometimes animal bodies and making careful observations of life in gardens and the countryside demonstrated to Darwin that, no matter how cultivated, outdoor places teamed with life in vigorous competition. Though intended to embody landscape beauty and to illustrate – and inform his studies of – Linnaean botany, the living botanic garden challenged and disturbed Enlightenment visions of scientific order and benign organic interdependency, as Darwin's *Phytologia* and *The Temple of Nature* expressed so clearly. As Roy Porter emphasised, as a naturalist Darwin's work demonstrates 'an unflagging curiosity towards the micro-economies of stamens and pistils, the techniques of grafting and breeding', insect behaviours and the 'character of domestic pets' – an 'inquisitiveness' that bore comparison with his contemporary the Rev. Gilbert White.[64] This is represented to some extent by the two parts of *The Botanic Garden*: one a popular exposition of the Linnaean system, the other a treatise on the 'economy' of vegetation, emphasising the nexus of interrelationships and interdependency across the natural world. Similarly, observations of the Hampshire countryside around Selbourne suggested to White that 'the most insignificant insects and reptiles' were of considerably greater 'consequence' and had 'much more influence on the economy of nature' than the 'uncurious' were 'aware of', and were 'mighty in their effect from their minuteness', teeming 'numbers' and 'fecundity'. They were the 'great promoters of vegetation', working and digesting the soil and mixing leaves and stalks within, and, without them, the earth would be sterile. Hence, though appearing to be a 'small and despicable link in the chain of nature', if 'lost', earthworms would 'make a lamentable chasm' and a 'good monography' of them would 'open a large and new field in natural history'. For White, the economy of nature operated, as Samuel Johnson said, like the 'management' or 'government' of a 'household', the 'disposition' or 'regulation' of things and the 'distribution of everything active or passive to its proper place' within God's creation.[65]

Darwin agreed with White on the significance of the apparently insignificant in the natural economy and the teeming tenacity of life, but emphasised organic and inorganic processes and interdependency rather than hierarchy, and competition instead of links in the divinely ordained chain of being. Like White and his grandson Charles Darwin, he was fascinated by insects, emphasising the 'truly astonishing' 'numbers and varieties of animated beings' living underground and 'sleeping' in winter, their bodies surviving beneath the frost. Once in his Lichfield botanic garden Darwin witnessed 'such immense numbers of small wing-sheathed insects' rising from ground near Sir John Floyer's cold bath – which he thought were fernchafers (*Scarabaeus solstitialis*) because they were much smaller than the May-Chafer (May bugs or cockchafers, *Scarabaeus melolontha*, known today as *Melolontha melolontha*), although of similar appearance – rising 'out of the ground near the cold bath'. There were so many that he estimated that 'one or two' had 'emerged from every square inch of many acres of land' within the garden. As a result he considered that the 'grubs or maggots' from which the 'wing sheathed flies' had apparently arisen, where situation and season was 'favourable to their production', were 'very destructive' to spring and early summer wheat fields, consuming stems near the ground surface 'at the joint, which is sweet, till it falls down or withers'. He believed this might explain why many crops had been 'nearly destroyed' in 1797 on lands he was told were 'previously well limed'. Insects were voracious in their internecine strife, even to the extent of cannibalising or attacking each other, as well as preying upon plants.[66] Darwin observed how 'uncommonly hardy or tenacious of life' the larvae, grubs and caterpillars of 'many insects' were; they could be destroyed without killing their plant hosts only with great difficulty by 'chemical mixtures'. He drew parallels with his experience of placing an 'ascaris' (roundworm) or invertebrate nematode from a human body 'above an inch long, and nearly as thick as a thin crow-quill' into a 'saturated solution' of lead acetate, where it had lived for 'many hours without apparent injury'.[67]

Darwin's arguments concerning animal intelligence, industry and communities also extended to the insect kind, although he recognised that even less was known about these than about other creatures (Figure 50). He was convinced that more 'histories' of insects 'formed into societies' would demonstrate their 'arts and improvements' to be less uniform than they hitherto appeared, having arisen 'in the same manner' as 'human arts', from 'experience and tradition', although reasoning 'from fewer ideas', 'busied about fewer objects' and 'exerted with less energy'.[68] He was convinced that, despite the 'habits of peace' or the 'stratagems of war' of insects' 'subterraneous nations' often remaining hidden, and the obvious physiological differences from humans and larger animals, the examples of insect communities that he had been able to observe or find out about demonstrated that, where their 'occupations' required, these beings demonstrated 'greater ... knowledge and ingenuity' than often accepted. According to Buffon's *Natural History*, insects were the most 'imperfectly formed' of all creatures, while for Goldsmith they were 'deservedly placed in the lowest rank of animated nature' and incapable of 'education' or 'arts' to 'turn' them from their 'instincts'. Yet Darwin believed that the inventiveness of birds and fish in forming nests and rearing their young was 'much inferior' to the 'arts exerted' by many 'insect tribes on similar occasions', including bees, ants and ichneumon flies, the last of which used live caterpillars as hosts for the 'incubation of their eggs', which was 'equal to any exertion of human science'. Like many naturalists, he was impressed with the 'mathematical exactness and ingenuity' of spider's webs, adapted for hunting and habitation, which he likened to the rope work employed at sea to support masts and ship sails. It may have been this that encouraged Darwin's philosophical friend the Wirksworth curate and electrical experimenter Abraham Bennet to utilise the astonishing 'tenuity' of spider's thread as the basis for his magnetometer, which was designed to detect the very smallest influences and which Bennet once observed being

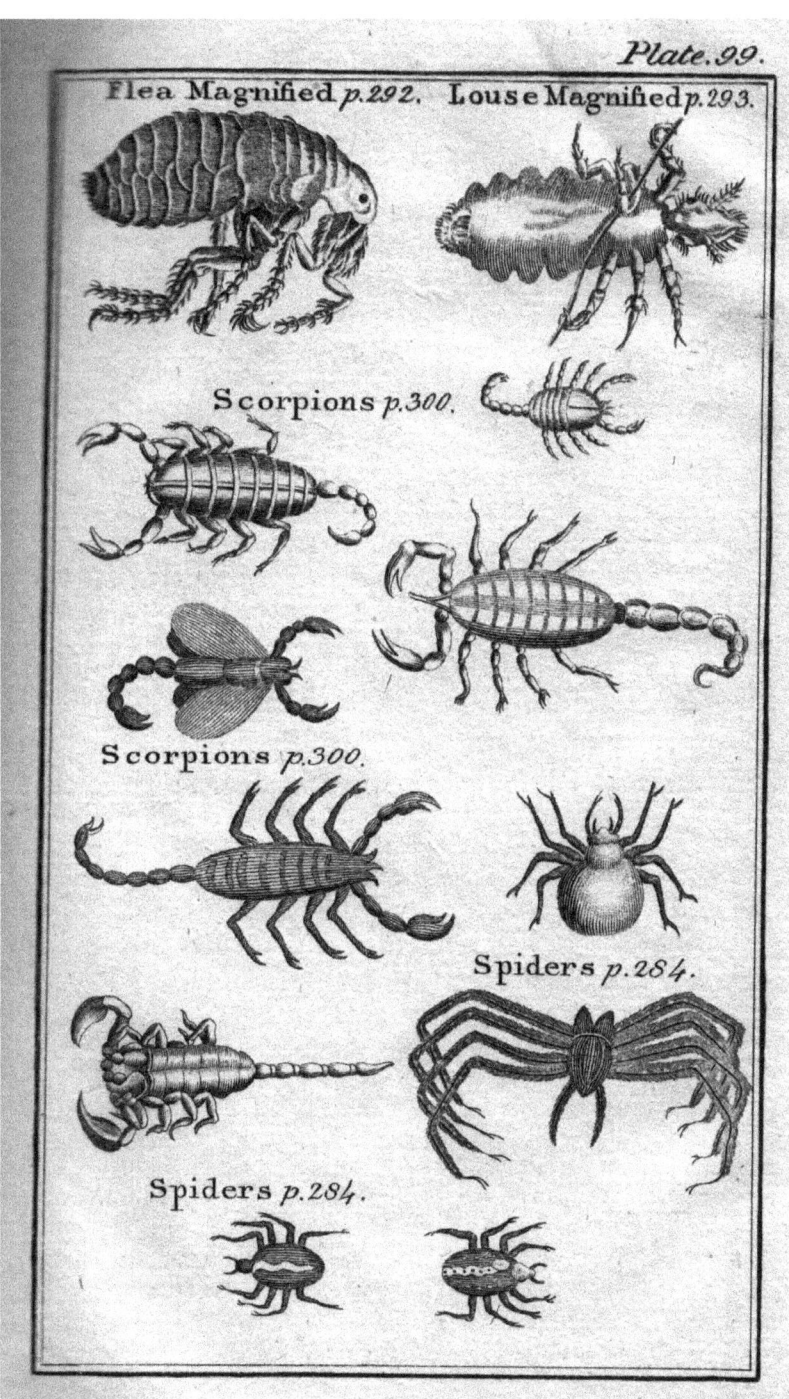

Figure 50.   Flea, scorpions and spiders, from *Buffon's Natural History Abridged*,
2 vols (1792), vol. 2.

twisted through 18,500 turns before snapping. According to Darwin, the manner in which spiders, widely seen during the Enlightenment as the chief order of insects, 'counterfeited death' when threatened and the habitations of bees, wasps, spiders and 'various coralline insects' all 'equally astonish us' for the 'materials' and 'architecture', which were often 'equal to any exertion of human science' and attained a 'great degree of perfection' in their sphere with 'as accurate knowledge, and as subtle invention, as the discoverers of human arts'.

Bees were seen as one of the most valuable insects for their production of honey and Darwin observed them closely in his gardens. He knew that 'bees of one society' often assailed those of another, plundering their honey and destroying most or all of their enemies 'in battle', so 'resembling the societies of mankind!' This 'war for plunder' had cost Darwin two hives in the previous two years, as he had been unable to intervene in time to prevent 'hostilities.' Seeing many bees about his 'only hive' again and witnessing a 'violent battle' resulting in the deaths of around a hundred, Darwin reduced the size of the hive mouth and placed a board over it to make access difficult for 'assailing bees'. But although fewer were 'slain in this day's battle' the 'war' did not cease, and he moved the hive in the early morning quiet to a 'distant part of the garden', facing eastwards, upon which the 'host of the enemy' did not follow and his bees 'resumed their work'; the few who returned to their 'old habitation' were 'carried early on the ensuing morning in their torpid state to their new situation'. Darwin successfully ended his bee war in the manner he hoped the great international revolutionary wars would end – 'without extermination' of each society.[69] Analogies between bee colonies and human nations were often drawn during the seventeenth and eighteenth centuries, particularly by monarchists, who celebrated their apparently structured, hierarchical and industrious ways of living – although this trope received a bit of a jolt when 'king bees' were recognised to be female in the 1740s. Bernard de Mandeville's *Fable of the Bees* (1724), which saw 'private vices' as essential for 'public virtues', was notorious for undermining traditional moral foundations, however, rather than for any apicultural insights.[70] Observing bee behaviours in his own hives, Darwin emphasised how much they co-operated together in 'societies' to feed their young and combat predators. Believing that, although bees could keep warm by moving their legs together, some hives died because resources were used up in warm weather as their residents revived, Darwin asked a friend to place two hives into a dry cellar for 'many weeks' and noted in consequence that the bees consumed none of their provisions and did not decrease in weight as they would have in the open air.[71]

CONCLUSION

As we have seen, the idea that animals shared many characteristics with humans and might even, therefore, be accorded greater protection was opposed by most in late Georgian society, although the rise of sensibility and moral feeling helped make actions of needless cruelty more socially unattractive. Yet still during the revolutionary wars from the 1790s the notion that beasts might enjoy rights or more moral and legal protections was considered sufficiently ridiculous to be employed by church and king loyalists to ridicule the arguments of supporters of political reform such as Darwin. In *A Vindication of the Rights of Brutes* (1792), for example, the classical scholar and expert on Greek philosophy Thomas Taylor (1758–1835) argued, in a parody of the 'wonderful productions of Mr. [Tom] Paine and Mrs [Mary] Wollstonecraft' and their supporters, that if women should have their own special rights then why not other entities in the chain of being or even physical objects, including animals, vegetables and minerals? Taking examples familiar to readers of Buffon and Goldsmith, Taylor asked why creatures such as lions, hares, flying birds and spiders, which had attributes such as strength, speed and ingenuity, shouldn't be accorded rights. He pointed out sarcastically that the 'sublime doctrine' of the 'equality of brutes to men' was as dear to the 'genuine modern' in the current 'amazing rage for liberty' as allowing the 'equality of mankind to each other', effacing distinctions between master and servant, abolishing all government or believing the soul and body to be 'essentially the same'.

According to Taylor, although things in nature might appear to be 'vile and contemptible' and brutes had bodily strength, they also possessed 'language' that humans might understand and 'reason in common with men, though not in quite so exquisite a degree'. Claiming to find roots for his ideas in Aristotelian and neo-Platonic philosophies, Taylor contended that 'brutes' communicated with each other and 'mankind', 'consulting' with each other's interests and 'providing for futurity'. Ancient fables and the divine status accorded to some animals by the Egyptians, Persians and Indians demonstrated how 'unjust and tyrannical' it was to destroy and eat animals that ought to be treated with 'kindness and familiarity'. 'Theatrical sports' and hunting were likewise unjust and cruel because of the level of animal reason and its similarities with 'human nature'. Taylor concluded his satirical pamphlet with ever more risible suggestions of how humans might benefit from beastly wisdom and example through provision of guidance from elephants on courtship, help from birds with music-making at fashionable Vauxhall Gardens and even assistance from dogs with discouraging the young from masturbation! The mere statement of arguments in favour of animal rights and intelligence was thought to be ludicrous enough to render the reformer's case ridiculous without further elaboration, but some of Taylor's points, especially in the earlier part of his *Vindication*, before he disappeared into the realms of far-fetched fantasy, are actually ones that Darwin might not have disgreed with.[72]

As 'markers of nature', taxonomic representations or material resource, animals helped define particular landscapes, from prize cattle on improved estates to their less distinguished cousins on common land, from hens and pigs on smallholdings to exotic beasts in exotic (or mundane) places, with horses almost everywhere and domestic pets living intimately in households. As John Aikin and Darwin emphasised, their presence and interaction with other living beings brought dynamic dimensions to farms, parks and other environments, and they became essential adjuncts to picturesque aesthetics and landscape gardening, by increasing productivity, transforming prospects and helping to manage the environment by their grazing and browsing. Historians of animal studies have sought to recover animal agency and intentionality, critiquing the assumption that non-human creatures were just passive subjects of the human or merely tame, idealised vehicles for sentimentalised moral or religious messages, but surprisingly little attention has been paid to Darwin's arguments concerning animal behaviour and cultures, despite it being one of his major concerns.

While Jeremy Bentham's brief suggestions concerning the moral status of animals, contained in one footnote of *Introduction to the Principles of Morals* (1789), are widely cited, Darwin's much more detailed analyses of animal physiology and behaviour in *Zoonomia* and *Phytologia* are relatively neglected, perhaps because of the attention devoted to his 'evolutionary' ideas and the fact that they were contained in lengthy treatises on agriculture, gardening and medicine.[73] Darwin's arguments that many creatures, including insects and birds, cooperated with each other in societies utilising sophisticated forms of communication and displayed many signs of sagacity or intelligence and self-identity have thus been passed over. But he was primarily a practising physician, and his medical experiences and observations of animal bodies and behaviours in varied environments led him to recognise the power of animal agency.

NOTES

[1]   A. Seward, *Memoirs of the Life of Dr. Darwin* (London, 1804), 96–105; S. Grimalkin (E. Darwin), letter to Po Felina (A. Seward), 6 November 1780 (originally 7 September 1780), in D. King-Hele ed., *The Collected Letters of Erasmus Darwin* (Cambridge, 2007), 177–9; T. Barnard, *Anna Seward: A Constructed Life: A Critical Biography* (Farnham, 2009), 113–18.

[2]   Seward, *Memoirs of Dr. Darwin*, 96–105; S. Grimalkin (E. Darwin), letter to Po Felina (A. Seward), 6 November 1780 (originally 7 September 1780), in King-Hele ed., *Collected Letters*, 177–9; Barnard, *Anna Seward*, 113–18; S. Bowerbank, *Speaking for Nature: Women and Ecologies of Early Modern England* (Baltimore, 2004), 137–8.

[3]   K. Thomas, *Man and The Natural World: Changing Attitudes in England, 1500–1800* (London, 1983), 17–41; H. Ritvo, *The Animal Estate: The English and other Creatures in the Victorian Age* (Harmondsworth, 1990), 1–3; E. Fudge, *Perceiving Animals: Humans and Beasts in Early-Modern English Culture* (Champaign, 2002); L. Kalof and B. Resl eds, *A Cultural History of Animals*, 6 vols (Oxford, 2007); J. B. Landes, P. Y. Lee and P. Youngquist eds, *Gorgeous Beasts: Animal Bodies in Historical Perspective* (University Park, 2012); H. Kean and P. Howell eds, *The Routledge Companion to Animal–Human History* (Abingdon, 2019).

[4]   S. M. M. Alberti, 'Introduction', in S. M. M. Alberti ed., *The Afterlives of Animals: A Museum Menagerie* (Charlottesville, 2011), 1–16, at 4, 38; L. Daston ed., *Biographies of Scientific Objects* (Chicago, 2000); M. Blackwell ed., *The Secret of Things: Animals, Objects and It-narratives in Eighteenth-Century England* (Lewisburg, 2007); A. MacGregor, *Curiosity and Enlightenment: Collectors and Collections from the Sixteenth to the Nineteenth Century* (New Haven, 2007).

[5]   J. Aikin, *An Essay on the Application of Natural History to Poetry* (Warrington, 1777), 33–4, 54, 110, 124.

[6]   E. Darwin, *Phytologia; or the Philosophy of Agriculture and Gardening* (London, 1800), 410.

[7]   R. E. Prothero (Lord Ernle), *English Farming, Past and Present*, 6th edn (London, 1961), 176–223; Ritvo, *Animal Estate*, 45–81; J. V. Beckett, *The Agricultural Revolution* (Oxford, 1990), 21–5; T. Williamson, *The Transformation of Rural England: Farming and the Landscape, 1700–1870* (Exeter, 2002), 39–41; N. Russell, *Like engendr'ing like: Heredity and Animal Breeding in Early-Modern England* (Cambridge, 1986).

[8]   P. Wakefield, *An Introduction to Botany, in a Series of Familiar Letters*, 5th edn (London, 1807), iii; T. Bewick and R. Beilby, *A General History of Quadrupeds*, 7th edn (Newcastle-upon-Tyne, 1820),

iii–iv, 8; Prothero, *English Farming*, 181–4; Ritvo, *Animal Estate*, 18–23; Bowerbank, *Speaking for Nature*, 141–60.

9   Prothero, *English Farming*, 176–9; R. Longrigg, *The English Squire and his Sport* (London, 1977), 114–66; Thomas, *Man and the Natural World*, 92–6; Ritvo, *Animal Estate*, 2–6, 15–21; Bowerbank, *Speaking for Nature*, 148–56.

10   G. Garrard, *Ecocriticism*, 2nd edn (Abingdon, 2012), 146–80; Landes, Young Lee and Younquist, *Gorgeous Beasts*, introduction, 1–11.

11   O. Goldsmith, *A History of the Earth and Animated Nature*, 4 vols, (York, 1804); Thomas, *Man and the Natural World*, 51–70; R. Wokler, 'Apes and races in the Scottish Enlightenment: Monboddo and Kames on the nature of man', in P. Jones ed., *Philosophy and Science in the Scottish Enlightenment* (Edinburgh, 1988), 145–68; Ritvo, *Animal Estate*, 10–15; H. Ritvo, *The Platypus and the Mermaid and other Figments of the Classifying Imagination* (Cambridge MA, 1997), 1–26; D. Knight, *Ordering the World: A History of Classifying Man* (London, 1981), 58–86.

12   G. L. Leclerc, comte de Buffon, *Buffon's Natural History Abridged*, translated by W. Smellie, 2 vols (London, 1792); Bewick and Beilby, *A General History of Quadrupeds* (1820 [1791]); E. Darwin, *Plan for the Conduct of Female Education* (Derby, 1797), 121–2; P. J. Marshall and G. Williams, *The Great Map of Mankind: British Perceptions of the World in the Age of Enlightenment* (London, 1982); Alberti, *The Afterlife of Animals*; C. Grigson, *Menagerie: The History of Exotic Animals in England, 1100–1837* (Oxford, 2016); C. Plumb, *The Georgian Menagerie: Exotic Animals in Eighteenth-Century London* (London, 2015).

13   *Derby Mercury*, 31 December 1789, 28 January 1790; Plumb, *Georgian Menagerie*, 36–40; Grigson, *Menagerie*, 96–121.

14   Ritvo, *Animal Estate*, 17–30.

15   J. Farey, *General View of the Agriculture and Minerals of Derbyshire*, 3 vols (London, 1811–17), vol. 3, 633, 642–8.

16   E. Darwin, letter to A. Reimarus, May 1756? in King-Hele ed., *Collected Letters*, 24–5; H. E. Hoff, 'Galvani and the pre-Galvanian electrophysiologists', *Annals of Science*, 1 (1936), 157–72; W. C. Walker, 'Animal electricity before Galvani', *Annals of Science*, 2 (1937), 84–113; R. W. Home, 'Electricity and the nervous fluid', *Journal of the History Biology*, 3 (1970), 235–51; R. Smith, 'The background of physiological psychology in natural philosophy', *History of Science*, 11 (1973), 75–123; K. M. Figlio, 'Theories of perception and the physiology of mind in the late eighteenth century', *History of Science*, 13 (1975), 177–212; G. N. Cantor, 'The theological significance of ethers', in G. N. Cantor and M. J. S. Hodge eds, *Conceptions of Ether: Studies in the History of Ether Theories, 1740–1900* (Cambridge, 1981), 135–56; P. Bertucci, 'Sparks of life: medical electricity and natural philosophy in England, c.1746–1792', DPhil thesis, University of Oxford (Oxford, 2001); P. Elliott, '"More subtle than the electric aura": Georgian medical electricity, the spirit of animation and the development of Erasmus Darwin's psychophysiology', *Medical History*, 52 (2008), 195–220.

17   J. Hunter, 'Anatomical observations on the Torpedo', *Philosophical Transactions*, 63 (1773), 481–9; J. Hunter, 'An account of the *gymnotus electricus*', *Philosophical Transactions*, 65 (1775), 395–407; T. Cavallo, *A Complete Treatise on Electricity in Theory and Practice*, 4th edn, 3 vols (London, 1795); vol. 3, 3–5.

18   W. Withering, *An Account of the Foxglove and some of its Medical Uses* (Birmingham, 1785), xv–xvi.

19   J. Priestley, *The History and Present State of Electricity*, 2nd edn, 2 vols (London, 1775), vol. 2, 253–9; Withering, *Account of the Foxglove*, xv–xvi; S. Finger, *Doctor Franklin's Medicine* (Philadelphia, 2006), 92–114.

20   Bowerbank, *Speaking for Nature*, 136–7.

21   E. Darwin, *Zoonomia; or the Laws of Organic Life*, 2nd edn, 2 vols (London, 1796), 3rd edn, 4 vols (London, 1801), vol. 1, 62, vol. 3, 79–85.

22 Cavallo, *Complete Treatise on Electricity*, 1–75; Hoff, 'Galvani and the pre-Galvanian electrophysiologists', 157–9; Walker, 'Animal electricity before Galvani', 109–11; M. Pera, *The Ambiguous Frog: the Galvani–Volta Controversy on Animal Electricity*, translated by J. Mandelbaum (Princeton, 1992), 64–6, 80–6, 123–31; S. L. Jacyna, 'Galvanic influences: themes in the early history of British animal electricity', *Bologna Studies in History of Science*, 7 (1999), 167–85; Elliott, 'More subtle than the electric aura', 208–12.

23 Elliott, 'More subtle than the electric aura', 208–12.

24 Darwin, *Zoonomia*, vol. 1, 264–5.

25 Darwin, *Zoonomia*, vol. 1, 202.

26 Darwin, *Zoonomia*, vol. 1, 186–265; G. White, *The Natural History and Antiquities of Selbourne* (1789), edited by E. T. Bennett and J. A. Harting (London, 1876), 288–9; W. Paley, *Natural Theology: or Evidences of the Existence and Attributes of the Deity* (1802) in *The Works of William Paley* (London, 1849), 119–24.

27 Darwin, *Zoonomia*, vol. 1, 187.

28 Darwin, *Zoonomia*, vol.1, 239.

29 Darwin, *Zoonomia*, vol. 1, 197–8.

30 Darwin, *Zoonomia*, vol. 1, 239, 237.

31 Darwin, *Zoonomia*, vol. 1, 212, 252; J. Bentham, *Introduction to the Principles of Morals and Legislation* (1789), 2 vols (London, 1823), vol. 2, 234–6.

32 G. Barker-Benfield, *The Culture of Sensibility: Sex and Society in Eighteenth-Century Britain* (Chicago, 1996).

33 Darwin, *Plan for the Conduct of Female Education*, 115–16.

34 Darwin, *Plan for the Conduct of Female Education*, 48.

35 Darwin, *Zoonomia*, vol. 1, 194–5, 223–4.

36 Darwin, *Zoonomia*, vol. 1, 200–4.

37 Darwin, *Zoonomia*, vol.1, 223, 224–5, 225.

38 Darwin, *Zoonomia*, vol. 1, 208, 205–6.

39 Darwin, *Zoonomia*, vol. 1, 208–9, 210; this idea was subsequently tested by his grandson, who published the results of his research in his fascinating *Expression of the Emotions in Man and Animals*, 2nd edn, edited by F. Darwin (London, 1889).

40 Darwin, *Zoonomia*, vol. 1, 212.

41 Leclerc, *Natural History*, abridged, translated by Smellie, vol. 1, 87–8; Goldsmith, *History of the Earth*, vol. 2, 5–7; Bewick and Beilby, *General History of Quadrupeds*, 1–2.

42 Darwin, *Zoonomia*, vol. 1, 226–7.

43 Darwin, *Zoonomia* vol. 1, 231, 237; Goldsmith, *History of the Earth*, vol. 3, 51–5.

44 Darwin, *Zoonomia*, vol. 1, 212–15, 219.

45 Darwin, *Zoonomia*, vol.1, 219–21.

46 Darwin, *Phytologia*, 335–7; Farey, *General View*, vol. 3, 646; Bewick and Beilby, *A General History of Quadrupeds*, 430–2.

47 Darwin, *Zoonomia*, vol. 1, 229, 222, 240–1.

48 Darwin, *Zoonomia*, vol. 1, 205–6, 222, 230–1, 234–5, 246–8, 252–3; on Butt: S. Kelly ed., *The Life of Mrs Sherwood* (London, 1847), 7–8.

49 Darwin, *Zoonomia*, vol. 1, 263.

50 Darwin, *Zoonomia*, vol. 1, 199; G. L. L. Buffon, *Buffon's Natural History Abridged*, 2 vols (London, 1792), vol. 1, 176; Goldsmith, *History of the Earth*, vol. 2, 338–43; Bewick and Beilby, *A General History of Quadrupeds*, 411–15.

51 Darwin, *Zoonomia*, vol. 1, 198–9; Alberti, *The Afterlife of Animals*; Grigson, *Menagerie*, 96–120; Plumb, *Georgian Menagerie*.

52 Prothero, *English Farming*, 139; R. Malcolmson and S. Mastoris, *The English Pig: A History* (London, 2001), 1–44.

53 Goldsmith, *History of the Earth*, vol. 2, 132–6.

54 Buffon, *Natural History Abridged*, vol. 1, 105–7.

55 Bewick and Beilby, *A General History of Quadrupeds*.

56 R. Bartel, 'Shelley and Burke's swinish multitude', *Keats-Shelley Journal*, 18 (1969), 4–9.

57 The 'Celestial Bard', 'The Quadrupeds, etc. or Four-footed Petitioners against the Sale of Nun's Green: A Terrestrial Poem', reproduced in J. Tilley, *The Ballads and Songs of Derbyshire* (London, 1867), 193–6.

58 J. Pilkington, *A View of the Present State of Derbyshire*, 2 vols (Derby, 1789), vol. 1, 315–16; Farey, *General View*, vol. 3, 164–76.

59 J. Barlow and P. Elliott, 'A brush with the doctor: Samuel James Arnold and Erasmus Darwin', *Notes and Records of the Royal Society*, forthcoming; E. Darwin, letter to S. J. Arnold, 20 December 1800, in King-Hele ed., *Collected Letters*, 562–3.

60 Darwin, *Phytologia*, 224.

61 Darwin, *Zoonomia*, vol. 1, 227–8.

62 *Derby Mercury*, 21 October 1784; the pair were back in Derbyshire in 1790: see *Derby Mercury*, 19 August 1790.

63 J. Boswell, *Life of Johnson*, edited by R. W. Chapman (Oxford, 1980 [1799]), 1357; A. Seward, letter to Miss Weston, 29 October 1784 in H. Pearson, *The Swan of Lichfield* (London, 1936), 71.

64 R. Porter, 'Erasmus Darwin: doctor of evolution?' in J. R. Moore ed., *History, Humanity and Evolution: Essays for John C. Greene* (Cambridge, 1989), 39–69, at 42.

65 White, *Natural History and Antiquities of Selbourne*, 234–5; S. Johnson, *Dictionary of the English Language* [1755], edited by J. Lynch (London, 2004), 155–6.

66 E. Darwin, *The Economy of Vegetation*, 4th edn (London, 1799), additional note xxxix; E. Darwin, *The Temple of Nature; or the Origin of Society* (London, 1803), 156–9, 166; Darwin, *Phytologia*, 328.

67 Darwin, *Phytologia*, 330–1.

68 Darwin, *Zoonomia*, vol. 1, 261, 252–6; A. Bennet, 'A new suspension of the magnetic needle, invented for the discovery of minute quantities of magnetic attraction ... communicated by the Rev. Sir Richard Kaye, Bart, FRS', *Philosophical Transactions*, 82 (1792), 81–98; Buffon, *Natural History Abridged*, vol. 2, 281–92; Goldsmith, *History of the Earth*, vol. 4, 157, 163–73.

69 Darwin, *Phytologia*, 332.

70 Buffon, *Natural History Abridged*, vol. 2, 345–57; Goldsmith, *History of the Earth*, vol. 4, 258–74; Thomas, *Man and the Natural World*, 62–3.

71 Darwin, *Zoonomia*, vol. 1, 258–60.

72 T. Taylor, *A Vindication of the Rights of Brutes* (London, 1792), advertisement, iii–vi, 10, 11–13, 18–19, 22–5, 29–30, 34–6, 47–72, 75–103; P. Fara, *Erasmus Darwin: Sex, Science and Serendipity* (Oxford, 2012), 208–9.

73 Bentham, *Principles of Morals and Legislation*, vol. 2, 234–6.

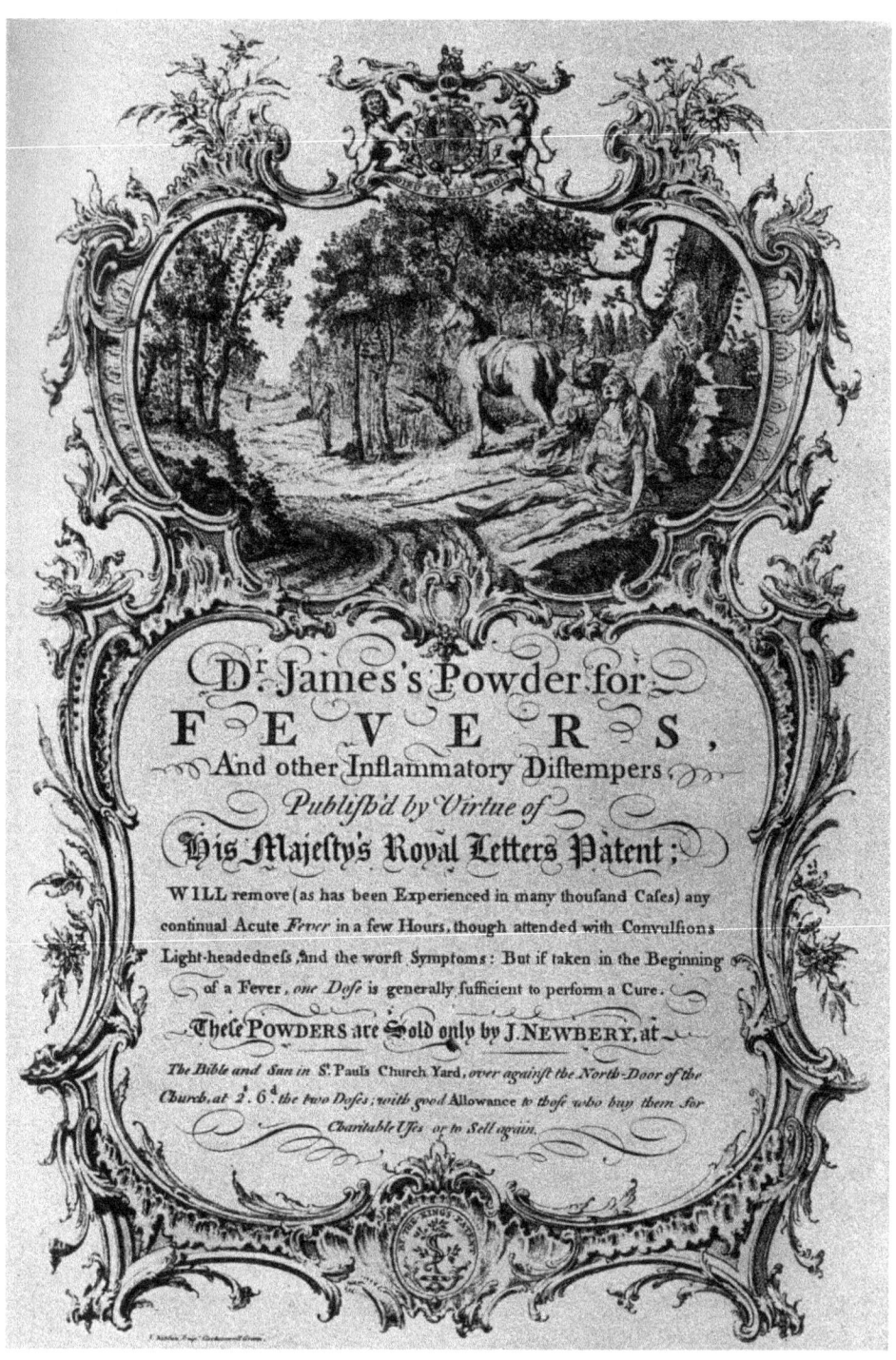

Figure 51. Advertisement for Dr James's Powder, reproduced from A. S. Turberville, *Johnson's England*, 2 vols (1933), vol. 2.

CHAPTER 7

ANIMAL DISEASES

Erasmus Darwin spoke of protecting animals and plants against disease in *Zoonomia* (1794/96) and *Phytologia* in a manner reminiscent of his analyses of human health, tending to emphasise inherent propensities to illness and disequilibrium within organisms above external putrefaction or miasmas. While remaining a strong advocate of traditional methods associated with neo-humoural medicine, such as bleeding and purging, as with his human patients, he was usually measured and careful in recommending these and his success as a physician probably arose in part from his preparedness to let ailments take their course and provide prescriptions designed to bolster the animal system, such as greater provision of fresh vegetables and fresh air. His understanding of animal physiology and development was closely informed by medical practice, and his recommendations for improving farming using natural philosophy encouraged greater attention to the problems of animal and plant diseases, competitive provision and promotion of new treatments in the medical marketplace, which paralleled those available for humans, and more critical assessments of these. As we have seen, larger farms and farming methods encouraged by the agricultural revolution inadvertently, perhaps, fostered increases in plant diseases, at least to begin with. However, outbreaks of illness among farm animals were one of the factors behind the decline in small farmers and the increase in larger farms and great estates during the eighteenth century, which (it was claimed) coped more efficiently with such problems. Diseases were linked to small-scale farming and the grazing of cattle on open and common land, allowing them to wander unimpeded and mingle with the livestock of other farmers. In addition to increasing the efficiency of farming, enclosure and improvement were therefore also associated with improving livestock health.[1]

Analogies between human and animal illnesses and the authority medical practitioners already had as experts encouraged farmers to consult them concerning their stock, as did the fact that some diseases, such as cowpox, were well known from 'time-immemorial' as spreading from animals to human.[2] These practises were reinforced by the use of animal versions of human medicines, such as James's Fever Powder, which was advertised and used 'very extensively' between 1750 and 1800. It was devised by the Staffordshire physician Robert James (1703–1776), a friend of Samuel Johnson and author of a *Medicinal Dictionary* and other works, including a study of canine madness, who took a keen interest in animal health (Figure 51). Originally a patent medicine, James's Fever Powder was promoted in his *Dissertation on Fevers* (1751) and marketed in an animal version for horses, horned cattle, dogs and other creatures. When subjected to chemical analysis by the physician George Pearson FRS (1751–1828) to 'ascertain the nature and manner of preparing' the medicine, he emphasised that it was 'well known' that it could not be prepared following the Court of Chancery specifications and concluded that there were 'no essential ingredients' except 'antimonial calces, phosphoric acid, and calcareous earth', the latter two being 'united together'. He also subjected the animal version to investigation, describing it as a 'light clay-coloured, gritty, tasteless substance' in which there were 'small *spicula*', and believed that this was 'nothing more than James's powder for fevers … made by calcining antimony and bone ashes together in open vessels'. Nevertheless, the powder was recommended by physicians such as Thomas Kirkland (1722–1798) for the treatment of cattle distemper.[3]

As various scholars have emphasised recently, despite the growing importance placed upon animal agency and the close and often intimate interconnections between bestial and human lives, animals are still comparatively absent from medical historical research. Where animals have been studied by medical historians this has been more from the perspective of their impact upon humans rather than a genuine effort to seek non-human perspectives. Animals need to be taken 'seriously as historical subjects' interfacing with humans, environments and each other rather than just being used to demonstrate anthropocentric changes in health and medicine, behaviours and thoughts, particularly given that animals and human/animal interactions have had a major impact upon many aspects of medicine, helping to define the scope and processes of human healthcare while also often challenging anthropocentrically defined boundaries between humans and other creatures.[4] By exploring the different ways that Darwin applied his medical treatments and ideas to animal ailments and the contexts in which this occurred, this chapter demonstrates how animals affected human healthcare and contrawise, challenging assumptions about bestial bodies just as Darwin had challenged assumptions about bestial behaviours.

ZOONOTIC DISEASE AND VACCINATION

The belief that some diseases could be spread between animals and humans and between different species of animals was quite widespread in the eighteenth century. Outbreaks of 'horse cold' or influenza and sometimes influenza among dogs, for example, were linked to attacks of the human version. In 1727 and 1733 horses in Staffordshire and Shropshire were recorded as 'seized with a cough and weakness', with others bleeding at the nose in parts of Ireland, and the same symptoms seized 'mankind' once the horses had appeared to get better. In 1750 'horse cold' struck parts of Ireland and in 1760 the *Annual Register* observed that an epidemic of the distemper 'rages amongst horses' around London and in Yorkshire, bringing 'sore throats and colds'. In 1775 influenza was observed by various physicians, including John Fothergill (1712–1780), John Haygarth (1740–1827) and Richard Pulteney (1730–1801), 'infesting' dogs and especially hounds; it also attacked horses and then humans, laying waste to whole hunting packs. In 1795 John Barker of Coleshill recorded that influenza had hit Birmingham, Lichfield and other midlands towns, bringing 'violent defluxions of the nose, throat and lungs', affecting some horses and causing dogs to experience symptoms of running eyes, a 'loss of the use of their hind legs' and in many cases death.[5]

In the first edition of *Zoonomia* (1794) Darwin emphasised that different animal genera did not usually 'naturally communicate infection to each other', which explains his caution in interpreting apparent evidence that diseases had spread between humans and other creatures, which he normally ascribed to a common cause such as miasmatic effluvias (clouds of tiny particles) in the case of '*catarrhus contagiosus*'. He had heard of experiments in which human diseases such as smallpox and measles were inoculated into quadrupeds, but these 'vain efforts' had failed. However, when 'parotitis or mumps' spread around the Haywood area of Staffordshire and was followed by an epidemic of '*parotitis suppurans* or mumps with irritated fever' among cats, he was 'inclined to believe that the cats received the infection from mankind'. The feline ailment, which Darwin christened '*parotitis felina*', caused swellings under the jaw which often suppurated or became infected, producing pus, and was usually 'very fatal'. An entire 'breed of Persian cats with long white hair' succumbed at Haywood, as did 'almost all the common cats of the neighbourhood'.[6]

The most famous – or notorious – instance of the spread of disease between humans and animals in the late eighteenth century, however, was cowpox and Darwin is usually presented as having welcomed the discovery of vaccination against smallpox using cowpox material (lymph fluid from skin pustules or lesions) by the physician Edward Jenner (1749–1823). However, he was actually rather cautious about it initially because of his belief that, with a few exceptions, diseases did not normally cross the interspecies barriers between humans and other animals.[7] Smallpox was a serious problem in Georgian society, killing around 15 per cent of the population and, as Darwin emphasised to his son Robert in 1791, in large cities some 35 per cent of children under the age of two died annually from the disease. The advent and refinement of inoculation, following a practice imported from the Ottoman Empire in the early eighteenth century and encouraged by Lady Mary Wortley Montagu (1689–1762), as Darwin emphasised, in which infected matter from smallpox pustules was deliberately placed in a cut upon the body, helped to seriously reduce mortality rates. Darwin's face bore the scars of a 'severe' smallpox attack when young and he made much effort as a medical practitioner to promote childhood inoculation even to the extent, according to his friend and biographer Anna Seward (1742–1809), of unsuccessfully trying similar methods to inoculate his children Robert and Elizabeth Darwin against measles.[8]

Darwin followed the physician Thomas Sydenham (1624–1689) in dividing smallpox into 'distinct and confluent' kinds, or '*variola discreta*', with 'distinct pustules', an often violent fever, haemorrhages, inflammation caused by 'absorption of variolous matter' over-exciting the 'cutaneous vessels' and other symptoms; and '*variolous confluens*', in which the pustules were combined. In 1789 he told Robert soon after his son had qualified as a physician that he supposed 'infectious miasmata' associated with scarlet fever and smallpox to consist of 'dry fine powder diffused in the air, not dissolved', which usually stuck 'on the mucus of the tonsils' and inflamed these parts 'like an inoculated arm', frequently with fatal consequences. In *Zoonomia* he again emphasised that the 'natural disease' was spread by dried dust particles of 'contagious matter' floating in the air and being breathed in; a single 'grain of variolous matter, inserted by inoculation' might in a fortnight 'produce ten thousand times the quantity of a similar material thrown out on the skin in pustules'. Darwin utilised various treatments to combat the symptoms, including blisters, cathartics, cordials, opiates, purges, mercury and cold air or bathing in some cases, with plasters or cerates to try to prevent disfiguring pock marks.[9]

Explaining his smallpox inoculation practice in some detail to Robert and subsequently in *Zoonomia*, Darwin inserted fresh matter from a pox into a child's arm with help from an apothecary and surgeon, following the practices of Thomas Dimsdale (1712–1800). This method required additional treatments and usually brought on a fever soon after infection, but one considerably milder than if the process had never been undertaken.[10] He believed that, while much remained unclear, inoculation worked so well because, by inserting a milder form of variolous contagion into the arm, it prevented the dangerous version from appearing in the tonsils and causing a swelling of the throat, resulting in far fewer pustules and associated fevers. Instances where it appeared to be ineffective and where pregnant women miscarried following inoculation probably occurred because of other factors, such as cross-contamination of patients from the 'natural' disease. Babies 'at the breast' were most vulnerable and therefore Darwin recommended that, except for crowded towns where there was a concentration of 'natural contagion', inoculation was best undertaken on children over the age of two using a 'small quantity' of fresh warm fluid matter from a diseased patient placed in as 'small and superficial' a scratch as possible to prevent ulcers. He argued that experiments concerning the vaccination process undertaken by Jenner, Charles Kite (1768–1811) and others supported his own approach to inoculation, and that Jenner's 'ingenious' observations concerning the much greater effectiveness of fresh variolous matter underlined the dangers of using older partially putrified matter, which risked causing other illnesses.[11]

The development of vaccination occurred because it had long been noticed in milking counties such as Gloucestershire that cowpox, which produced 'an eruption' on the udders of cows and could be 'communicated' to the arms or hands of their milkers, producing an ulcer and fever, 'secured those who had been infected with it from afterwards being liable' to smallpox (Figure 52). This discovery was publicised by Jenner and his supporters, who urged that programmes of vaccination against smallpox using milder cowpox material be implemented.[12] However, when he first read about Jenner's work and received further information and an 'infected thread' from the physician George Pearson (1751–1828), who was conducting extensive trials with it in London, Darwin was somewhat sceptical. This was on the grounds that, as he told Robert in 1799, the physician John Ferrier (1761–1815) at Manchester had reported a case in which an individual vaccinated with cowpox had subsequently caught smallpox, while inoculation conducted 'in the best manner' with 'diluted' smallpox

re 52. Portrait of Edward Jenner, reproduced from *The History of Inoculation and Vaccination for the ...ention and Treatment of Disease* (1913).

'matter' on young children produced only a very 'mild' form of the disease, suggesting there might be no need to vaccinate with cowpox material. He believed that childhood fatalities from inoculation resulted from the spread of the 'natural disease' by 'careless surgeons' visiting before the process had begun and even suggested, perhaps half jokingly, that Robert inoculate cowpox patients with smallpox and that Pearson might have sent smallpox labelled as cowpox 'to deceive you and the world'. What he may have had in mind here, as later emphasised in *Zoonomia*, was the dangers presented by using smallpox material to vaccinate that was not fresh and had undergone some putrefaction and so contained both the variolous matter and 'typhus contagion', thereby resulting in typhus or putrid fever.[13] In December 1800 Darwin was still cautious, recommending Jenner and William Woodville's latest work on the cowpox to Robert in preference to Charles Aikin's book on the subject and emphasising the fundamental importance of distinguishing 'the genuine cowpox from that which is not genuine', because patients who had received only the 'spurious cowpox' were 'liable to the variolous infection' (Figure 53).[14]

A crucial aspect of Jenner's discovery, in Darwin's view, was that, as the 'vaccine-pox' was not 'infectious', through 'careful inoculation' members of families might be inoculated without endangering pregnant women and others, which was 'of great consequence to the public' because smallpox inoculation was often 'known' to 'propagate' that disease and cause miscarriages and death. The cowpox vaccine, by comparison, was considerably less hazardous to infants than was smallpox. There were some concerns, however, one being that, in Darwin's opinion, because cowpox was 'much less infectious' than smallpox, so greater care was needed to carry out the inoculation correctly and effectively; inflammations caused by the lancet puncture had sometimes been mistaken for the symptoms of the vaccination process itself, which he believed had led to at least four cases of people being struck down by smallpox in Derbyshire when they had supposedly already had the inoculation. However, such instances ought most emphatically not to cause anyone to 'discredit' this otherwise

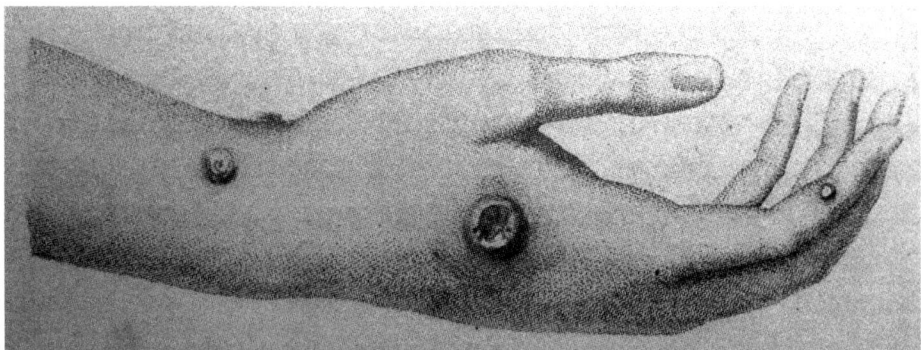

Figure 53.   Engraving of the hand of a dairy maid infected with cow-pox from a cow's udder, from E. Jenner, *Inquiry into the Causes and Effects of the Variolae vaccinae* (London, 1798), reproduced from C. Singer, *A Short History of Medicine* (Oxford, 1928).

'fortunate and wonderful discovery'. Darwin believed that Jenner's argument that cowpox was communicated to the udders by manservants (who were largely responsible for milking in the Gloucestershire area) who had 'previously acquired the infectious matter' from exposure to the 'acrid sanies' discharged by horses suffering from 'the grease' was worth 'further investigation'. In some striking and profound reflections, Darwin also speculated that because of the 'preservation' of so many from smallpox from their previous bout of cowpox, it might be that the former was 'originally acquired' from the latter. This was 'so wonderful a phenomenon, so contrary' to previous assumptions concerning analogies between human and animal infectious diseases, that it might even mean that smallpox had originally come from cowpox. If so, 'by process of time', the 'much older' cowpox might have become 'much milder' than its progeny, just as smallpox appeared to have become 'milder than formerly', probably because the 'incapacity of receiving it' (or resistance) existing in those who had experienced the disease had through 'process of time become hereditary'. Yet, while moving towards a theory of the hereditary development of resistance or immunity to contagious disease, Darwin still sought to explain this by the principles of his neo-humoural medical theory, arguing that the growing 'incapacity of receiving' the smallpox for a second time might be explained by the 'general law of animation', because 'stimuli greater than natural lose their effect by habit, or from their being too violently or too frequently applied'.[15]

Smallpox vaccination faced much opposition, partly because of revulsion felt at the idea of taking infected material from the skin pustules of patients suffering from a disease caught from cows and placing it within healthy bodies as means of supposedly preventing future illness. Cartoons appeared ridiculing the idea, with human heads replaced with those of cattle, and Jenner himself and his supporters attracted considerable opprobrium (Figure 54).[16] Many medical practitioners, such as the apothecary Charles Cooke of Gloucester, were sceptical of vaccination's benefits precisely because it entailed 'introducing animal disease to the human system', one being merely a 'local disease in the brute' while smallpox was a 'constitutional disease in man'. Likewise, the leading 'cow doctor' in Gloucestershire, Henry Clayton, who also attended horses for the 'grease', believed cowpox to be a 'local disease ... invariably cured by local remedies'.[17] Nevertheless, in a letter to Jenner of February 1802 Darwin called his discovery a means of 'preventing the dreadful havoc made among mankind by the small-pox' and proclaimed himself hopeful that, 'by introducing into the system so mild a disease as the vaccine inoculation produces', smallpox would 'in time' be eradicated 'from all civilised countries'. 'Testimony of innumerable instances' had demonstrated that the 'vaccine disease' was 'so favourable to young children' that in time it might 'occur that the christening and vaccination of children may always be performed on the same day'.[18] As he emphasised the previous year in the third edition of Zoonomia (1801), as the cowpox was 'said to be so favourable to infants, great benefit might accrue to mankind by their early inoculation', which could eventually 'exterminate' smallpox. Darwin suggested that dispensaries might be established 'in towns and even villages' to administer inoculations to the poor, providing them with 'a premium of a few shillings' and 'their daily sustenance for eight or ten days' so

The Cow Pock __ or __ the Wonderful Effects of the New Inoculation.

Figure 54. G. Cruikshank, 'The Cow Pock – or – the wonderful effects of the new inoculation' (1802), from *The History of Inoculation and Vaccination for the Prevention and Treatment of Disease* (1913).

that the process could be closely observed by surgeons to check that they genuinely caught the cowpox.[19]

COMBATING ANIMAL DISEASES

Darwin and other members of the Derby Philosophical Society, such as the surgeon John Hunt (1743–1813), had a professional interest in veterinary medicine, and it looks likely that this was shared by at least some of their colleagues, given that around half the resident and over half of the non-resident members of the Society were medical practitioners. As Society president, Darwin ordered various works on zoology and animal medicine for the Society library, including the royal Scottish farrier James Clarke's treatise on the diseases of horses (1790) and *Lectures of the Elements of Farriery* (1793) by veterinary surgeon Charles Vial de Sainbel (1753–1793), as well as agricultural works such as Richard Parkinson's *Experienced Farmer* (1798) and transactions of medical and agricultural societies that included studies of the subject. Parkinson's *Experienced Farmer*, which was dedicated to the celebrated Thomas

William Coke of Norfolk, included a chapter on curing corns in horses' feet, while Clarke's study of equine diseases included recommendations on preventing problems by better management of stables, water, diet, air and exercise, as well as advice on bleeding, purging, moulting, lameness and 'observations on some of the surgical and medical branches of farriery'. The purchase of Sainbel's work is significant because it examined various aspects of equine management and diseases and, at a time before the Society acquired a museum or laboratory apparatus, it was provided with 'a box with plaister models of horses' feet and brass models of horseshoes', which were stored with the books. Sainbel was a professor at the new London veterinary college and his book included analysis of the anatomy and physiology of equine legs and feet, consideration of how careful shoeing could help forestall problems and analyses of equine disorders such as corns, sand cracks, canker, swellings, bone fractures and strains to muscles, tendons and ligaments.[20]

According to Hunt, who knew him well, the 'breeding of sheep' was 'one great object of Darwin's attention' and in *Phytologia* he considered treatments for 'sheep rot' and attacks by 'flewk worm' (the common liver fluke or sheep liver fluke, a parasitic flatworm that infects animal and human livers), which was often countered using weak brine. Darwin agreed that this might be effective, but believed that, as with many human and animal diseases, an improved diet helped and suggested this would work better if 'hay was moistened with the solution' so providing 'better nourishment' than generally obtained by diseased sheep. Instead of blaming the fluke worm he saw it as a symptom and believed that, as with human intestinal worms, the proximate cause was 'inactivity of the absorbent vessels' in the sick sheep's liver which made the bile duct 'too dilute', particularly in 'moist seasons'. It was this that encouraged worm growth, which 'eroded the liver', caused ulcers and, 'from the sympathy of the lungs with the liver', resulted in coughing and 'hectic fever' from 'absorption of the matter'. The brine probably operated, he thought, by making the bile duct 'less dilute' through 'promoting a greater absorption of its aqueous parts' and more secretion, although 'sixty grains of iron-filings' rolled into a flour ball with some salt and fed to sheep every morning might prove 'more efficacious'.[21]

As we have seen, Darwin attributed some diseases to miasmas but believed that internal characteristics determined the susceptibility of animal bodies; thus morbidity did not always transfer easily because it manifested differently in different frames. He maintained that '*Catarrhus contagiosus*' (influenza) in animals originated, like the human version, from miasmas 'diffused in the atmosphere' and communicated from individual dogs or horses to wider populations by 'contagion', as 'ancient philosophers' had argued, particularly in warm climates, where it was spread among the 'heavy inflammable air or carbonated hydrogen of putrid marshes', reaching beasts first because their heads were closer to the ground that those of humans. However, although the cause was the same, this did not mean humans and other creatures contaminated each other because, although the latter could infect other animals 'of the same genus', the 'contagious matter generated' from 'their own bodies' might not be the same as that received, as the number of deaths that occurred during an outbreak at Oxford from 'jail fever' (epidemic typhus) without infecting others suggested. The higher prevalence of influenza among horses and dogs might also be due to 'the greater extension and sensibility' of their nasal 'mucous membrane', which spread 'over their wide nostrils and their large maxillary and frontal cavities'.[22]

Darwin tried to reflect these differences in the impact of miasmas upon various creatures in his treatment recommendations. As equine influenza was of the 'sensitive, irritated or inflammatory kind', bleeding was not to be employed excessively and mild purgatives were more effective. He recommended that, unless too cold, horses should be 'turned out to grass both day and night' for the 'pure air' or kept untied in an 'open airy' stable with plenty of fresh vegetables, carrots, potatoes, 'mashes of malt' or oats and 'plenty of fresh warm or cold water' given frequently during the day, while a pint of ale combined with 'half an ounce of tincture of opium' provided every six hours was likely to be effective. Dogs needed to be placed in the 'open air' with 'constant access to fresh water' because all the air saturated with 'contagious particles' they breathed passed 'twice over the 'putrid sloughs of the mortified parts of the membrane' lining the nostrils. They must be fed fresh milk and broth often and allowed to wander near running water and through fields to dissipate any falling 'contagious matter' and enable use of the grass as an emetic. Raw meat was more effective than cooked flesh, and 'five to ten drops of opium tincture', depending upon the size of the dog, could be 'given with advantage every six hours. If 'sloughs' were visible in the nostrils of either horses or dogs, they ought to be moistened twice each day with a solution of lead or alum formed by dissolving half an ounce of lead acetate or alum into a pint of water using a sponge on a stick or syringe.[23]

While some animal diseases were caused or exacerbated by external factors such as miasma or rotting materials, Darwin's main treatments, like his prescriptions for human conditions, focused upon internal bodily state and how it could be improved through, for example, more nutritious diet or exercise. His belief in an active 'power impressed on organised bodies by the great author of all things', which enabled them to grow in 'size and strength' to maturity, also extended to an assumption that this could sometimes 'cure their accidental diseases, and repair their accidental injuries', and so bodies were not merely passive victims of disease but could actively counter it by various means. This power even on occasion allowed them to form 'armour to prevent those more violent injuries which would otherwise destroy them'.[24]

Darwin was able to put his theories concerning animal disease and the best methods of combating it into practice when a 'most alarming' cattle 'disorder' occurred in south Derbyshire in autumn 1783, which was reported in the newspapers. By early September this had proved fatal to a number of cattle; twenty-five of Sir Harry Harpur's 'fine beasts' at Calke House (now Abbey) had died 'in a very short time' – and, despite farmers, gentry and 'inhabitants of the neighbouring villages' using 'all proper precautions' to stifle the 'dreadful malady', cows died at Melbourne, Chellaston and Repton, apparently from the same distemper. In 1750, during the minority of Sir Harry, his guardians had been obliged to remit the rents of many estate tenants who lost the largest portion 'of their stock of cattle by the contagious [cattle] distemper' then raging across the area, which explains his concerns that history might be repeating itself. Darwin became involved as a friend of Sir Henry and Lady Frances Harper, who were his patients during the 1770s and 1780s, and he was a frequent visitor to Calke. In March 1788 Darwin remarked to Josiah Wedgwood in a letter written from Calke that he passed 'almost all' his 'evenings' there 'now … with Sir Harry' and it is therefore not surprising that Harpur asked Darwin's advice.[25]

On 3 September 1783 Darwin sent a letter to *Derby Mercury* proprietor John Drewry – which did not arrive in time to be included in the new edition, so appeared the week after (and was also carried by other papers and nationally in the *Weekly Entertainer*) – in which he explained how the cattle distemper had struck and provided recommendations to prevent its reoccurrence; this, however, put him in direct conflict with another medical practitioner he knew well, Thomas Kirkland, who had advised rigorous bleeding and purging.[26] According to Darwin, the outbreak had begun about two weeks before with the death of a cow owned by Francis Kinsey of Melbourne 'as was supposed, by a quinsey' (a complication or abscess linked to tonsillitis). The cow's carcass was taken to the Calke Abbey kennels minus the head, and a dog carried a large piece of it among the other cows, which 'were seen to smell of it in a circle, as is the custom of those animals, when they see raw flesh, or blood on the ground'. Subsequently two or three of these 'became ill in three or four days and died in about 24 or 27 hours' after they had been seen to 'abate in the quantity of their milk, or to appear drooping'. Harpur was 'unfortunately advised to have all his other cattle blooded and purged' and these had 'dropp'd off day after day since that time', with sixteen deaths, the last being buried the previous Tuesday morning.

According to Darwin, the hides from two of the first to die 'before the disease was suspected to be infectious' were carried to a Repton tan-yard. Unfortunately, two cows in a field adjoining the tan-yard owned by William Bryant of Ticknell were taken ill, with one dying, while two others owned by 'Mr. Taylor' of Repton became sick, one dying. The owner of the tan-yard also lost a cow 'in the close adjoining the vats'. Meanwhile, the head of Kinsey's cow was carried in a lime-cart to a Chellaston orchard owned by Richard Foreman, where four calves were 'observed to smell of it, and even to lick it', along with several pigs. Three of the former died, while three pigs and cows were lost to Mr Woodward of Repton, Mr Robinson and [Samuel] Erpe [Earp] of Melbourne and others, although Darwin did not believe that 'these were infected from the same source'. Of the sixteen cows lost at Calke, Darwin thought nine were 'blooded by the same fleam' that had been used to bleed the first two or three, 'as hard swellings appear'd on all these about the orifice', which was believed by some to have spread the 'distemper'. Fleams were bloodletting instruments with triangular-shaped blades designed to cause the least damage to surrounding tissue that were placed over veins and struck with a fleam stick. Once sufficient blood was taken, incisions were sealed with a pin, horse tail hair or thread. Darwin considered it unlikely that the use of the same fleam had caused inoculation (the spread of the disease) because the cattle had fallen 'sick at very different times', from the first to the ninth day, after being blooded. Other sick cows bled by an uninfected fleam also had the swellings and he thought the fleam 'must have lost its infection' after being used upon two or three beasts.[27]

In response to reports of the 'contagious disorder' and fears that it would spread without 'speedy and effectual methods' to prevent this, the Fox–North coalition government moved swiftly to pass an order of council on 10 September with strict measures to be enforced by county justices and constables.[28] On the advice of the Privy Council, the king ordered that on the appearance of 'marks of the said distemper', all 'cowkeepers, farmers, and owners of any oxen, bulls, cows, calves, steers, and heffers' had to 'immediately remove such cattle' to a distant place and 'cause the same to be shot dead or otherwise killed', spreading as little blood as possible and burying the bodies with skins and horns attached within one day. Having 'cut and slashed the hides from head to tail' along with the rest of the bodies to render them 'of no use', they were buried in soil at a depth of at least four feet. Cattle-driving within a mile of infected places was also forbidden, except for trips for water, pasture or for slaughter, and if this exclusion zone was breached without permission of the justices the cattle were stopped until 'full proof' of their health was provided. If these instructions were disobeyed the township constable was empowered to 'pursue the said cattle with all expedition' and if he could not overtake them had to notify the constable in the next constablewick to do the same until they were caught. Farmers and cowkeepers were also forbidden from taking cattle to markets in the parishes of Calke, Repton, Ticknall and Chellaston or any place within ten miles of these (which therefore included Derby itself and parts of Leicestershire and Staffordshire), while hay that had been in contact with infected cattle was to be immediately burnt (Figure 55).

The movement of people in contact with the beasts was also constrained under the government measures, and no person who attended 'any infected cattle' could 'go near 'sound' ones in the 'same clothes'. Buildings in which infected cattle had stood had to be fully cleared of dung and smoked with burning 'wet gunpowder, pitch, tar, or brimstone' and washed in vinegar and warm water, and no cattle could be brought back in for at least two months. All these measures were to be enforced by JPs and commissioners of the land tax, who were empowered to appoint salaried inspectors and other officials to inspect houses, farm buildings and cattle in infected parishes within the ten-mile radius and to 'seize and kill' cattle within and burn hay, straw and other 'litter' to make sure these were 'purified'. The severity of these measures followed a pattern formed by experience of earlier epidemics and reflected fears that outbreaks could spread rapidly, with devastating economic and social consequences. Some relief for farmers and cowkeepers affected was available, as they could claim compensation 'from the Commissioners of His Majesty's Treasury' for 'one full moiety of the value of such cattle' as assessed by the magistrates if they were certified by them to have complied with the regulations in full.[29] There were, however, early signs of improvement and on 9 September 1783 Thomas Grimes, Sir Henry Harpur's bailiff, was able to reassure Darwin 'with pleasure' that:

> no more cattle have dropt in this neighbourhood since I saw you, except a yearling of Sir Henry's at Swarkestone, which was taken ill on Saturday, and died in the night, and a cow of – Snow's of Barrow, which likewise died on Sunday morning, and were both buried in their hides. – Our cattle at Calke all hold very well, and those that were recovering continue to do so.

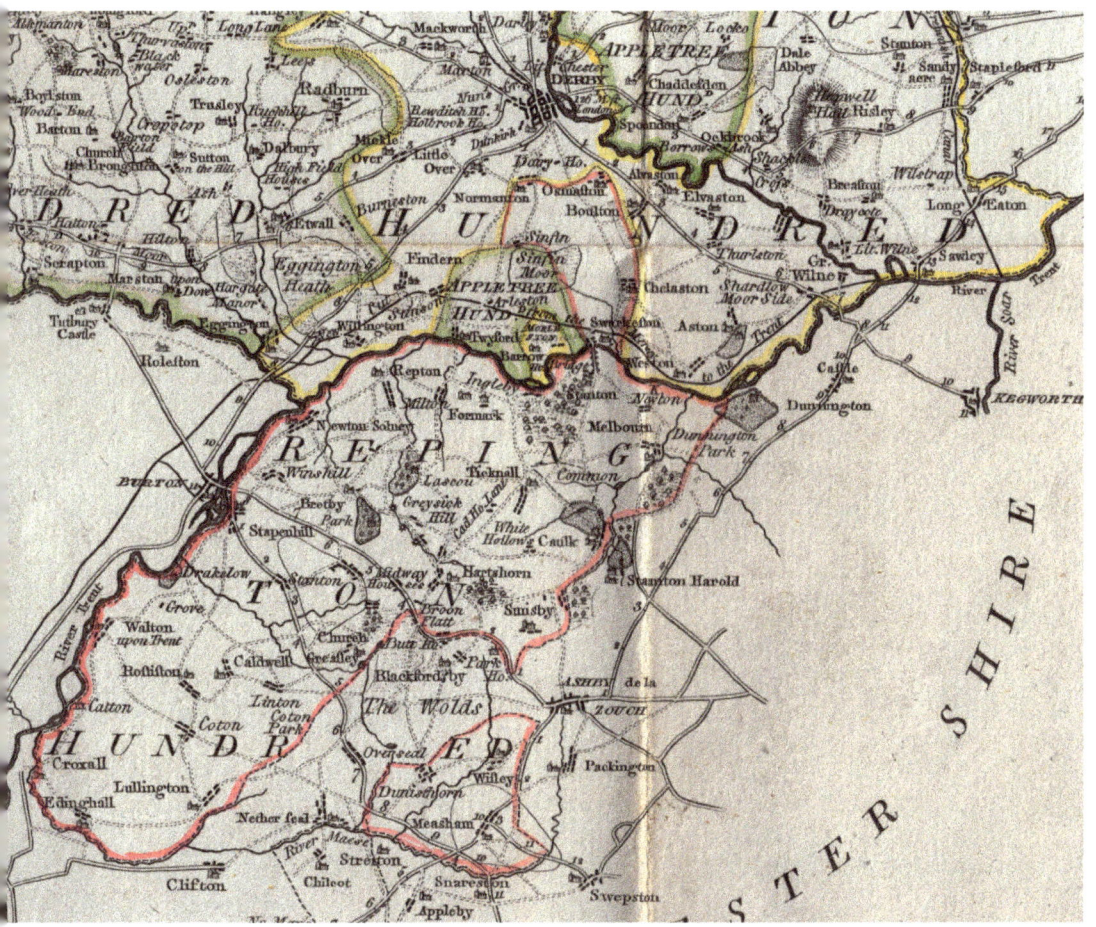

Figure 55.   Calke and area of cattle distemper of 1783, detail from P. P. Burdett, map of Derbyshire with 'corrections and improvements', J. Pilkington, *A View of the Present State of Derbyshire*, 2 vols (1789), vol. 1.

Grimes and Harpur therefore had 'great hopes' that the distemper outbreak might have passed the worst, and their valuable surviving beasts might now be safe.[30]

One reason for the authorities' heavy intervention was that medical opinion about the cattle distemper was divided, with practitioners such as Darwin and Kirkland advocating different treatments and preventive measures. As well as capitalising upon his prestige as a physician, Darwin's recommendations relied upon his belief in the essential similarities between human and animal physiology and anatomy, which he often emphasised and which provided authority to make recommendations to farmers. From descriptions of the disease's progress and his own observations, Darwin concluded that the 'putrid fever' originated from 'a highly putrid carcass because it destroy'd swine as well as cattle', and there was therefore 'reason to hope' that, with vigilance, the infection might be stopped quickly; hence he supported the 'wise measures' already introduced of destroying and deeply burying the infected cattle. Care needed to be taken when 'burying all other carrion' in 'this putrid season of foggy weather' because the horned cattle seemed 'particularly liable' of infection. Following the advice of some medical writers, such as the physician John Barker (1708–1748), Darwin advised that 'blood-letting and purging' should be avoided because the cattle needed to be as strong as possible to fend off infection; instead, feeding with 'a quarter of a peck of malt, oats, barley, or other grain, either whole, or in meal, or in mashes' twice daily and open stabling was preferable. If the danger of infection remained great, a pint of 'decoction of oak-bark' created by boiling two pounds of powdered oak-bark for a quarter of an hour in two gallons of water should be given each morning, while cabbage, turnips and carrot were also ideal because they were 'known to counteract putridity'. Informed by the experience of the mid-century outbreak and Barker's recommendations, Darwin also advocated 'ten grains of opium dissolved in a pint of ale' to be given night and morning. At the same time, where the stomach was 'bound' half a pound of 'white bryony root boiled to a pulp in a quart of water' had, it was claimed, acted as an effective aperient to relieve constipation. On the other hand, it was clear that confining the cattle to stables and keeping them hot 'to promote sweating', 'scarifying the affected part', 'boring the horns' and using different rowels and setons had all done no good.[31]

After examining the cattle plague or '*Pestis vaccina*' again in *Zoonomia*, which had much 'afflicted' the country half a century previously, Darwin advocated treatments similar to those he had recommended for the horned cattle distemper of 1783. He believed that, like the disorder of 1783, this was a 'contagious fever with great arterial debility', which resulted in emphysema being formed in some places, the 'latter stage of the disease' showing how far the 'progress of gangrene' had gone. For treatment of the 'sensitive inirritated fevers' of cattle he recommended about 'sixty grains of opium with two ounces of extract of oak-bark' to be administered every six hours, along with 'thirty grains of vitriol of iron' should a 'tendency to bloody urine' become evident, to which cattle were 'liable'. As in 1783, the government would need to issue an order to prevent movement of cattle within five miles of the place of infection, until it could be certain that the 'pestilence' had abated as judged by a 'committee of medical people', and if this was agreed slaughtering and disposal of hides in lime water would again become necessary before inspections took place.[32]

The physician Joseph Denman of Buxton, who held estates in Derbyshire and Norfolk, argued that as the 1783 cattle distemper apparently spread so rapidly it would 'admit of little, if any, assistance from medicine', but did call for a fuller account of its diffusion and accompanying symptoms, suggesting that a medical practitioner needed to observe the dissection of some infected carcasses to see the 'morbid appearances after death'. Presuming that the disorder was largely 'analogous' to previous outbreaks of horned cattle disease, he believed it would be better to 'kill every beast' suffering from 'violent symptoms of infection' and agreed that burial and hide slashing were essential. Furthermore, like Darwin, Denman maintained that there could be no doubt that 'in so very putrid a disease' bleeding and purging were 'highly wrong in every respect', as they would reduce strength and help foster further infection, although he recommended that a quarter pint of 'ale-wort just beginning to ferment' might 'prove beneficial', rather than Darwin's selection of vegetables. Denman suggested that moving cattle into open air on higher ground away from stables or close buildings might also help, along with the provision of plentiful cold water and much clean straw. Cattle should be washed and dung removed, while servants should avoid attending in the same clothes they had worn while looking after the healthy cattle, and dogs should not be permitted to follow them to the beasts. Slaughter needed to be undertaken with as much care as possible so no blood was spread, burning the straw and firing pitch over the cattle burial sites.[33]

The terms of the order of council in 1783 demonstrated acceptance that the disease was 'communicated' by dead hides and implementation of the measures to manage these and restrict the movement of cattle was believed to have helped hasten its eradication. However, stung by Darwin's letter, which appeared in local and national papers, Thomas Kirkland – who had originally been consulted by Harpur, whereupon he had recommended vigorous bleeding and purging – published a detailed pamphlet justifying his behaviour on the basis of his interpretation of humoural medicine.[34] Kirkland emphasised that he had best knowledge of the distemper as the only 'person of the [medical] faculty' who saw the dead cattle or their original symptoms and stages of decline. Without naming Darwin directly, he argued that, even though the distemper was apparently abating, 'accounts may be given of it which have a tendency to mislead'. As a surgeon as well as a physician, he detailed 'the appearances upon dissection [which Darwin had not done], the progress of the disease, and the events of different methods of treatment'.[35] Kirkland described how, after being called out by Harpur, he found the 'membranes (or inward coat) which lined the windpipe' were 'amazingly inflamed and thickened', which was 'extended to the lungs in a violent degree' and that 'all the membranes and spaces in and about the lungs, were loaded abundantly with a yellow lymph' and that the chest interior was in a 'similar situation', along with other symptoms. Given the hot weather, and having no 'suspicion of fever' or infections, as the beast seemed 'strong and vigorous', Kirkland diagnosed a 'gangrenous inflammation of the lungs' that had caused strangulation, believing it to have arisen from particular habits or behaviours of the cattle and an unusual 'state of the air', with seasonal changes producing differences even in the same disease. He maintained that 'a general alarm might be avoided' by 'preventing communication betwixt' diseased cattle and recommended for treatment, as dictated by 'common sense', 'bleeding, purging, nitre, diluting liquors, and screening the beasts from the heat of the sun' to reduce that 'extreme inflammation' that he believed caused the gangrene. Thirty-one surviving cows were bled and purged using glauber's salt to evacuate the 'morbid bile' from the 'gall bladder and its ducts' and Kirkland claimed that 'those who had the management of the cattle' were 'fully persuaded' that this was the best course of action, which was, in any case, in 'no way injurious' and had even prolonged some beasts' lives.[36]

The death of some of the cattle was admitted by Kirkland, although he argued that this was due to an alteration 'in the seat of the malady' rather than the vigorous programme of bleeding and purging he had advocated. In these 'the wound made by the fleam in bleeding' had become inflamed, swollen and 'in some cases turned to a gangrene'. Despite the application of a poultice of bark, sea salt and a 'decoction of herbs' to 'resist putrefaction', the cow's neck swelled to an 'enormous size' with a 'gangrenous emphysema' full of a 'yellow gangrenous bile' and, although some cows survived, many died. Yet Kirkland claimed that no deaths had resulted from bleeding but that the 'violence of the swelling' of the neck had 'diverted the disease from falling on the lungs' to the bodily exterior, relieving the 'vital parts' and retarding its progress.[37] If the disease were caught early enough, Kirkland thought that James's Fever Powder for cattle might be efficacious, along with 'pegging' to 'discharge' the noxious matter from the diseased part using rowels or setons, which consisted of silk threads, horsehairs or a linen strip passed beneath the skin to form an 'issue' using a knife or needle. Likewise, 'pledgets' or wads of lint or other soft material containing a softened poultice with extract of lead and other 'preventatives' might be applied to the wound to prevent the disorder 'from falling upon the lungs' by irritating the part. Kirkland also recommended that incisions be made if the 'external gangrene' came on, in which 'fomentations of wormwood, century, sage, water and vinegar' might be employed with 'spirit of sea salt and water', over which a 'stale beer poultice' might be applied 'till the digestion comes on'.[38] As well as these external remedies, some internal medicines might be useful, including the spirit of sea salt in a 'strong decoction of rue and chamomile flowers' with bark and a purgative of 'lenetive electuary' (a medicated paste prepared with honey or other sweet substances to alleviate pain) and Epsom salts in gruel to open the bowels.[39] Like Darwin, he believed that water or gruel with verjuice or vinegar and a rich vegetable diet of 'mashes of malt, oats', cabbage, turnips and carrot would help, but because Kirkland was convinced it was not a 'putrid contagion' he saw bleeding and purging as easily the best remedies.[40]

Darwin's recommendations for treatment of the diseased cattle demonstrate again how closely he applied the same medical interventions with beasts as with humans and how finely he drew his human–animal parallels. The disagreements between medical practitioners over treatment methods also show how important the medical dimensions of agricultural improvement had become. Prized livestock were valuable commodities and there were major fears that disease could spread across the countryside, infecting other cattle and potentially other species. Cattle health was almost as important as the health of servants or estate workers and therefore, in a period before the development of specialist veterinary medicine, the most prestigious medical practitioners in the region were employed, who, in turn, were prepared to stake their considerable reputations upon finding a successful treatment and to publicise and defend their courses of action with letters to newspapers and publications.

INSECT ATTACKS

Darwin's insectivorous investigations were driven partly by the need to counter their depredations against plant and animal bodies but also partly, as we have seen, by wonder at their prominence in the economy of nature, and he opposed cruelty against them to a degree far beyond fashionable sensibility. While insects were perceived as pests in particular contexts and might destroy crops, they were also mainstays of natural systems with impressive prodigality and animate tenacity, which might be harnessed for the benefit of agriculture, horticulture, gardening, industry or manufactures, just as honey was collected from bees, and, as we will see, Darwin even thought that some insects might be better exploited as a new foodstuff. The size of the eighteenth-century Derby silk industry, for instance, depended upon how much raw silk could be imported of sufficient quality to produce clothing and other items. Darwin and his philosophical associates, some of whom had close interests in the county textile industries, therefore tried to breed silkworms, believing that if this could be successfully accomplished it might transform local silk manufactures.[41] At the same time, the problems that insects caused in agriculture and horticulture were fully evident, although some plants had devised various means of combating insect depredations, as detailed above.[42] Around 1760, when travelling between Chesterfield and Pleasley in Derbyshire, Darwin remembered seeing 'every' hedge leaf eaten by May-chafers, who flew 'like bees in a swarm' for 'two or three miles together'. He also recollected finding a 'true locust like a very large grass-hopper with very long and broad wings' lying dead in a field near Chesterfield, which he 'preserved in spirits', when many others had been discovered across the country the same year.[43]

Darwin followed Linnaeus in his *Philosophia Botanica* (1751) in detailing various 'excrescences' caused in plants by the deposition of eggs, including oak galls and those found in ground-ivy, cistus, trembling poplar, willow and hawk-weed. Insects also caused unusual duplications or proliferations of flowers, such as on those of *Carduus caule crispo* (*Carduus acanthoides*, welted thistle), where 'wounds' inflicted by them resulted in larger florets 'with pistils growing into leaves'. Excrescences on plant leaves and vegetable growth mutations were caused by the deposition of eggs, larvae or puncture wounds from insects, which often stimulated their hosts into 'unnatural motions' and growth akin to 'inflammation' and 'new granulations of flesh in the wounds of animal bodies', which, if skin could not regrow, would 'rise into large substances' of 'fungous flesh' on the surface or subcutaneously.[44] Various insects caused problems for trees and plants, including caterpillars that attacked fruit trees, which Darwin frequently observed in his Derby orchard and gardens, although the leaves of most trees were 'renewed' following almost total destruction in the early season. He suggested that caterpillar populations might be limited by burning leaves where their eggs were laid, noting that tying fringes or wrapping tarred paper around stems discouraged them from climbing up for food (Figure 56). Selecting apple trees that flowered early made them 'less liable' to insect depredation, but late flowering kinds were 'less liable to the injuries of frost'. Cabbage white butterflies were another pest that should be caught or poisoned if possible, but could be combated using their natural predator, an ichneumon fly which laid eggs on their backs and eventually killed the insect at chrysalis stage, before it could change into a butterfly. Slugs were likewise, of course, injurious to crops and flowers, and Darwin saw 'many artichoke stems above a foot high' being consumed near wet Derwent-side ground in his garden until they collapsed 'like trees felled by the axe'; they might be killed or controlled using a heavy roller before sun rise, tar water, lime, salt or bran.[45] Other insects were more directly dangerous to humans, such as some toxic hairy caterpillars (perhaps those of the brown-tail moth) which caused an itch and left pointed bristles in the skin, while others were apparently so poisonous that they were used as a weapon, according to the French naturalist Francois Levaillant (1753–1824) in his *Travels into the Interior Parts of Africa* (1796).[46]

Aphids and other insects could destroy plants or curl leaves but seemed not to lay eggs in 'putrifying' animal or vegetable bodies, although Darwin sometimes came across 'microscopic animals' in the 'stagnating juices of animal bodies', pustules, faeces or even semen, but not in blood or other secretions. He was fascinated with aphid lifecycles and reproductive methods, which were so 'extensively injurious not only to gardens and hot-houses' but to half of all plants, and feared that with their 'innumerable progeny' they might 'in process of time destroy the vegetable world'. Darwin watched them closely in his gardens using magnifying glasses as they propagated on trees and shrubs, arguing that they reproduced prodigiously with 'wonderful increase' in a manner similar to vegetables, with their eggs resembling plant seeds. In 1798 he observed his garden trees suffering from innumerable aphids who fed on 'sap juice' even after washing with a 'strong stream from a forcible water-engine'. He tried using different substances to destroy them without injuring the plants, such as diluted acids, essential oils, smoke, tobacco, lime water, tar water and soap suds, but only really had success on nut tree leaves with tobacco from bellows and 'Scotch snuff', which the gardener John Kennedy had used to good effect on pineapple plants.[47] Just as he advocated new vegetable foods, Darwin also suggested that insects might be eaten by humans as they were by birds and some peoples around the world, which would turn menaces or pests into useful sustenance. They could therefore become a 'grateful food, if properly cooked, as the locusts or termites of the east' were, while the 'large grub' picked up by rooks after ploughing was probably 'as delicious as the grub called groogroo [Fat palm grubs] and a large caterpillar which feeds on the palm', both of which, when roasted, were West Indian staples. The Chinese silkworm aurelia, or chrysalis (after the silk was wound off), the 'white earth-grub' and the 'larva of the sphinx moth' were all supposed to be 'delicious' too.[48]

Encouraging natural predators, such a 'beautiful small spotted beetle called a lady-bird by the people', was one way of battling aphids. Darwin believed that other insect populations might be controlled by welcoming and breeding birds such as sparrows, larks and rooks, rather than by 'taking their nests'.[49] The 'most ingenious' way of destroying aphids was to propagate its 'greatest enemy, the larva of the aphidivorous fly', so counteracting this 'plague' by 'natural means'. Darwin worked with two Derby friends to test this: John Horrocks, an attorney, and Thomas Swanwick, who, as we have seen, was a mechanic, surveyor and proprietor of a commercial academy who became secretary of the Derby Literary and Philosophical Society, whom he asked to observe and draw both the larva and fly (Figure 57, drawings 3 and 4). Swanwick explained how he had the 'pleasure' of seeing the aphidivorous fly larva devouring aphids on a leaf supplied by Horrocks. The larva was 'like the sloth in his disposition' and did not 'ramble about', but lifted up his head and 'extended it in various directions';

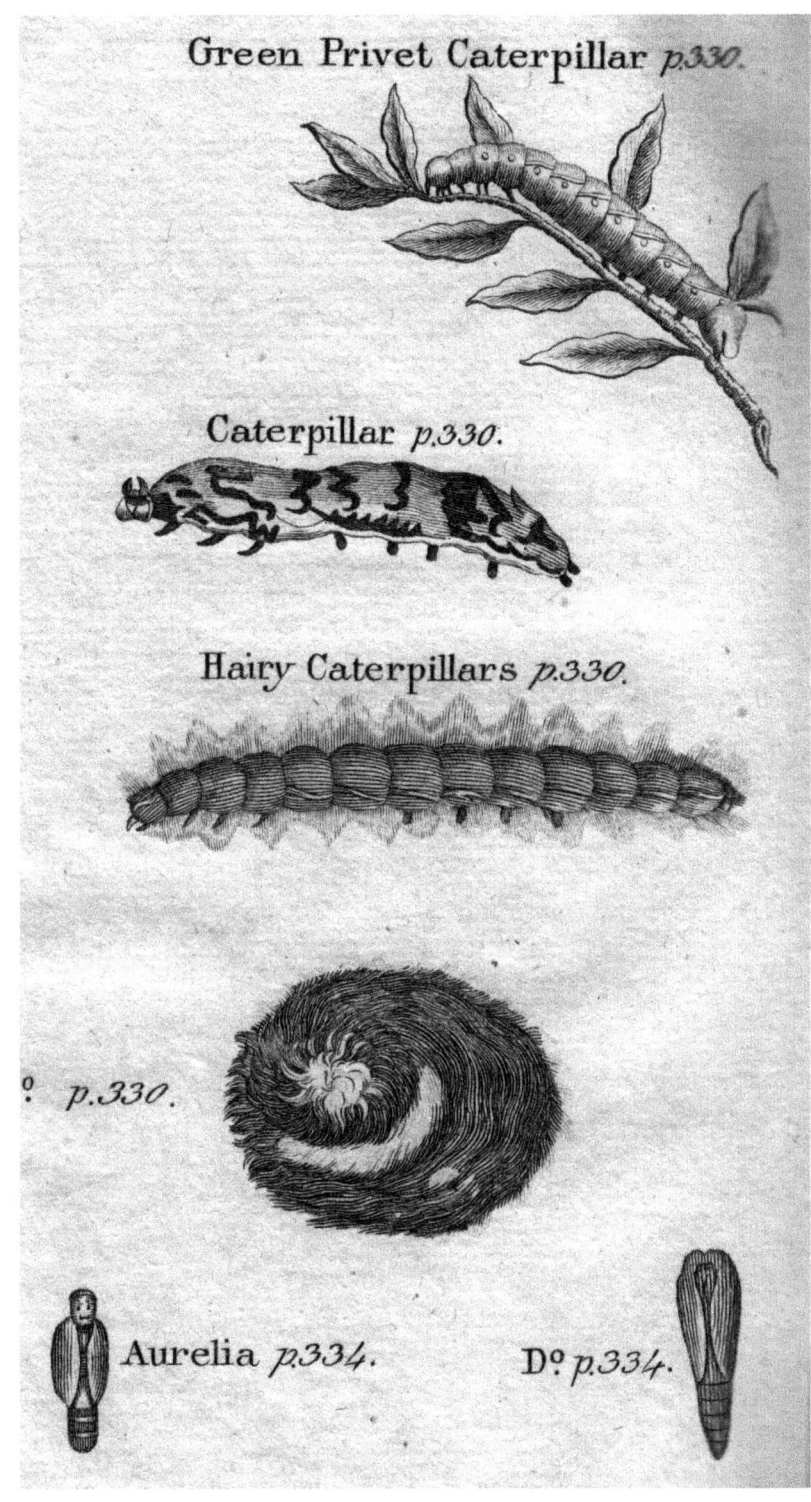

Green Privet Caterpillar *p.330.*

Caterpillar *p.330.*

Hairy Caterpillars *p.330.*

p.330.

Aurelia *p.334.* D? *p.334.*

re 56. Caterpillars,
1 *Buffon's Natural*
ory Abridged, 2 vols
2), vol. 2.

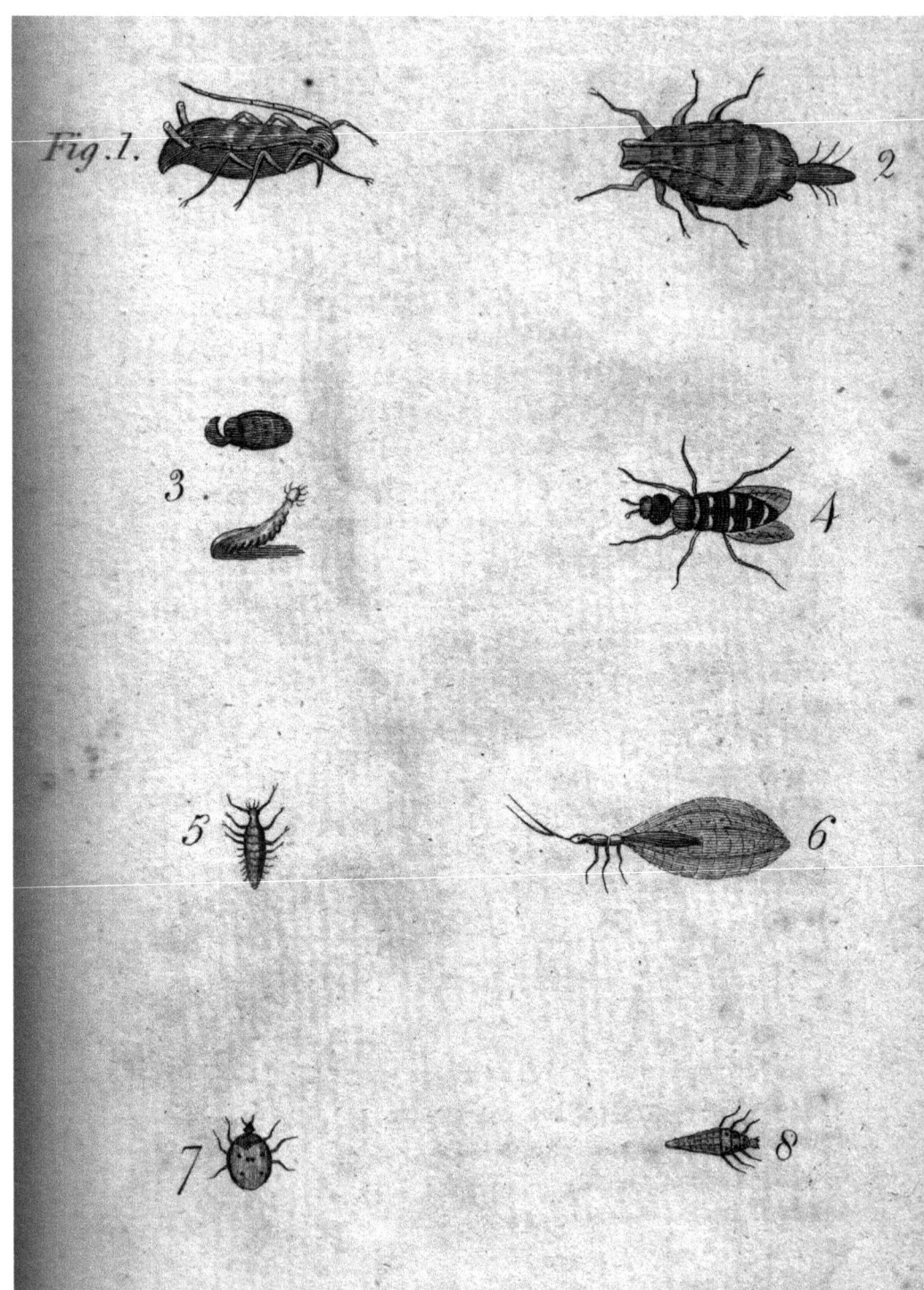

Figure 57. Aphids and their predators, from E. Darwin, *Phytologia* (1800).

when feeling an aphid it seized it 'by the back', lifted it up and positioned it in the air as if seeking 'to prevent it from liberating itself by its struggles against the surface of the leaf' or to allow it inside 'the cavity of his mouth' more easily. The larva then pierced the aphid, sucked 'the juice out of the body', dropped the skin and licked 'his lips round with his little black tongue' before repeating the operation when another aphid came into its orbit. Curious to see what the fly looked like, Swanwick continued feeding it until it turned into a chrysalis and after 'ten or eleven days' it burst out of the cell as a 'beautiful butterfly'.[50]

CONCLUSION

This examination of Darwin's methods of treating sick animals and advising upon their healthcare reveals much about his understanding of disease and medical treatment and demonstrates how and why some medical practitioners came to incorporate veterinary medicine into their practice in the context of the agricultural revolution and the medical marketplace. It has demonstrated how and why he tried to apply treatments adapted from humans to animals, arguing that the latter impacted upon human healthcare and contrawise, which subverted expectations concerning bestial bodies, behaviours, cultures and intelligence. The chapter has detailed how an eminent and highly successful physician came to take a close interest in animal medicine, which might have been considered a risky enterprise given that it stretched his knowledge, skills and experience, sometimes drawing him into disagreements with other practitioners over treatments while extending medical practice from domestic consultations into the farm, field, stable and allotment.

Although veterinary science is often believed to have been placed upon a systematic footing during the nineteenth century, Darwin's veterinary interventions show that, encouraged by the twin pressures of the competitive Georgian medical marketplace and the increase in landed wealth driven by enclosure and agricultural revolution, it became worthwhile for medical practitioners to seek to apply their expertise to farm animals in particular. Darwin shaped and applied his theories of physiology, anatomy, aetiology and nosology in the context of veterinary science, drawing upon his practices with humans and close observations of domestic animals. Despite the constant parallels and analogies that Darwin drew between humans and animals, he resisted the idea that some diseases spread between them commonly, while emphasising that all creatures were susceptible to diseases from similar sources such as miasmatic contagions. One of the animals who accompanied him most closely in his later years was a veteran horse aptly called 'Doctor', who walked freely behind his carriage with a saddle on, enabling him to ride to patients when the roads became impassable. According to Samuel Galton's daughter Mary Anne Schimmelpenninck (1778–1856) he would have been happy, as 'lashed on the place' normally 'appropriated to the boot' on Darwin's carriage was a 'large pail' used for 'watering the horses, together with some hay and oats beside it'. 'Doctor' survived to a 'great age' and was buried at the Priory. While this 'Doctor' really was an animal, it is true to say that his master definitely was a doctor of animals.[51]

NOTES

[1]   J. D. Chambers and G. E. Mingay, *The Agricultural Revolution: 1750–1880* (London, 1966), 42, 49; J. C. Loudon, *An Encyclopaedia of Agriculture*, 5th edn (London, 1844), 258–63; L. Wilkinson, *Animals and Disease: An Introduction to the History of Comparative Medicine* (Cambridge, 2005); R. A. Dunlop and D. J. Williams, *Veterinary Medicine: An Illustrated History* (Mosby, 1995); J. Swabe, *Animals, Disease and Human Society: Human–Animal Relations and the Rise of Veterinary Medicine* (London, 1998), especially 61–80, 99–101; M. Henninger-Voss, ed., *Animals in Human Histories: The Mirror of Nature and Culture* (Rochester NY, 2002); A. Guerrini, *Experimenting with Humans and Animals: From Galen to Animal Rights* (London, 2003); Wilkinson, *Animals and Disease*, especially 35–64, which examines cattle plague in eighteenth-century England; L. H. Curth, *The Care of Brute Beasts: A Social and Cultural Study of Veterinary Medicine in Early Modern England* (Leiden, 2009); K. Brown and D. Gilfoyle, eds, *Healing the Herd: Disease, Landscape Economies and the Globalisation of Veterinary Medicine* (Athens GA, 2010), especially P. A. Koolmees, 'Epizootic diseases in the Netherlands, 1713–2002: veterinary science, agricultural policy, and public response', 19–41 and D. Hunniger, 'Legislation and administration during outbreaks of cattle plague in eighteenth-century Northern Germany as continuous crisis management', 76–91; V. Anderson, *Creatures of Empire? How Domestic Animals Transformed Early America* (Oxford, 2004); D. Donald, *Picturing Animals in Britain, 1750–1850* (New Haven, 2007); D. Brantz, *Beastly Natures: Animals, Humans and the Study of History* (London, 2010); L. Cole, 'Introduction: human–animal studies and the eighteenth century', *The Eighteenth Century*, 52 (2011), 1–10; R. Kirk and M. Worboys, 'Medicine and species: one medicine, one history?' in M. Jackson ed., *The Oxford Handbook of the History of Medicine* (Oxford, 2011), 561–77; S. Nance, *The Historical Animal* (New York, 2015); A. Cassidy, R. M. Dentinger, K. Scoefert and A. Woods, 'Animals roles and traces in the history of medicine', in A. Rees ed., 'Animal agents: the non-human in the history of science', *British Journal for the History of Science*, themes 2 (2017), 11–33; A. Woods, M. Bresalier, A. Cassidy, R. M. Dentinger, *Animals and the Shaping of Modern Medicine: One Health and its Histories* (Cham, 2017); H. Kean and P. Howell eds, *The Routledge Companion to Animal–Human History* (London, 2019); A. Woods, 'From one medicine to two: the evolving relationship between human and veterinary medicine in England, 1791–1835', *Bulletin of the History of Medicine*, 91 (2017), 494–523.

[2]   G. Pearson, *An Enquiry Concerning the History of the Cow-Pox* (London, 1798), 1.

[3]   R. James, *Dr. Robert James's Powder for Fevers* (London, 1748); R. James, *A Vindication of the Fever Powder* (London, 1776); G. Pearson, 'Experiments and observations to investigate the composition of James's Powder', *Philosophical Transactions*, 81 (1791), 317–67.

[4]   Woods et al., *Animals and the Shaping of Modern Medicine*, 1–16; Kirk and Worboys, 'Medicine and species', 561–77.

[5]   C. Creighton, *The History of Epidemics in Britain*, 2 vols (London, 1894); vol. 2, 345–6, 355, 361, 370–1; J. Barker, *Epidemicks, or General Observations on the Air and Diseases* (Birmingham, 1795).

[6]   E. Darwin, *Zoonomia; or the Laws of Organic Life*, 3rd edn (London, 1801), vol. 3, 366–70.

[7]   Darwin, *Zoonomia*, vol. 3, 369.

[8]   A. Seward, *Memoirs of the Life of Dr. Darwin* (London, 1804), 60–1; E. Darwin, letter to R. Darwin, 4 September 1791, in D. King-Hele ed., *The Collected Letters of Erasmus Darwin*

(Cambridge, 2007), 389–90; Creighton, *History of Epidemics*, vol. 2, 434–597; Burroughs, Wellcome and Co, *The History of Inoculation and Vaccination for the Prevention and Treatment of Disease* (London, 1913), 27–92; D. King-Hele, *Erasmus Darwin: A Life of Unequalled Achievement* (London, 1999), 8; C. Darwin, *The Life of Erasmus Darwin*, edited by D. King-Hele (Cambridge, 2003), 109; P. Razzell, *The Conquest of Smallpox: The Impact of Inoculation on Smallpox Mortality in Eighteenth-century Britain* (Lewes, 1977).

9   E. Darwin, letter to R. Darwin, 18 March 1789, in King-Hele ed., *Collected Letters*, 334–5; Darwin, *Zoonomia*, vol.1, 380–4.

10   E. Darwin, letter to R. Darwin, 4 September 1791, in King-Hele ed., *Collected Letters*, 389–90; T. Dimsdale, *The Present Method of Inoculation for the Small Pox* (London, 1767).

11   Darwin, *Zoonomia*, vol. 1, 384–9.

12   Darwin, *Zoonomia*, vol. 1, 389–91.

13   E. Darwin, letter to R. Darwin, 10 May 1799, in King-Hele ed., *Collected Letters*, 527–9; *Zoonomia*, 388–9.

14   E. Darwin, letter to R. Darwin 24 December 1800, in King-Hele ed., *Collected Letters*, 563–4; E. Jenner and W. Woodville, *A Comparative Statement of Facts and Observations relative to the Cow Pox* (London, 1800); C. Aikin, *A Concise View of all the most Important Facts … respecting the Cow Pox* (London, 1800).

15   Darwin, *Zoonomia*, vol. 3, 389–92.

16   E. Jenner, *An Inquiry into the Cause and Effects of the Variolae Vaccinae* (London, 1798); E. Jenner, *A Complete Statement of Facts and Observations relative to the Cow-pock* (London, 1800); Creighton, *History of Epidemics*, vol. 2, 558–67; A. J. Harding Rains, *Edward Jenner and Vaccination* (London, 1974); G. Miller ed., *The Letters of Edward Jenner and other Documents Concerning the History of Vaccination* (Baltimore, 1983); H. Bazin, *The Eradication of Smallpox* (San Diego, 2000).

17   T. Beddoes, *Contributions to Physical and Medical Knowledge Principally from the West of England* (Bristol, 1799), 387, 391–2, 394.

18   E. Darwin, letter to E. Jenner, 24 February 1802, extract in J. Baron, *Life of Edward Jenner*, 2 vols (Colburn, 1838), vol. 1, 541; King-Hele ed., *Collected Letters*, 576.

19   Darwin, *Zoonomia*, vol. 1, 92.

20   Derby Philosophical Society, *Rules and Catalogue of the Derby Philosophical Society*, with supplements of 1795 and 1798 (Derby, 1793–8), 9, 18; Derby Philosophical Society, *Rules and Catalogue of the Derby Philosophical Society* (Derby, 1815), 21; J. Clarke, *Treatise on the Prevention of Diseases Incidental to Horses*, 2nd edn (Edinburgh, 1790); C. V. de Sainbel, *Lectures on the Elements of Farriery; or the Art of Horse-Shoeing, and on the Diseases of the Foot* (London, 1793); C. V. de Sainbel, *The Posthumous Works of Charles Vial de Sainbel, Late Equery to the King* (London, 1795); R. Parkinson,*The Experienced Farmer*, 2 vols (London, 1798), vol. 2, 305–6; P. A. Elliott, *The Derby Philosophers: Science and Culture in British Urban Society, 1700–1850* (Manchester, 2009), 75.

21   E. Darwin, *Phytologia; or the Philosophy of Agriculture and Gardening* (London, 1800), 305–6; Darwin, *Zoonomia*, vol. 2, 495–6.

22   Darwin, *Zoonomia*, vol. 3, 372–5; F. Sargent II, *Hippocratic Heritage: History of Ideas about Weather and Human Health* (New York, 1982), 191–210.

23   Darwin, *Zoonomia*, vol. 3, 372–5.

24   Darwin, *Phytologia*, 317.

25   E. Darwin, letter to J. Wedgwood, 8 March 1788, in King-Hele ed., *Collected Letters*, 311–12.

26   *Derby Mercury*, 4 September 1783; *The Weekly Entertainer*, 19 September 1783, 301–3.

27   *Derby Mercury*, 4 September 1783.

28   *Derby Mercury*, 18 September 1783.

29   *Derby Mercury*, 18 September 1783; J. Broad, 'Cattle plague in eighteenth-century England', *The Agricultural History Review*, 31 (1983), 104–15; C. A. Spinage, *Cattle Plague: A History* (New

York, 2003); M. DeLacy, *Contagionism Catches on: Medical Ideology in Britain, 1730–1800* (Cham, 2017), 89–124.

[30] Obituary of Thomas Grimes, *Gentlemen's Magazine*, 75 (1805), part i, 86, where he is described as the retired house steward to the late Sir Harry Harpur, residing at Swarkestone, Derbyshire.

[31] J. Barker, *An Inquiry into the Nature, Cause, and Cure of the Present Epidemick Fever* (London, 1742); *Derby Mercury*, 11 September 1783; *Shrewsbury Chronicle*, 27 September 1783; King-Hele ed. *Collected Letters*, 217–18; Creighton, *History of Epidemics*, vol. 2, 79–83, 90–8; DeLacy, *Contagionism Catches on*, 24–30.

[32] Darwin, *Zoonomia*, vol. 3, 403–4; Creighton, *History of Epidemics*, vol. 2, 558–64.

[33] *Derby Mercury*, 18 September 1783; for Denman, *Gentlemen's Magazine*, 82 (1812), part ii, 93; King-Hele ed., *Collected Letters*, 262; later J. Lawrence, *A General Treatise on Cattle* (London, 1805), 581–2, argued that large-scale slaughters were 'needlessly harsh' and suggested more targeted killing and careful medical attention instead.

[34] T. Kirkland, *An Account of the Distemper Among the Horned cattle in Derbyshire in 1783* (Ashby-de-la-Zouch, 1783).

[35] Kirkland, *Account of the Distemper*, 1–3.

[36] Kirkland, *Account of the Distemper*, 4–7, 16, 11–12.

[37] Kirkland, *Account of the Distemper*, 7–10.

[38] Kirkland, *Account of the Distemper*, 19–23.

[39] Kirkland, *Account of the Distemper*, 23–4.

[40] Kirkland, *Account of the Distemper*, 24–5, 28.

[41] Darwin, *Phytologia*, 330–1.

[42] Darwin, *Phytologia*, 315, 317–18.

[43] Darwin, *Phytologia*, 329.

[44] Darwin, *Phytologia*, 308.

[45] Darwin, *Phytologia*, 324–6.

[46] Darwin, *Phytologia*, 330.

[47] Darwin, *Phytologia*, 309–16, 318–22; J. Kennedy, *Treatise upon Planting, Gardening and the Management of the Hot House*, 2nd edn, 2 vols (London, 1777), vol. 2, 151–81.

[48] Darwin, *Phytologia*, 330.

[49] Darwin, *Phytologia*, 329–30.

[50] Darwin, *Phytologia*, 312–13.

[51] C. C. Hankin ed., *Life of Mary Anne Shimmelpenninck*, 2nd edn (London, 1858), 127; Darwin, *Life of Erasmus Darwin*, 61.

CHAPTER 8

'EATING OF THE TREE OF KNOWLEDGE': FORESTRY, ARBORICULTURE AND MEDICINE

Despite the profound importance of Erasmus Darwin's many different encounters with trees and the numerous scholarly studies of his botany and his literary portrayals of the natural world, evolutionary ideas and the midlands social, agricultural and industrial milieu, there has never been a study of his work on trees and the impact of arboriculture upon his ideas.[1] In many ways this is equally true for his grandson Charles Darwin (1809–1882), whose analyses of trees have also been neglected, even in studies of the place of botany in his evolutionary theories, despite the long section on trees and fruits in *The Variation of Animals and Plants under Domestication* (1875).[2] Tom Williamson has emphasised how much work has been undertaken on aristocratic perspectives on arboriculture in the 'long' eighteenth century compared with those of the 'wider population', and a consideration of Darwin's approach to silviculture and that of his close circle of friends and tree admirers will help to address this gap in our understanding.[3] Like his Nottinghamshire contemporary Major Hayman Rooke (1723–1806), many of Erasmus Darwin's tree encounters were inspired by the economic utility of agricultural and horticultural improvement as well as a love of arboriculture; however, Darwin was able to unite, as his friend Anna Seward claimed, practical botany with picturesque landscapes by creating and experiencing both planted and poetical botanic gardens.[4] Darwin's romantic appreciation of landscape beauty was founded upon the interrelationship between a picturesque understanding of trees and plants and his botanical, psychological and physiological theories, which he utilised to weave value and meaning into local environments. Hence it was local tree places (and amorous joy) that inspired him to express himself poetically for the first time since his student days as he personified the spirits of a grove in his garden under threat from the axe. Just as Seward was motivated by threats to Lichfield trees to intervene to try and save them, so similar circumstances provided an opportunity and impetus for her friend Darwin to assert their centrality in defining layers of landscape beauty, value and meaning as animate beings of fixity and longevity. However, Darwin's arbophilia also stemmed from a general interest in tree botany and physiology within his literary circle, represented by the work of close friends such as Brooke Boothby, Thomas Gisborne and Francis Noel Clarke Mundy. The latter's poem on 'Needwood Forest' in Staffordshire, for instance, inspired Darwin to write his own poem on the Swilcar oak, which he inserted into the chapter on the production of leaves and wood in *Phytologia*, and likewise prompted Seward to respond with two poetical tributes, which praised it as one of the most effective and beautiful 'local' landscape poems and lamented the destruction of the forest.

The trees in Darwin's orchards and gardens, from Lichfield to Derbyshire, and his knowledge and experience of local woodland strongly informed his analyses of vegetable physiology and anatomy, which, along with the personified trees of his poetry, emphasised the significance of arboricultural agency as well as human–tree relationships.[5] The size and longevity of trees provided the scale necessary for Darwin to observe vegetable physiology in detail, encouraging parallels with animal physiology. Observations of local estate and forest trees informed his suggestions for the improvement of timber production, which he believed was vital to strengthen the defence of the country against foreign invasion. His visits to local estate plantations, such as those at Shugborough Hall, and important woodland, such as Needwood Forest in Staffordshire and Sherwood Forest in Nottinghamshire, demonstrated where trees flourished best and underscored their cultural status as 'monarchs of the forest' in the natural economy. Observations of estate woodlands on the edge of the Staffordshire moors suggested that large coniferous plantations might be nurtured on exposed mountainous and boggy moors such as the Pennines for the benefit of economy, industry and naval usage. In *Phytologia*, his general study of gardening, horticulture and agriculture, Darwin provided detailed recommendations concerning the most efficacious means of growing, nurturing, straightening, curing, transplanting and felling trees.

Darwin encountered trees in many different contexts, from ancient midlands forests to the extensive plantations of improving landowning friends and patients, and believed that the 'whole habitable world' had once been covered with woods, until humans had destroyed so many of them 'by fire and steel'. His various uses of the 'tree of life' metaphor suggested that the tree was also, for Darwin, the pre-eminent symbol of the progress and determined audacity of human knowledge as well as its challenges and ambiguities. The next two chapters argue that, for Darwin, trees, more than other plants, bridged the divide between humans and nature, between medicine and botany and between poetry and natural philosophy. They inspired him to compose poetry when faced with their destruction and he celebrated them as the closest vegetable cousins of humanity, monarchs of the forest and miniature worlds in the economy of nature to be nurtured and protected, rather than destroyed. We will examine some of Darwin's arboricultural encounters and poetic personifications of trees and consider why trees were so crucial to his medicine before detailing the importance of analogies between trees and animals in his philosophical studies.

MIDLAND TREES

The most important stimulation for Darwin's poetical and scientific arboricultural studies was encounters with trees close by in the midlands and northern England, and his own experience planting and growing trees in his various gardens and orchards. Living in Nottinghamshire, Staffordshire and Derbyshire, Darwin knew some surviving forest areas well, notably Sherwood Forest, Nottinghamshire, Charnwood Forest, Leicestershire and Needwood Forest, Staffordshire, in addition to the many woodlands of the post-Restoration 'great replanting', connected with estate improvement, patriotism, recreation and fashions in landscape gardening, which he passed through on tours and when visiting patients. Sherwood in Darwin's native Nottinghamshire was, of course, probably the most famous partially surviving ancient British forest and, though much depleted by the eighteenth century, it still included scores of veteran oaks among the wood pasture, especially in the Birklands area of former royal forest, many of which had become familiar as local landmarks and were recorded on contemporary maps and drawn, described and measured, as Emily Sloan has shown, by naturalists such as Major Hayman Rooke of Mansfield Woodhouse, in the county (Figure 58).[6]

Figure 58. Sherwood Forest, detail from map of Nottinghamshire engraved by J. Roper from drawing by G. Cole, F. C. Laird, *Topographical and Historical Description of the County of Nottingham* (1820).

Other striking and ancient trees in the region were also very familiar to Darwin. Derbyshire included both extensive areas of upland and lowland agricultural land, a varied topography that helped to determine a very mixed distribution of trees and woodland. Although there was not much natural woodland remaining across much of the county by the eighteenth century, particularly on the uplands, woodland existed in river valleys, notably those of the Derwent and the Dove, and trees grew in hedgerows and copses in the rich agricultural land to the south. There were historically two royal forest regions in the county with deer, the Peak Forest in the north-west and Duffield Frith, just to the north of Derby, although the former had never been much wooded in historical times, and disafforestation and enclosure removed most remaining trees by the seventeenth century. Duffield Frith, in contrast, was probably originally covered by trees, but much woodland was felled during the sixteenth and seventeenth centuries and the land was formally disafforested and declared common between 1643 and 1786, although some areas of woodland in close proximity to parks such as those of Kedleston and Alderwasley appear to have survived into the eighteenth century. Towns such as Derby and Nottingham were surrounded by orchards and small woods, while extensive tree planting was undertaken on the great estates of Derbyshire, Staffordshire and Nottinghamshire for a mixture of cultural, aesthetic and economic reasons. Thomas Brown and John Farey, in their reports on Derbyshire agriculture (1794, 1811–17), believed that the amount of timber and wood was decreasing, with coal- and lead-mining and other industries, such as iron-working, requiring supplies for fuel, props, buildings and other purposes.[7]

Observations of trees in estate plantations, woodlands and forest informed Darwin's ideas concerning the improvement of timber production. He suggested that large coniferous plantations might be nurtured on exposed mountainous and boggy moors such as the Pennines for the benefit of economy, industry and the navy, and in *Phytologia* provided detailed recommendations on the best means of growing, nurturing, straightening, curing, transplanting and felling trees. As we have seen, Anna Seward described how Darwin adorned 'the borders of the fountain, the brook, and the lakes' in his Lichfield botanic garden with 'trees of various growth', some of which are depicted on the 1884 Ordance Survey map, surveyed in 1882. No list of the trees in this garden seems to survive, although we can see from the map that a few coniferous kinds were planted with a far larger number of deciduous types, but the preserved notebook or catalogue relating to Darwin's Derby orchard and garden in the Cambridge University Library Darwin Collection lists over 200 trees, identified with the assistance of Dr William Jackson (1734/5–1798) of the Lichfield botanical society, describing locations and the qualities of fruit produced by each.[8]

Darwin advocated tree planting not just for economic reasons but also because of the benefits that woodland brought in landscape gardening and estate improvement. As large plantations and specimen outliers, trees were also pivotal to changes in landscape gardening aesthetics in the period from Lancelot 'Capability' Brown to Humphry Repton, in terms of both new planting and the careful preservation and reconfiguration of existing features such as older pollards, banks, streams or hedgerows, and became badges of honour for the gentry and aristocracy, underlining the extent of their estates and symbolising their patriotism, rootedness, endurance and investment in the landscape through the generations (Figure 59). Led by a few extremely wealthy aristocratic families, landowners and their agents took advantage of the changing social and political opportunities to improve their estates by tree planting. Trees and woods beautified parks and could draw the eye towards features beyond or within if necessary, masking and demarcating boundaries where necessary (often being recorded on estate maps, for instance) and enhancing the apparent extent of holdings while serving as cover for game (Figure 60). They also helped to obscure or demarcate sections and divisions within gardens and parks, such as those between park and pleasure grounds, and emphasise the centrality of the house, which was often redesigned or reconstructed in fashionable Palladian and, later, gothic styles; and, like fencing or hedging, they could help to facilitate social inclusion or exclusion by, for instance, screening villages or farm buildings.

Afforestation also provided a means for managers and farmers to maximise estate income by making relatively marginal areas such as stony or boggy land productive by growing trees, and the Scottish physician Alexander Hunter (1729–1809) of York argued that if all landowners and their managers sought out every 'useless bog'and

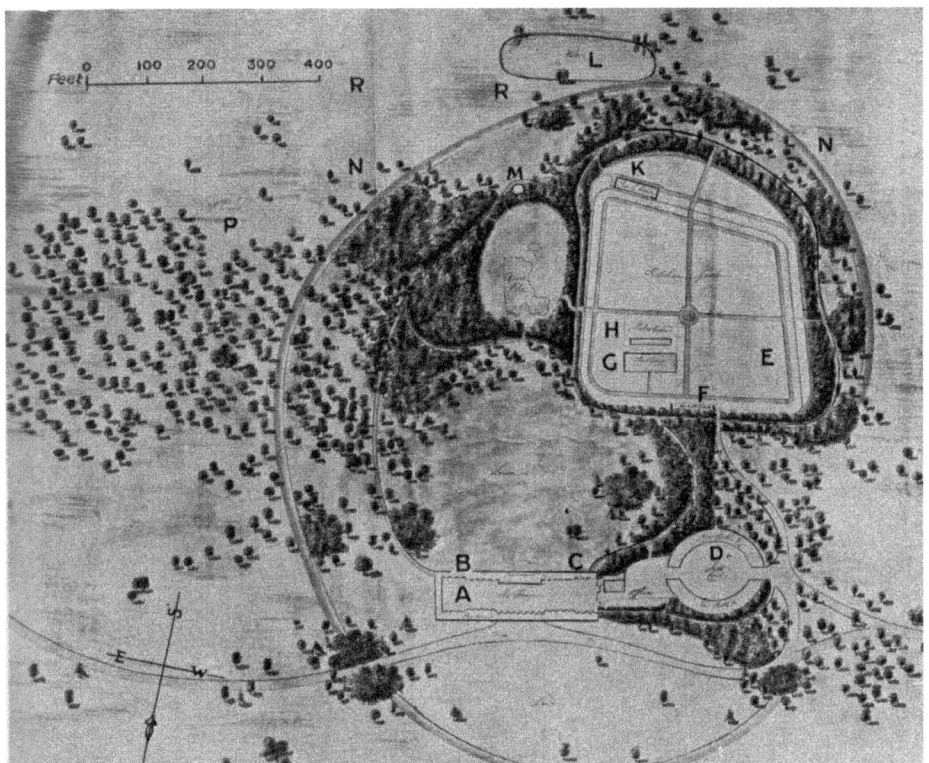

Figure 59. Detail of plan for the improvement of Hevingham, showing the configuration of plantations (1781), reproduced from A. S. Turberville ed., *Johnson's England* (1933), vol. 2.

An irregular base forming bays & promontories.

The summit regularly varied.

Figure 60. Effects of woodland summits designed to heighten landscape beauty, from W. Gilpin, *Remarks o Forest Scenery and other Woodland Views*, 3rd edn, 2 vols (1808), vol. 1.

planted it with poplars or aquatics, planted out all 'waste grounds' and allocated land for cottagers there would be 'few persons complaining that their ancient common-rights are invaded' or that they had needed to emigrate. At the same time, despite pressures from enclosure, depopulation and landed consolidation, and ensuing conflicts, traditional forms of wood management and community usage continued during the eighteenth century within and without estates alongside efforts to enclose, empark and improve, although pollarding declined and the length of coppice rotations increased, as did the value and extent of timber plantations owing to the replacement of wood by coal as fuel and other factors.[9]

Tree planting was strongly encouraged by philosophers such as John Evelyn, whose *Sylva* (1664) provided the main model for an arboricultural treatise; it went through a number of editions in the subsequent decades and was republished in 1776 in a new edition with extensive notes by Hunter.[10] The list of subscribers of Hunter's first edition provides a good summary of those in Georgian society most interested in arboriculture, which included some of the wealthiest landed gentry and aristocracy, natural philosophers such as Joseph Priestley and professionals such as clergy and medical men.[11] Likewise, encouraged by Darwin, the interest of members of the Derby Philosophical Society in arboriculture, horticulture and tree planting is evident from their acquisition of works such as Benjamin Rush's *An Account of the Sugar Maple Tree* (1792), Aylmer Bourke Lambert's *A Description of the Genus Cinchona* (1797) and William Marshall's *On Planting and Rural Ornament* (1796), as well as catalogues of tree and plant collections, such as William Aiton's *Hortus Kewensis* (1789), and books that examined arboriculture and forestry from an aesthetic and landscape-gardening perspective, such as Uvedale Price's *Essay on the Picturesque* (1794). The Society also obtained volumes on natural history that made reference to trees and arboriculture around the world, such as an edition of Antonio de Ulloa's *A Voyage to South America* (1758) and Carl Peter Thunberg's *Travels in Europe, Africa and Asia* (1793), as well as transactions of learned associations that did likewise, such as the *Memoirs of the Bath Agricultural Society*, *Transactions of the Linnean Society* and *Memoirs of the Parisian Agricultural Society*.[12]

As well as the work of Lancelot 'Capability' Brown at Chatsworth and Humphry Repton at Welbeck Abbey, Nottinghamshire, and various other places, Darwin was particularly familiar with the designs of Derbyshire landscape gardener William Emes, of Bowbridgefield near Derby. He worked extensively across the midlands and Wales, including at Markeaton Hall, Kedleston Hall, Locko Park, Foremark Hall, St Helen's House, Etruria and Radburne Hall (c. 1790), seat of the Pole family and former residence of Darwin and his wife Elizabeth, and by the 1790s had 'obtained great reputation for his taste in ornamental gardening'. Emes's style was similar to Brown's and he 'frequently assured' the Staffordshire historian the Rev. Stebbing Shaw (1762–1802) that his methods were inspired by the relationship between natural and cultivated woodlands and landscapes in the Needwood Forest vicinity, where he lived for some of the time and undertook various commissions. Emes generally removed or adapted formal features, planted trees extensively in the form of clumps and larger plantations to frame views within and beyond park boundaries, used earthworks, effaced field boundaries and incorporated ha-has and curved drives to create a more naturalistic appearance.[13] These new arboricultural attitudes are also evident in changing representations of trees in landscape art, which impacted upon Darwin's understanding of the subject, as we will see in the final chapter (Figure 61).[14]

Within the midland and northern English counties with which Darwin was most familiar there were various country estates with extensive or historically significant plantations formed and maintained for either aesthetic or economic purposes, demonstrating the contexts in which trees flourished best and reinforcing a perception of their status as 'monarchs of the forest' in the economy of nature. These included Kedleston Hall, Chatsworth House and Calke Abbey in Derbyshire; Staunton Harold, Leicestershire; Shugborough Hall, Staffordshire; and Welbeck Park in Sherwood, Nottinghamshire. Major improvements undertaken by the Curzon family at Kedleston during the second half of the eighteenth century included much tree planting (and also reconfiguration and felling), especially under Nathaniel Curzon, first baron Scarsdale, who had the house reconstructed from 1759 and new farm buildings, stables and gardens formed to designs by the architects James Paine, Matthew Brettingham and

Figure 61.   North view of High Tor at Matlock (Bath) by J. Farrington, engraved by F. R. Hay, from D. and S. Lysons, *Magna Britannia: Derbyshire* (1817), extra-illustrated edition.

Robert Adam, with assistance from William Emes. Arthur Young praised the landscape park at Kedleston for its beautiful 'winding vale', lawns and pleasure grounds 'bounded by woods of noble oaks', as well as the carefully contrived views of the house gained on the approach from Derby through the 'spreading plantations', which, when matured, would be 'a noble effect' (Figure 62). The line of approach towards the house via the 'fine bridge of three arches' was 'exceedingly well varied without betraying any marked design of pursuing fashion at the expense of everything else'.[15] Young thought that Lady Caroline Curzon had traced 'with great taste' the pleasure grounds through a 'winding lawn decorated with trees, shrubs and knots of wood' and along a gravel path through the woods, rising up the hill; a vantage point commanded 'very picturesque views of the lake and adjoining wood' and 'noble prospects' of the surrounding countryside 'broken into inclosures'. The path would, it was intended, carry on around the south of the park along the brow of the hill with commanding views of the vale and town of Derby to the south-east (as it indeed did).[16]

ure 62. Kedleston Hall and park, from J. Britton and E. W. Brayley, *The Beauties of England and Wales*, vol. 3, *mberland*, Isle of Man and Derbyshire (1802).

At Welbeck Park in the 'Dukeries', Nottinghamshire, the seat of William Cavendish Bentinck, third duke of Portland and prime minister in 1783 and 1807–9, large parts of Sherwood Forest were added to the estate by enclosure; extensive improvement, planting and remanagement of the woodlands was undertaken by Francis Richardson from the 1740s, work continued by the head gardener William Speechly (1735–1819) and Humphry Repton with assistance from Hayman Rooke, who recorded and studied many of the oldest oaks. Thousands of acres were planted with specimen and forest trees, including pine, firs, beech, birch and Cedar of Lebanon, to form or enhance vistas, and dense timber plantations were created to increase estate productivity, while old oaks were felled if in terminal decline, retained in situ or moved to prominent positions.[17] Sir Harry Harpur at Calke, whom, as we saw, Darwin often visited for medical and social reasons, also undertook major estate planting, sweeping aside earlier formal gardens and avenues and adding to the many fine mature oaks and other trees in his park, and redesigned his gardens during the 1770s to accentuate the picturesque possibilities of his sloping hills, valleys and lake. Taking advice from Emes, who prepared a plan for the gardens as part of a programme of enclosures and agricultural improvements across the family estates, which helped to significantly grow rental incomes, Harpur spent £3,600 on the nursery, levelling ground adjacent to the house and planting trees, while physic, kitchen and flower gardens were created in the walled section and a conservatory was constructed. Likewise, trees were a major part of estate improvement at Chatsworth and a key attraction for the growing number of visitors. As John Barnatt and Tom Williamson have demonstrated, the formal gardens were replaced by ostensibly more natural gardens from the 1720s under the influence of William Kent and more directly Brown from the 1760s. These improvements, which complemented new features such as James Paine's stables, included the enlargement of the park, the rearrangements of roads, the beginning of the reconfiguration of Edensor village and major afforestation across the park, with plantations and clumps employed to enhance views from the house and emphasise distant prospects such as rocky outcrops.[18] After he became baronet, Brooke Boothby planted many trees on his Ashbourne estates, including exotics in the gardens (Figure 63).

Darwin's industrial and manufacturing friends and acquaintances also acquired and improved country estates as they became wealthy and aspired to join the landed classes, with arboriculture and forestry being seen as crucial and indeed as patriotic. One of Matthew Boulton's first actions after he acquired Soho House in Staffordshire was to improve the gardens, taking much inspiration from a close study of Charles Hamilton's garden at Painshill, Surrey; Boulton planted a thick belt of conifers as shelter from the wind and subsequently many exotic varieties including 'Neapolitan evergreen cytissus', Swedish and English junipers, variegated hollies, Scots pine and three 'shining leav'd laurestinum', purchased from the nurseryman

Figure 63. Ashbourne Hall by S. Raynor and P. Gresley, from S. Dawson, *The History and Topography of Ashbourne* (1839).

Roger Eykyns.[19] Likewise, the Derwent valley area was transformed by the picturesque planting and forestry schemes of the Arkwright, Strutt and Evans families. As part of the development of their park at Willersley Castle, near Cromford, overseen by the architect and landscape gardener John Webb, Richard Arkwright and his son Richard junior planted some 350,000 trees by 1802.[20] Equally, the Strutts eagerly engaged in afforestation on their Derbyshire and Nottinghamshire land holdings, including Strutt's Park, considerable estates around Belper and the 1,300-acre estate and woodlands at Kingston-Upon-Soar, Nottinghamshire, which William Strutt jointly purchased in 1796 with his father-in-law Thomas Evans from Francis Godolphin Osborne, fifth duke of Leeds (1751–1799). These planting schemes were informed by William Strutt's reading, which included works such as Humphry Repton's *Observations on the Theory and Practice of Landscape Gardening* (1803), William Forsyth's *Treatise on the Culture and Management of Fruit Trees* (1802) and Huddersfield nurseryman and forester William Pontey's *Forest Pruner* (1805) and *Profitable Planter* (1808), which, as president in succession to Darwin, Strutt had purchased for the Derby Philosophical Society library. According to John Farey, the Strutts and their gardeners and foresters 'very laudably' applied themselves to planting some 100,000 larch, Scots pine and other trees on their Belper estates and oversaw the 'pruning and management' of extensive plantations, keeping 'accurate and systematic accounts of the expense and time of planting', pruning and thinning to facilitate more effective growth and increase the value of cut timber, while carefully measuring and recording the results of their endeavours.[21]

One means to quickly change the character of pleasure gardens or parks and to provide instant maturity was to plant trees nurtured in pots or to transplant old or striking examples, a process Darwin described in *Phytologia*. Some landscape gardeners, landowners and arboriculturists, such as Sir Henry Steuart (1759–1836) in Scotland, achieved this by moving trees in which the soil had been removed from the roots, which meant that they were much lighter and easier to manoeuvre (Figure 64). Darwin emphasised, however, that many roots might be damaged or 'torn off' by transplanting in this way, while roots from trees and shrubs that had been 'compressed in a garden pot' tended to produce fewer leaves. For trees to thrive in their new locations he believed that they ought not to be planted too deeply in soil because 'the most nutritive or salubrious parts of the earth' were those affected by the 'reach of the sun's warmth', by 'descending moisture' and by the atmosphere's oxygen. As the tree's root fibres usually grew 'towards the moistest part of the soil' and 'young shoots and leaves' grew towards the 'purest air and brightest light', so roots would not tend to rise higher than the original level they had been planted at and not spread out in their normal way across the soil horizontally. This meant that it might not be sufficiently anchored in the ground to survive heavy winds or storms or grow with sufficient 'healthy vigour', and might produce insufficient flower or leaf buds.[22] Darwin recommended digging a circular trench two or three feet deep around tall trees intended for transplantation in early spring, which would encourage numerous 'new roots' to 'shoot from those' that had had their ends severed, binding the ball of earth more strongly together when the tree was moved to its new location and giving it a head start in the new location.

Figure 64. Tree transplanting using machine designed by Sir Henry Steuart, J. C. Loudon, *Encyclopaedia of Agriculture*, 5th edn (1844), 643.

The likelihood of success was also increased by attaching the tree to 'wooden props' to prevent overturning in high winds before root systems were fully re-established. He recognised, however, that the bark might receive damage from 'contusion by the pressure of props against it' or 'strangulation' from the 'bandage' tethering the tree to them. Darwin therefore recommended fastening smaller transplanted trees directly to props by hammering a nail through them into the trunk, which he tried with fruit trees in his own Derby orchard, finding that this provided better support with 'much less apparent injury' than occurred using the usual technique of tying the young trees between three props by 'cordage'.[23]

IMPROVING NATIONAL TIMBER SUPPLIES

Georgian tree planting was stimulated by various factors, including a strong but perhaps incorrect perception that woodland had declined to its lowest ever extent, most strongly articulated during major wars with the French, when fears concerning timber supplies for the construction of naval and merchant vessels were heightened. In his influential *Sylva*, for instance, putting much blame upon Oliver Cromwell and the Parliamentarians, Evelyn called for 'nothing less than a universal plantation of all sorts of trees' to restore British woodlands, 'the waste and destruction' of which was 'so universal'.[24]

As well as a form of fuel often harvested through coppicing, wood remained, of course, the staple material for a wide range of processes and products in Georgian society, from tanning, dyeing and colouring to tools, walking sticks, coffins, fishing rods and furniture. Although in many ways the 'long' eighteenth-century urban transformation was achieved through the substitution of traditional regional vernacular building methods, such as cobbles, pebbles, turf, flint, thatch and timber-framed wattle and daub with brick, slate, tile and ashlar stone, wood remained crucial as a domestic and industrial building material and older buildings were often refronted with brick while original structures remained disguised to the rear.[25]

Darwin shared the concerns over timber shortages and at various times emphasised the importance pf timber for the national economy, urging that afforestation was needed as part of agricultural improvement. In 1756 Darwin's fellow midlander Samuel Johnson worried about the impact upon British naval and commercial hegemony after passing through tracts with the 'preserved names of forests' but now 'barren and useless', which had been cleared of trees because 'our industry has for many ages been employed in destroying the woods which our ancestors have planted'. Similarly, in 1755, Edward Wade argued that major planting for timber needed to be undertaken on forests, commons, heaths and waste grounds with nurseries in each parish to supply the navy, provide employment for the poor and beautify the nation. In 1756 a House of Commons enquiry into the problem of the decline of timber production and its impact upon international trade and naval power led to an act to further planting on common lands. The traditional practices of pollarding and coppicing continued in some parts of the midlands, only declining slowly through enclosure and other factors. The provision of new plantations was encouraged as part of estate improvement and as an economic resource, with the Society of Arts offering prizes for tree planting as a patriotic as well as an economic duty.[26]

It was in this context that Darwin and his friends the business partners Thomas Bentley (1730–1780) and Josiah Wedgwood argued that timber production would benefit from a new network of waterways in their pamphlet promoting the Grand Trunk Canal (1765). Referencing the arguments in Roger Fisher's 'very interesting' recent book *Heart of Oak: the British Bulwark* (1763), they argued that the timber industry was one of 'great importance' and would greatly benefit from the canal, especially through the supply of additional oak, as there were 'many large woods near the course of the intended canal' that, because conveyance to the coast for ship building was difficult, had to be sold at low prices locally. Fisher was a Liverpool shipwright who followed Wade and others in arguing that there had been a serious decline in the quality and quantity of timber for naval and merchant vessels over the previous four decades. If not rectified by major afforestation on the part of the 'landed interest' he argued that this would damage trade and debilitate the nation so much it would be 'crushed by its enemies'; pointedly, one of his claims was that the naval timber supplies of the French, Britain's greatest international rival, would outlast those of the British partly because of the superiority of their river and canal networks. Darwin, Bentley and Wedgwood contended that, by facilitating the cheaper and easier conveyance of large quantities of wood, new canals would 'greatly encourage the growth of it and help to repair' the 'alarming' national 'decrease of ship timber' which had damaged a nation 'whose riches and power depend so greatly upon navigation'.[27]

The number of trees being grown and the quality and size of timber might be maximised by various means, which Darwin summarised in *Phytologia*, building upon recommendations made by British and French arboriculturists such as Evelyn and Henri-Louis Duhamel de Monceau.[28] To hasten the growth of wood, nourish the bark – which Darwin believed was the only true living part of the tree and gradually turned into 'alburnum or sap-wood' before becoming 'heart wood' – and bring forward the development of leaf-buds, trees needed more water than they received in their 'natural state'. To provide this, regular sprays to the roots from a 'water engine' could be administered, which probably worked because it 'lessened the cohesion' of leaf-bud cuticles and was akin to what Darwin thought was the healthy impact upon skin of soaking hands in water. Removing flowers might help enlarge the proportion of tree bark and wood by 'increasing the number and vigour of leaf-buds' which were essential for the 'nourishment' of the tree and reducing the energy expended upon flowers, fruits and seeds. Close planting resulted in straighter timber trees, because of the 'powerful contest with each other for light and air', which propelled them upwards and discouraged wasteful lateral branches. This was also hastened by judicious thinning and intermixing more economically valuable or slower-growing trees such as oak with more rapidly maturing and less valuable ones such as pine, which could be removed or lopped if they impeded the former too much.[29] While individual or freestanding trees might have special value in landscape gardens or pleasure parks, these were not necessary or desirable for timber production because they did not face competition and were thus usually smaller and more branched, and suffered from increased exposure to wind, as the stunted trees growing on mountain sides or in coastal areas demonstrated.[30] If plantation timber trees needed to be grown in a curved shape for construction or shipbuilding purposes this could be accomplished by 'annually scratching the external bark or cuticle' either across or downwards on the south side of the trunk, which was 'known to grow faster' than the north side, as cross sections from felled trees showing the concentric growth rings demonstrated. Where timber trees that were supposed to be straight curved because of movement towards light and air, this could likewise be corrected by similar means or by redirecting their growth towards each other using 'cordage'.[31]

Tree felling, according to Samuel Pepys (1633–1703), as Darwin emphasised, was not best done in winter, as is often still supposed, but should be done, in the case of oak for ship-building, after the removal of bark in early spring and the subsequent death of early foliage, because the new buds had drawn the 'saccharine matter' from the sapwood, which therefore became 'as hard and durable as the heart wood', rendering it less liable to decay and insect attack.[32] Darwin emphasised that 'barren-commons' such as Cannock Heath in Staffordshire might be fruitfully planted with trees such as sugar maples, while pines such as 'Scotch-fir' would 'succeed astonishingly' in such places, or on 'barren mountains', as the estate plantations of Thomas Anson, first Viscount Anson (1767–1818), of Shugborough Hall in Staffordshire demonstrated. The large-scale planting of Scots pine on the Longleat estate of Thomas Thynne, the first marquis of Bath (1734–1796), in Wiltshire, also showed what could be achieved. The latter was managed by the steward and agriculturist Thomas Davis (c. 1749–1807), author of the *General View of the Agriculture of the County of Wiltshire* (1794), who was associated with the Bath Agricultural Society.[33] The extensive 'summit' of England, which principally consisted of 'a ridge of mountains extending from south to north' along the Pennines from the Staffordshire Moorlands and Derbyshire Peak, was 'so bleak or so barren' that it was 'totally unfit for the plough or for pasturage' and might as well be covered with 'extensive' plantations. As well as providing crucial timber for ships, the afforesting of 'unfertile mountains' from Cornwall to the tip of Scotland would transform them into major woodland resources for the national good.[34]

Here Darwin may have been inspired by John Kennedy's *Treatise upon Planting* (1777), which argued that, following the example of 'nature' and the sight of trees growing in some very rocky areas, 'rocky, hilly, waste and heath lands' and 'apparent barren mountains' could be planted with 'thriving' forest trees for the benefit of the landowner, the rural economy, the nation and 'posterity'. Gardener to Sir Thomas Gascoigne of Parlington Hall, Yorkshire, from 1771, Kennedy claimed that, while the afforestation of 'naked and disagreeable' 'poor wastes, moorlands' and mountains had been 'but seldom treated of' or 'attempted', he had enjoyed considerable 'success' in planting such places, even in northern Scotland, so 'extensive tracts' now apparently useless might become both 'ornamental and profitable'. Kennedy recommended that Scotch fir and larch were suitable for higher ground and noted that many of the firs and pines imported from the American colonies thrived upon 'poor' land too, while beech, elm and even oak would grow perfectly well lower down hills with 'clumps of sycamore' and mountain ash 'for ornament' and 'beauty'.[35]

Afforestation of relatively marginal places such as stony hillsides and mountains, however, while furthering estate productivity, also increased the likelihood of damage from the elements as these were often on exposed areas. Landowners and tree enthusiasts such as Hayman Rooke and Darwin were thus concerned to create a scientific forestry that might protect valuable and expensive plantations from weather and disease. Rooke carefully recorded the impact of severe weather such as frost, wind and lightning strikes on plantations in particular conditions in his diaries, and considered how best to mitigate these, while, having experienced damage to his fruit trees, as we have seen, Darwin believed that lightning strikes were 'more frequent than is generally supposed', as the 'great injury' to many timber trees split by large 'longitudinal cracks' only discovered during felling demonstrated. His solution was to position 'pointed wires, as thick as a goose quill' as lightning conductors on some of the tallest examples. Other bark damage was caused by woodpeckers and, although these were a 'beautiful and ingenious bird', could only be prevented by killing them.[36]

The continuing reliance upon wood as a construction material meant that means of preserving within structures were at a premium, while the number of buildings destroyed by fire, the amount of damage to timber caused by water and insects, and costs of insurance were also serious problems. This accounts for the interest among Darwin's textile-manufacturing friends in fire-resistant structural materials and the techniques pioneered by William Strutt, such as the use of plaster and metal sheathing to encase timber beams, hollow earthenware pots within floor and ceiling structures and iron beams, because expensive mill buildings with warm dry atmospheres were at particular risk. One problem was chemical reactions between construction materials: lead sheets were sometimes corroded by the action of gallic acid and sap from oak timbers when moisture was present, which might also cause 'fermentation' or 'dry-rot' in the sapwood. This might be prevented by careful selection of building materials and making holes in walls on the sides of buildings under the roof to keep them dry and ventilated.[37] One way of reducing timber rot was to soak dry wood in lime water until as much had been absorbed as possible, before immersing it in a 'weak solution of vitriolic acid' once dry, which would 'unite with the lime already deposited' in the timber 'pores' and, he believed, would convert it intogypsum. This would act almost like its own coating of plaster, potentially preserving the wood for 'many centuries' if dry and also render it more fire-resistant. A further benefit was that wood thus impregnated would be 'less liable to swag' or warp, with benefits for industries such as mining, textile mills or domestic construction.[38]

Some timbers, such as 'ironwood' (*Sideroxylum*), cypress (*Cupressus*), cedar (*Cedrus*), mahogany (*Swietenia*) and Bermuda cedar (*Juniperous bermudiana*), were particularly durable and highly resistant to insects, either because of their 'hardness or the general cohesion of their particles', while box (*Buxus sempervirens*) was especially valuable because of its combination of 'hardness and smoothness'.[39] Some timbers were particularly moisture-resistant as a result of the characteristics necessary for survival in their environments, such as alder (*Alnus glutinosa*), usually found on river banks, and Darwin was assured by James Brindley (1716–1772), the great canal engineer, that 'red Riga-deal, or pine wood would endure as long as oak in all situations', probably because of its resinous quality.[40] Woods also varied naturally in terms of their specific gravities, meaning that while some, such as oak, would sink, others, such as deal (fir or pine boards), floated easily, making them an ideal construction material for rafts and landing ships. Darwin suggested that the buoyancy of these vessels might be much improved by hollowing the timbers by boring and plugging the ends to make them watertight, or creating 'long square' air-filled 'wooden troughs' that could be fastened together to produce boats that could carry much greater weight and not 'so easily to be destroyed by storms' or sunk by weaponry. These proposals were similar to Lionel Lukin's 'unimmergible', boats based upon Norwegian yawls, of which Darwin may have been aware, which utilised air pockets in watertight bulkheads, along with the early lifeboat designs made by William Wouldhave and Henry Francis Greathead during the 1780s and 1790s.[41]

But increases in timber production were not only useful for fighting wars and boosting the economy. Darwin also believed that trees could help feed the population and supply medical products, which, as we have seen, was one of the main reasons for the establishment of physic and botanical gardens and scientific associations such as Linnean Society in Britain and its colonies.[42] He was led towards this notion partly because of his experiences tasting different tree parts, a method he employed to provide initial information about their likely chemical composition but also because of the apparent multiple opportunities for novel foodstuffs and commercial crops presented by trees such as the sugar maple (*Acer saccharum*) and the fact that some mucilage derived from the pith of trees and shrubs, such as sagoe, derived from the palm tree (*Cycas circinalis*), were already sold in shops for consumption. The pith from the stalks of the globe artichoke (*Cinara scolymus*) could be eaten if prepared correctly, while young shoots of elder (*Sambucus nigra*) were likewise promising if 'agitated in cold water' beforehand to remove 'acrid material', just as starch was prepared.[43] Similarly, nutritious foodstuffs might be obtained by 'grating', 'rasping' or 'pounding' the sapwood of 'most trees', especially in winter when it was at its most wholesome, and boiling the resulting powder or dust.[44] The taste of young ash or oak leaves suggested to Darwin that rising vernal sap juices in deciduous trees must contain not only 'sugar and mucilage' but also 'other ingredients as yet undetermined by investigators'. These 'nutritious juices', which were so useful for medicine, dyeing and tanning, were partly 'expended on the young leaves', which explained why they possessed something of the 'taste and qualities' of bark. The fact that the young leaves on both decorticated and undecorticated trees had the same 'bitter flavour' suggested that much of the astringent material continued to reside in the sapwood and roots.[45] Interior tree barks, roots and sapwood might 'probably all be used as food for ourselves or other animals in years of scarcity' because they frequently contained so much 'mucilaginous

ure 65.   Deer eating holly branches, drawn by S. Gilpin and engraved by T. Medland, from T. Gisborne, *lks in a Forest*, 3rd edn (1797).

or nutritious matter'; potentially useful species in this regard included elm, holly and 'probably' all trees and shrubs with thorns or bitter-tasting bark, such as hawthorn, gooseberry and gorse, which had evolved to defend themselves against animal or insect damage. This was demonstrated by the eagerness with which deer in Needwood Forest 'greedily' consumed holly cut from the tops of the trees, where the leaves did not develop spikes (Figure 65), or the way in which horses were 'well nourished by gorse' if the plants were first damaged by being crushed by stones – some horses at liberty, indeed, apparently deliberately trampled gorse in order to eat it.[46]

TREE POETRY AND PRESERVATION

Darwin's encounters with midlands trees also encouraged him to celebrate them in poetry, and his poetical portraits of trees helped to reinforce his analogies between animals and vegetables. Like the poems of his friend Seward, his poetical and prose portraits of trees mined common neo-classical tropes such as pastoral imagery and idealism to demonstrate how tree places were invested with special meaning, but they also supported his belief that they were crucial elements of the natural economy as defining denizens of woodland ecosystems and rich habitats in microcosm.[47] As previously emphasised, Darwin believed that his epic poems were an ideal vehicle for presenting scientific ideas, but used the 'interlude' in *The Loves of the Plants* to contrast pleasing but more superficial poetical scene-painting with the presentation of serious philosophical subjects in prose, employing 'stricter' reasoning and analogy rather than playful metaphors and personifications.[48] In *Zoonomia* Darwin argued that the 'Creator' had 'stamped a certain similitude on the features of nature' which demonstrated that the 'whole is one family of one parent' and upon this 'similitude is founded all rational analogy'. This encouraged philosophical comparisons concerning the 'essential properties of bodies' founded upon 'rational' analogies that led to 'many and important discoveries' and might 'collect ornaments for wit and poetry'.[49] As *The Temple of Nature* makes clear and Ashley Marshall has claimed, Darwin 'reclaims analogy as a valid epistemological tool, as no less defensible than the individual perception of physical objects'. His epic poetical works were therefore a 'union of scientific observation' and poetry' and a 'calculated combination of strict and "looser" analogies' bringing meaning beyond cold rationalism to nature and the universe.[50]

Darwin's belief in the special qualities of trees is reflected in his poetical and scientific treatment of them. They were often personified in poetry and literature because of their size and other traits, perhaps the most significant of which was their longevity, in which they resembled – or outstripped – humans more than did the annual, biennial or perennial non-ligneous plants. Following a well-established neo-classical tradition in which trees and groves were accorded life spirits, as in the mock-epics of Alexander Pope (1688–1744), the poetic personification of trees suggested a close relationship with humans denied other plants, helping to illustrate why some late Georgian Whigs and reformers promoted an iconography of woodland as part of their political message. Tree lifecycles, too, resembled the stages of human development, growth and decline, taking place over many years. Trees were one of Darwin's favourite and most long-established metaphors for those dear to him. For example, in a poetical rendering of the fifth satire of Roman poet Persius Flaccus (AD 34–62) sent to his old Chesterfield schoolmaster William Burrow in 1750, when he was nineteen, Darwin likened him to a 'stately pine' about which his fellow classmates, the 'young scions', were nurtured as 'pupil-plants' to become the new 'Burrows of the rising Age'. Decades later he used a great maturing tree with spreading branches and wholesome fruit as the metaphor for intellectual progress over the centuries, while maintaining that the world of antiquity had knowledge of some things awaiting rediscovery.[51]

In his foundation address to the members of the Derby Philosophical Society in 1784 Darwin employed the tree of knowledge metaphor to inspire enthusiasm for his vision of their contribution to Enlightenment science. Together they would help to nurture this precious tree, 'whose fruit, forbidden to the brute creation' had been 'plucked by the daring hand of experimental philosophy'. This was the tree whose seed had been 'sowed in Egypt' from where 'the very name of chemistry' had apparently been derived and which had 'put forth buds and branches afterwards in Arabia'. The tree of learning that the Derby philosophers honoured through their studies had then, through the 'abundance of its flowers', 'exuberance of its fruit', 'salubrious shade' and 'honeyed dews', enriched the 'whole terraqueous globe' by the innumerable contributions to the 'arts and sciences' and the 'necessities' and 'embellishments' of life that it had made.[52] The tree of knowledge metaphor was employed more extensively in *The Temple of Nature*, when Darwin described the rise of the 'reflective faculties of Man':

> In Eden's groves, the cradle of the world,
> Bloom'd a fair tree with mystic flowers unfurl'd;
> On bending branches, as aloft it sprung,
> Forbid to taste, the fruit of KNOWLEDGE hung;
> Flow'd with sweet Innocence the tranquil hours,
> And love and beauty warm'd the blissful bowers.
> Till our deluded parents pluck'd, ere long,
> The tempting fruit, and gather'd right and wrong;
> Whence good and evil, as in trains they pass,
> Reflection imag'd on her polish'd glass;
> And conscience felt, for blood by hunger spilt,
> The pains of shame, of sympathy, and guilt.[53]

In a note below he countered the argument that 'the acquisition of knowledge' reduced 'the happiness of the possessor' because Adam and Eve were supposed to have been made 'miserable' by eating an apple from this tree. In fact, the 'foresight and the power of mankind' had been 'much increased by their voluntary exertions in the acquirement of knowledge' and they had 'undoubtedly' avoided numerous evils and discovered many sources of good while possessing sensory and imaginative delights 'as extensively as the brute or the savage'.[54] Just as Darwin saw trees as one of the most successful life forms, which might even be capable of limitless growth, so they served as an apt metaphor for the almost boundless potential for human improvement in the arts and sciences. 'Eating of the tree of knowledge' revealed, among many other things, the manifold potential of arboriculture as a science.

Poetic personifications elided easily between science and poetry in Darwin's work and, as underlined earlier, he adopted what he believed to be a Linnaean scheme combined with a Newtonian framework for the new medical system of *Zoonomia*. The special status of trees in botanical terms was reinforced by difficulties accommodating them within the 'artificial', abstract Linnaean system, which scattered them across twenty-four classes and among varied non-ligneous plants. This meant that, especially for horticultural, economic or medical purposes, they were still often placed together in a pre-Linnaean style division, a grouping that fostered the adoption of more 'natural' taxonomies which tried to use a wider range of characteristics, such as that of the French botanist Antoine Laurent de Jussieu (1748–1836). This tension between conventional classifications of trees in natural history and those of Linnaeus and more 'natural' taxonomies is evident at times in *The Loves of the Plants*, *The Economy of Vegetation* and the Lichfield society's translation of Linnaeus's *Systema Vegetabilium* (1783), which described trees as 'highnesses, of the order of nobility' that were surrounded by the woodland 'servants' – the grasses (the 'plebeians'), ferns ('new colonists'), mosses ('servants'), flags ('slaves') and funguses ('vagabonds'). According to the Lichfield botanical society's translation, trees 'erect their heads, resist the winds, overshadow the sultry, moisten the dry with invisible showers, allure the birds, and perfect their fruit like so many orchards of nature.'[55]

The personification of trees also reinforced the feeling of humanity's kinship with the natural world, encouraged Darwin to appreciate the picturesque qualities of plants and confirmed his notion of the interdependency of animal and vegetable worlds, while dramatising the perniciousness of their destruction. The threat of annihilation faced by some trees stimulated him, like his Lichfield companion Anna Seward, to plead on their behalf using the language of poetic personification, and to express his love for his future wife Elizabeth Pole too. As Darwin stated to the actor, dramatist and writer Joseph Cradock (1742–1826), in 1775 he 'interceded' with her to prevent the lopping of a grove of trees, which 'occasioned' him 'from inspiration' to 'try again the long neglected art of verse-making'. Giving voice to the tree, Darwin implored Elizabeth not to 'Lop my green arms, my leafy tresses tear' for within the grove there slept in every tree 'A nymph, embalm'd by some poetic spell, who once had beauty, wit and life like thee'. If she spared the trees her name would be celebrated forever and be 'carved on every smooth green rind', while the grove would survive through 'summer's heat' and 'winter's blasts' as a place for the 'earliest fruits', the 'sweetest blossoms' and the 'love-struck swain' to walk with his sweetheart beneath the shades. Although usually expressed in a humorous manner, a desire to condemn the destruction of trees occasioned by enclosure or improvement was an important motivating factor in the arboricultural poetry of Darwin and friends such as Francis Noel Clark Mundy, Rev. Thomas Gisborne, Gisborne's younger brother John and Anna Seward, who implored the dean and chapter not to cut down some of the lime trees along the Dean's Walk, Lichfield, on practical and aesthetic grounds.[56]

The threat to trees and woodlands presented by enclosure and disafforestation may have partly prompted Darwin's acquisition and development of the Lichfield botanic garden from 1776, as well as his composition of a series of poems employing silvicultural personification to engender sympathy for their plight. As we have seen, his romance with the garden-loving Elizabeth also encouraged him in both ventures. When their friend Francis Noel Clark Mundy composed a poem which lamented the destruction of Needwood Forest in Staffordshire by the Duchy of Lancaster, Darwin and Seward contributed verses of their own. Mundy's work seems to have been more of a collaborative effort than has previously been realised, and one which his friends Darwin and Seward enthusiastically encouraged and promoted, the latter holding it to be a 'prime' example of 'local verse' which invested the beautiful trees and landscapes of Needwood with layers of meaning and rich associations. Mundy had some copies privately printed, but opposed wider distribution, even to the extent of buying up copies of an unauthorised version printed by William Jackson of Lichfield, although after substantial extracts were published in the *History of Staffordshire* (1798) by the Rev. Stebbing Shaw (1762–1802) the poem became more widely known.[57]

Despite Darwin's general support for many enclosure measures, probably encouraged by the affability of age, he believed, like the Herefordshire landowner, writer and antiquarian Richard Payne Knight, that, as he emphasised in the *Temple of Nature* (1803), landscape beauty was also evident in the 'rural charms' of 'unchastised' nature as well as in highly designed and managed parks and gardens. Darwin therefore lamented how much 'nature' had been 'chastised' in 'cities or their vicinity and even in the cultivated parts of the country', where it was rare to see it in an 'undisguised' state. The fields were ploughed, 'the meadows mown, the shrubs planted in rows for hedges, the trees deprived of their lower branches', while animals such as horses, dogs and sheep were 'mutilated in respect to their tails or ears'. These actions often resulted from the 'useful or ill-employed activity of mankind!' All these 'alterations' added to the 'formality of the soil, plants, trees, or animals', so that, when 'natural objects such as an uncultivated forest and its wild inhabitants' were encountered, they provided a source of charm and beauty. Darwin's concern at the chastising of nature was most evident and clearly articulated in relation to trees, but as these observations demonstrate, extended to the mistreatment by humans of all animals and plants in both urban and rural areas.[58]

Threats to trees and woodland from enclosure and the depredations of landlords and their agents were a source of considerable controversy during the eighteenth century, and conflicts were caused by perceived challenges to local rights and wood customs. Gilbert White described the impact of arguments that erupted over local timber resources in his *History of Selbourne* (1789). One nearby wood, the Holt, abounded with 'oaks that grow to be larger timber' and when the local landowner Henry Bilson-Legge, second Baron Stawell (1757–1820) had about a thousand oak trees felled in 1784 and claimed the 'lop and top', the poor from local parishes asserted their ownership and assembled 'in a riotous manner', taking it away with them, although Stawell served forty-five individuals with legal actions.[59]

THE BEAUTY AND DESTRUCTION OF NEEDWOOD FOREST

Needwood Forest in Staffordshire was a substantial and ancient midlands forest which in the medieval period had four wards and four keepers, each with their own lodges, although these were eventually purchased by local gentry (Figure 66). In the seventeenth century Needwood was about twenty-four miles in circumference and covered some 92,000 acres. A total of 47,000 trees was listed in 1684, along with 10,000 cords of hollies and underwood with a value of over £30,000. The forest had been a chase of the Duchy of Lancaster until the reign of Henry IV, when it became crown property. Needwood contained 'some of the finest oaks in England' and the 'wildest and most romantic spot' within the forest was held to be the Park of Chartley, which had once been owned by the De Ferrers family. Some of the trees had great local cultural significance and parts of the forest were associated with the late fourteenth-century chivalric poem *Sir Gawain and the Green Knight*. Efforts were made to enclose the forest in the late seventeenth century and again in 1778, but it was only after an enclosure act was successfully passed in 1803 that enclosure gathered pace. However, disafforestation was already occurring under the auspices of the duchy court and was substantially completed by 1811. By the late eighteenth century, while the extent of the forest still technically amounted to 9,220 acres, only about a thousand acres remained covered with woodland – all, it was claimed, still lying in a 'state of nature, wild and romantic' and 'beautiful to the fox hunter and the sportsman'. Despite acknowledging that it was full of the most beautiful birds and had a 'great and beautiful variety of aspect with "meandering rills" and deep glens', William Pitt (1749–1823), the Board of Agriculture's county investigator, argued that it was 'a fit subject' for 'improvement by human art and industry', which would increase the value of the land by enabling it to be effectively drained and sown with crops and vegetables, although, following a suggestion originally from George Venables-Vernon, Lord Vernon (1709–1780), whose main seat was at nearby Sudbury Hall, Derbyshire, he allowed that 1,000 acres should be preserved in 'clumps and coppices' as an 'ornament to the country' and a 'nursery for stout oak timber'. According to Pitt, this would put the Forest in 'an improved state', making it 'one of the most delightful spots' in the country, with matchless variety and management for sports and agriculture, while the 'wildest and most romantic part' would be preserved.[60]

The Needwood enclosure schemes attracted some opposition, however, and Stebbing Shaw argued that, while enclosures were of 'great public utility, to 'the eye of an admirer of picturesque scenery in all its wild and natural beauties' they were an 'insufferable innovation'. The Yorkshire artist Harriet Green (c. 1751–1821) and her husband Amos (1735–1807), also an artist who had lived at Halesowen in Worcestershire for many years and worked with the printer and publisher John Baskerville, who had close ties to the Lunar circle, visited Needwood quite often during the 1790s and early 1800s. Harriet took a close interest in its fate, noticed that some of the woods near Holly-bush were already 'despoiled of their largest oaks' and 'but a shadow' of what they once were, worrying that these 'depredations' were as 'nothing' compared with what was likely to ensue in the future. After the couple visited again in 1802, Harriet

ure 66. Detail from 'A New Map of the County of Stafford' by J. Smith and engraved by C. Smith, showing
:dwood Forest and vicinity, from W. Pitt, *Topographical History of Staffordshire* (1817).

complained that there were many who would wreak such 'irreparable devastation'
upon all such beautiful places across the country where 'the sweet enthusiasm, which
nature inspires, can be excited'.[61]

Darwin's friend John Gisborne was also a close associate of the Greens and author of
the *Vales of Wever* (1797), and lived at Holly-bush in the forest between 1795 and 1806.
He hoped that, though directed by 'motives of self interest', the Needwood enclosures
might bring some social improvements, but made 'great efforts to save single trees'.
He told the Greens that the problem was that, although farmers found trees useful

for sheltering their cattle, the 'bargain between commissioners and woodcutters was easiest made', resulting in swathes of woodland being swept away because of the 'expense of time and great difficulty' in selecting individual trees for preservation, despite the Gisborne brothers' efforts. At the same time, the surrounding hollies were 'bought up with great avidity' by Manchester cotton traders because of the usefulness of their by-products for textile printing. John Gisborne was sufficiently concerned about the preservation of the forest to ask Darwin to intervene with Matthew Boulton and request that he offer him his allotment if the full enclosures took place. However, the destruction continued and, after a visit to Gisborne at Holly-bush in 1802, Harriet and Amos Green likened walking through the 'devastation' of 'oaks and hollies ... piled upon the ground' to 'crossing a field of battle'. When from 1806 general enclosure did come Gisborne sold his estate and moved away because he was unable to afford the purchase of the 'beautifully wooded' forest 'elevations' near his house, the destruction of which he believed would 'seriously injure the beautiful scenery' and destroy the 'idea' of the 'place' to which he had 'become so ardently attached'. But even Stebbing Shaw believed that Bertram's Dingle, to the north of the Forest, though a 'singularly romantic valley or dell', could only be made 'uncommonly beautiful' if a 'gentleman of taste' improved its 'natural charms by a judicious display of art'.[62]

After hymning the beauties of Needwood in his poem, Mundy explained how 'a deeper gloom the wood receives, and horror shivers on the leaves' as the depredations of the axeman tore their way through the ancient oaks. Mundy – like Darwin, a supporter of moderate political reforms who opposed the American revolutionary war – saw the ancient oaks as symbols of British identity and precious freedoms sanctioned through the centuries. Seward later lauded Mundy's efforts to celebrate and preserve the 'ravaged groves' for posterity and designated *Needwood Forest* 'one of the sweetest local poems in our language', suggesting that 'such a beautiful work cannot die' and would be 'given to future times', like the Vale of Needwood itself. Four short poems signed by the initials E.D., A.S., B.B. and E.D. jun. were appended as a tribute to Mundy's efforts. The poem was privately printed with a small circulation among friends in 1776 (Seward was given four copies, for instance) and later reprinted by John Drewry at Derby in 1811. Few of those who purchased the later edition would have known that the initials stood for Darwin, Seward, Brooke Boothby and Erasmus Darwin junior, and still less that, as Seward emphasised, Darwin senior was the author of all the short poems apart from that by Boothby. Most readers would likewise not have been aware that Darwin and Seward assisted Mundy with the composition of *Needwood Forest*. This secrecy seems to have been partly a joke (which irritated Seward), but may also have reflected Darwin's concern for his professional reputation and perhaps a fear that, despite Mundy's status as a local landowner and magistrate, questioning enclosure and disafforestation for estate improvement, even in poetic form, might dent his popularity with patients and discourage potential gentlemanly benefactors. Nevertheless, the extent of Darwin's contribution to the Needwood poems again demonstrates the depth of his interest in arboriculture as well as his delight in sociable versifying.[63]

Mundy's *Needwood Forest* inspired Darwin to compose an 'Address to Swilcar Oak' in the forest, which he later inserted into the chapter on the production of leaves and wood in *Phytologia*, underscoring the association between poetic personification and arboricultural physiology and offering some light relief for readers rather in the same way that poems were inserted into Evelyn's *Sylva* – probably Darwin's main model. Personified as the 'Father of the Forest', the Swilcar oak, near Ealand Lodge, Marchington, towards the north-west portion of Needwood, stood within the royal forest and on the boundary of the land of the freeholders. It was of 'stupendous size, of noble form' and 'high antiquity', both 'buried in solitude' yet also 'surrounded by his congenial forest' and a guard of 'filial oaks', though 'separated from them' as 'one of the rarest sights in nature'.[64] According to Darwin, this venerable oak measured 'thirteen yards round at its base and eleven yards round at four feet from the ground', and was believed to be 600 years old (Figure 67). His poem on the Swilcar Oak was composed in response to Mundy's work, according to Darwin, 'on leaving' Needwood'. A number of themes developed by Darwin in his 'Address' were to reappear in – and inspire – his future poetic and scientific work. Inspired by Mundy's composition, Darwin was attracted by the size, venerable antiquity and appearance of the ancient oak, which he immediately personified according to poetic convention:

Figure 67. Swilcar Oak, drawn by J. Smith and engraved by H. Moore (1807), originally for J. Nightingale, *The Beauties of England and Wales*, vol. 13, part II, Somersetshire and Staffordshire (1813).

> Gigantic Oak! Whose wrinkled form hath stood,
> Age after age, the patriarch of the wood!-
> Thou, who hast seen a thousand springs unfold
> Their rave'd buds, and dip their flowers in gold;
> Ten thousand times yon moon relight her horn,
> And that night star of evening gild the morn!-

The cyclical nature of life and the oak's life experience were also venerated, with allusions to the significance of trees in the pagan religion and mythology of the ancient Britons:

> Erst, when the druid bards with silver hair
> Pour'd round they trunk the melody of prayer;
> When chiefs and heroes join'd the kneeling throng,
> And choral virgins trill'd the adoring song;
> While harps responsive sung amidst the glad,
> And holy echoes thrill'd thy vaulted shade;
> Say, did such dulcet tones arrest thy gales,

This kind of poetic portrayal of the personified oak mirrored Darwin's scientific investigations and observations of tree bodies, poetical and botanical, which were informed by close observation, and both, in a sense, preserved the memory of the tree:

> As Mundy pours along the listening gales?
> Gigantic oak! – they hoary head sublime
> Erewhile must perish in the wrecks of time;
> Should round they brow innocuous lightning shoot,
> And to fierce whirlwinds shake they steadfast root;
> Yet shalt thou fall!-they leafy tresses fade,
> And those bare shatter'd antlers strew the glade;
> Arm after arm shall leave the mouldering dust,
> And they firm fibres crumble into dust!-
> But Mundy's verse shall consecrate they name,
> And rising forests envy Swilcar's fame;
> Green shall they gems expand, they branches play,
> And bloom for ever in the immortal lay.[65]

Darwin wanted to indicate how much there was to be learned from the study and contemplation of trees, as the reference to the oak as a 'hoary head sublime' suggests. Although trees' natural demise was inevitable eventually, as with all living creatures (though organic matter was transmogrified into other living beings through the generations), it also underscored the reasons for his concern at the unnecessary destruction of trees.[66]

Shortly after Mundy's *Needwood Forest* and Darwin's 'Address to Swilcar Oak' were written in 1783, Thomas Gisborne – close friend of the artist Joseph Wright, who had painted his portrait in 1777, older brother of John and a founder member of the Derby Philosophical Society – moved to Yoxall Lodge in the heart of Needwood to hold the perpetual curacy of Barton-Under-Needwood, although he continued to stay in Derby quite frequently until the 1790s. Thomas Gisborne's Derby residence, inherited from his father John Gisborne senior was St Helen's House, designed by Joseph Pickford, which had a considerable park to the rear running down the hill towards the Derwent that was extensively planted with trees. His larger estate, which crossed between Derbyshire and Staffordshire, included part of Needwood and 'its unenclosed woods of oak and holly, with their glades and dells, diversified the more cultivated landscape of English husbandry'.[67]

Yoxall Lodge was a 'delightful spot' upon a stream called the Linbrook and was 'pleasantly situated in a recluse valley, well suited to the placid and studious mind[s]' of Thomas and his wife Mary, who 'enlarged and greatly improved the house' and turned it into their 'principal residence in preference to the superb mansion' in Derby.[68] Darwin, Joseph Wright and William Wilberforce stayed at Yoxall quite often, Wilberforce emphasising the 'riches' on display in 'overflowing profusion' with 'the first foliage of the magnificent oak' contrasting 'with the dark holly, the flowering gorse, and the horse chestnut'.[69] Darwin lodged at Yoxhall Lodge in 1802 and his host was one of the last patients he treated before his death shortly afterwards. As a naturalist and accomplished artist, Gisborne was captivated by Needwood and to a large extent his fascination paralleled White's love for Selborne. Gisborne made close observations of the natural history of the forest and his library was 'like his mind, full of curious things stored up – birds stuffed, and birds alive, creatures he had caught and tamed'; 'there was not a flower within the circuit of the Forest, or an insect which flitted its brief life on the wing, that he had not noticed and studied'. As 'his eye was also a painter's eye', and encouraged by Wright, he made 'sketches from nature [which] were remarkable both for fidelity and power', and demonstrated his keen awareness of the changing picturesque qualities of the forest. Encouraged by Gisborne, Mundy and Darwin, Wright painted at least nine versions of a cottage on fire in Needwood during the 1790s. He also painted a portrait of Thomas and his wife Mary Gisborne (1760–1848) together in a wooded setting, probably in the forest near Yoxhall, in 1786 (Figure 68). A series of landscape studies of trees standing alone or with rocks in ink, pencil, charcoal and other mediums completed during the 1780s and 1790s also seems to have been inspired by Gisborne and Needwood.[70]

Like Mundy, Darwin, Seward and Boothby, and inspired by their arboricultural poetry, Gisborne too chose to express his love for the forest in verse and composed *Walks in a Forest* in 1794, partly modelled on William Cowper's poetry. Consisting of six blank verse compositions ostensibly occasioned by walks in the woods at various times of the year, *Walks in a Forest* was one of Gisborne's most popular works and ran to ten editions by 1814. Like Mundy, Darwin and their friends, Gisborne extolled the virtues of Needwood and regretted its disafforestation, emphasising the changing picturesque qualities of the woodland through the seasons, a portrayal that was

Figure 68. J. Wright, *Rev. [Thomas] and Mrs [Mary] Gisborne* (1786), courtesy of the Paul Mellon Collection, Yale Centre for British Art.

reinforced by the six engravings that accompanied the text after William Gilpin, with whom he had formed a close friendship from 1792. However, in contrast to the worldviews of Darwin and Mundy (but not Seward), in keeping with his understanding of moral philosophy and conduct, Gisborne intended that by examining 'the face of nature' *Walks in a Forest* would 'inculcate those moral truths, which the contemplation of the works of God suggests'. Gisborne's perception of the natural world as part of the divine economy as portrayed in *Walks in a Forest* and in his geological theories was informed by the moral principles detailed in a series of other major works during his long lifetime and in his sermons.[71]

His brother John Gisborne – who also spent his younger years at St Helen's House, married Elizabeth Darwin's daughter Millicent Pole and as we saw, often visited the Darwin family at Full Street and The Priory – was likewise inspired by Needwood and Darwin, 'whose poetry' he believed, would 'perish only' with language itself', and he engaged in his 'favourite pursuit' of botany with 'unabated zeal' throughout his life, keeping a 'Hortus-siccus' of dried plants that he carried around with him everywhere and corresponding with 'most of the leading botanists of his day'. John Gisborne bought Holly-bush in Needwood, near his brother's house, where he resided all year, to write and relish the surrounding 'sylvan scenery', while beautifying the grounds 'with excellent taste', forming a walk beside a sheet of water so he could enjoy both the 'delights' and 'security' of his garden and the 'grandeur and extent of the forest'. The inspiration that he gained from Darwin (and, of course, Thomas) is strongly evident in the language and scenes of the *Vales of Wever* (1797), which is suffused with the 'delight he experienced' walking through the 'varied beauties of Needwood' while staying at Yoxall Lodge. He 'often spoke' of the 'majestic oaks', 'dark groves of holly', 'golden furze', 'murmuring brooks' and 'tangled glades' of the forest.[72]

The impact of the sylvan poetical world of Needwood created by Mundy, Darwin, Seward and the Gisborne brothers is fully evident in the way that visitors to the forest perceived it, with poetry, art and place informing and colouring perceptions of each other. The artists Amos and Harriet Green were both already very familiar with Mundy's 'genuine and delightful' poem when they visited during the 1790s and early 1800s, the former having been presented with a copy of the book by its author 'many years' previously, while the latter received her copy from a friend of Seward, who emphasised how Darwin and Seward had assisted the author. Their admiration for the poem 'continually increased' as they more directly experienced the 'forest beauties' aided by their friend John Gisborne, and each one of Mundy's descriptions had 'all the faithfulness of a copy with all the freedom of an original' as the 'natural product of the scene'. Harriet emphasised the large extent of the forest, with its varied terrain of lawns and patches of woodland interspersed with dales and 'perpetual undulation', contrasting the valley sides 'crowned with trees' and intermingled with the 'turf, bushes, furze, and heath' to a nearby comparatively tame landscaped park, which provided a 'meagre' prospect with its 'trees, abruptly separated from smooth lawn' providing the 'only variety'. The endless varieties in tree forms and sizes, punctuated by the huge hoary oaks with their blasted trunks and 'giant arms' stretching through the 'dark foliage', were special to Needwood, as were the 'wild and fantastic' hollies encircling each tree like 'guards' surrounding a 'monarch'. The Greens experienced the 'wild picturesque' Darley oaks in Needwood directly, guided by John Gisborne; in the scenic descriptions of Mundy's and Thomas Gisborne's poems; and through a picture drawn by the latter that they saw hanging in his study at Yoxall Lodge. All these experiences were mediated through the 'historic and poetic ground' of the geniuses of Spencer, Milton and Shakespeare, as well as the poetry of the Needwood circle, the ancient oaks with their wild forms and 'gigantic size' evoking the 'days of the Druids' for them as they had for Mundy, Darwin, Seward and Thomas Gisborne. They were alarmed that Darwin's favourite Swilcar oak was being left exposed by felling on the edge of the royal domain, the forest glades receding and exposing his 'bare arms', but found some solace in the fact that 'an order came just in time to spare a few guards to his venerable age'.[73]

CONCLUSION

This chapter has argued that Erasmus Darwin was a strong advocate of afforestation as part of agricultural improvement, developing his arguments most fully in *Phytologia*. He contended that large-scale tree-planting schemes were required for naval timber supplies for national defence and also to ensure that Britain's lifeblood of international maritime trade was maintained through the construction of merchant vessels, especially in time of war. Trees were required for agricultural improvement too, making estates more profitable by rendering difficult ground potentially useful, enhancing landscape beauty and providing shelter from rain and sunshine. To this extent his ideas were not especially original and were shared by a large number of his contemporaries.

Where Darwin was more original was that he had a keen sense of the physiology and anatomy of trees as individual beings and, as we have seen with plants in general, pressed the analogies with other living beings, including animals and humans, further than most of his contemporaries. His medical perspective encouraged him to see tree-products as both a potentially valuable source of foodstuffs and forms of treatment – with the copious medical usage of Peruvian bark demonstrating how many hitherto undiscovered virtues might exist in trees from around the globe (at least from a European perspective). As far as food was concerned, he believed there to be enormous potential opportunities for materials such as some sapwoods to supply a nutritious meal if properly prepared, potentially preventing food shortages and famines, which were exacerbated by over-reliance upon a few plants such as wheat and potatoes or costly and expensive meat supplies. In the case of medical treatments, as with plants and animals more generally, he believed that the context and environment in which trees grew, including the threats they faced from insects, animals and other plants in the competitive economy of nature, dynamically determined their qualities and medical virtues. Analysis of these dimensions held the key to a potentially vast natural storehouse of potential remedies that might stock future pharmacy cupboards and alleviate a thousand ills with which life was plagued. For Darwin, trees were therefore not merely pretty plants to adorn a lawn or rocky gorge nor even a useful supply of timber, but a life-giving opportunity.

NOTES

1 R. Schofield, *The Lunar Society of Birmingham: A Social History of Provincial Science and Industry in Eighteenth-century England* (Oxford, 1963); M. McNeil, *Under the Banner of Science: Erasmus Darwin and his Age* (Manchester, 1987); J. Browne, 'Botany for gentlemen: Erasmus Darwin and the Loves of the Plants', *Isis*, 80 (1989), 593–620; R. Porter, 'Erasmus Darwin: doctor of evolution?' in J. Moore ed., *History, Humanity and Evolution, Essays for John C. Greene* (London, 1989), 39–69; L. Schiebinger, 'The private life of plants: sexual politics in Carl Linnaeus and Erasmus Darwin', in M. Benjamin ed., *Science and Sensibility: Gender and Scientific Enquiry, 1780–1945* (Oxford, 1991), 121–43; T. Fulford, 'Coleridge, Darwin, Linnaeus: the sexual politics of botany', *The Wordsworth Circle*, 28 (1997), 124–30; D. King-Hele, *Erasmus Darwin: A Life of Unequalled Achievement* (London, 1999); D. Coffey, 'Protecting the botanic garden: Seward, Darwin and Coalbrookdale', *Women's Studies*, 31 (2002), 141–64; J. Uglow, *The Lunar Men: The Friends who Made the Future, 1730–1810* (London, 2002); C. Darwin, *The Life of Erasmus Darwin*, ed. by D. King-Hele (Cambridge, 2003); C. U. M. Smith and R. Arnott eds, *The Genius of Erasmus Darwin* (Aldershot, 2005); G. Rousseau, '"Brainomania": brain, mind and soul in the long eighteenth century', M. Green, 'Blake, Darwin and the promiscuity of knowing: rethinking Blake's relationship to the Midlands Enlightenment', T. H. Levere, 'Dr Thomas Beddoes (1760–18108) and the Lunar Society of Birmingham: collaborations in medicine and science', and G. Budge 'Erasmus Darwin and the poetics of William Wordsworth: "excitement without the application of gross and violent stimulants"', all in G. Budge ed., 'Science and the Midlands Enlightenment', special issue of *British Journal for Eighteenth-Century Studies*, 30 (2007); P. Ayres, *The Aliveness of Plants: The Darwins at the Dawn of Plant Science* (London, 2008), P. Jones, *Industrial Enlightenment in Birmingham and the West Midlands, 1760–1820* (Manchester, 2008); P. A. Elliott, *The Derby Philosophers: British Urban Scientific Culture c1700–1850* (Manchester, 2009); A. Seward, *Anna Seward's Life of Erasmus Darwin*, ed. by P. K. Wilson, E. A. Dolan and M. Dick (Studley, 2010); P. Fara, *Erasmus Darwin: Sex, Science and Serendipity* (Oxford, 2012); M. Priestman, *The Poetry of Erasmus Darwin: Enlightened Spaces, Romantic Times* (Farnham, 2013).

2 C. Darwin, *The Variation of Animals and Plants under Domestication*, 2nd edn, 2 vols (London, 1899 [1875]), vol. 1, 352–88; M. Allan, *Darwin and his Flowers: The Key to Natural Selection* (London, 1977), 226; K. Thompson, *Darwin's Most Wonderful Plants: A Tour of His Botanical Legacy* (London, 2019). Ayres, *Aliveness of Plants*, does not really cover any of the Darwins' analyses of trees.

3 T. Williamson, 'The management of trees and woods in eighteenth-century England', in L. Auricchio, E. H. Cook and G. Pacini eds, *Invaluable Trees: Cultures of Nature, 1660–1830* (Oxford, 2012), 221–35, at 221.

4 Lichfield Botanical Society, *A System of Vegetables ... translated from the thirteenth edition of the Systema Vegetabilium of the late Professor Linneus*, 2 vols (Lichfield, 1783); Lichfield Botanical Society, *The Families of Plants ... Translated from the Last Edition of the 'Genera Plantarum'*, 2 vols (Lichfield, 1787); E. Darwin, *The Loves of the Plants*, 4th edn (London, 1799); E. Darwin, *The Economy of Vegetation*, 4th edn (London, 1799); E. Darwin, *Zoonomia; or the Laws of Organic Life*, 3rd edn, 4 vols (London, 1801); E. Darwin, *Phytologia; or, the Philosophy of Agriculture and Gardening* (London, 1800); E. Darwin, *The Temple of Nature; or the Origin of Society* (London, 1803); E. Sloan, *The Landscape Studies of Hayman Rooke (1723–1806): Antiquarianism, Archaeology and Natural History in the Eighteenth Century* (Woodbridge, 2019), 65–84.

5 O. Jones and P. Cloke, *Tree Cultures: The Place of Trees and Trees in their Place* (Oxford, 2002); S. Bowerbank, *Speaking for Nature: Women and Ecologies of Early-Modern England* (Baltimore, 2004), 177–80; L. Auricchio, E. H. Cook and G. Pacini eds, *Invaluable Trees: Cultures of Nature, 1660–1830* (Oxford, 2012).

6   H. Rooke, *Descriptions and Sketches of Some Remarkable Oaks in the Park at Welbeck, in the County of Nottingham* (London, 1790); R. Lowe, *General View of the Agriculture of the County of Nottingham* (London, 1794), 19–20, 63–4, 71–88; H. Rooke, *A Sketch of the Ancient and Present State of Sherwood Forest in the County of Nottingham* (Nottingham, 1799); W. Gilpin, *Remarks on Forest Scenery and other Woodland Views*, 3rd edn, 2 vols (London, 1808), vol. 1, 324–9; F. C. Laird, *A Topographical and Historical Description of the County of Nottingham* (London, 1820), 48–76; R. Marquiss ed., *The Nature of Nottinghamshire* (Buckingham, 1987), 19–44; O. Rackham, *Trees and Woodland in the British Landscape*, rev. edn (London, 2001); O. Rackham, *Woodlands* (London, 2012); P. A. Elliott, C. Watkins and S. Daniels, *The British Arboretum: Trees, Science and Culture in the Nineteenth Century* (London, 2011); Williamson, 'The management of trees and woods', 221–35; C. Watkins, *Trees, Woods and Forests: A Social and Cultural History* (London, 2014), 140–74; Sloan, *Landscape Studies of Hayman Rooke*, 65–84.

7   T. Brown, *General View of the Agriculture of the County of Derby* (London, 1794), 42–3; J. Farey, *General View of the Agriculture and Minerals of Derbyshire*, 3 vols (London, 1811–17), vol. 1, 381, vol. 2, 215–16, 244–324; J. C. Cox, 'Forestry' in W. Page ed., *The Victoria County History of Derbyshire*, vol. 1 (London, 1905), 397–425; K. C. Edwards, *The Peak District* (London, 1973), 77–83.

8   Cambridge University Library, E. Darwin, Notebook (DAR, 227.2:11); Darwin, *Phytologia*, 449–54; D. King-Hele ed., *The Collected Letters of Erasmus Darwin* (Cambridge, 2007); Laird, *Topographical and Historical Description of the County of Nottingham*, 48–76, 349–52; Darwin, *Life of Erasmus Darwin*, 31; J. C. Brown, *The Forests of England and the Management of Them in Bye-Gone Times* (London, 1883); S. Daniels, 'The political iconography of woodland in later Georgian England', in D. Cosgrove and S. Daniels eds, *Iconography of Landscape* (Cambridge, 1988), 43–82; J. Zonneveld, *Sir Brooke Boothby* (Voorburg, 2003); C. Jarvis, *Order out of Chaos: Linnaean Plant Names and their Types* (London, 2007); T. Barnard, *Anna Seward: A Constructed Life* (Farnham, 2009); Elliott, Watkins and Daniels, *British Arboretum*, 1–26; Watkins, *Trees, Woods and Forests*, 104–18, 131–6.

9   J. Evelyn, *Sylva: or a Discourse of Forest-Trees*, with notes by A. Hunter, 3rd edn, 2 vols (York, 1801), vol. 1, 3, 75–96; D. Hudson and K. W. Luckhurst, *The Royal Society of Arts, 1754–1954* (London, 1954), 86–9; G. E. Mingay, *English Landed Society in the Eighteenth Century* (London, 1963), 50–79, 163–88; J. V. Beckett, *The Aristocracy in England, 1660–1914*, rev. edn (Oxford, 1989), 43–90, 157–205; Daniels, 'Iconography of woodland'; S. Daniels, *Humphry Repton: Landscape Gardening and the Geography of Georgian England* (New Haven, 1999), 52–3, 91–6; Elliott, Watkins and Daniels, *British Arboretum*, 11–26; Williamson, 'The management of trees and woods', 221–35; C. Watkins and B. Cowell, *Uvedale Price (1747–1829): Decoding the Picturesque* (Woodbridge, 2012), 46–58; D. Brown and T. Williamson, *Lancelot Brown and the Capability Men: Landscape Revolution in Eighteenth-Century England* (London, 2016), 103–8.

10 J. Evelyn, *Sylva; or a Discourse of Forest Trees* (London, 1664), '5th' edn 1729 (actually a reprint of the 4th edn); Evelyn, *Sylva*, with notes by Hunter, 3rd edn.

11 J. Evelyn, *Sylva*, with notes by Hunter, 1st edn, 2 vols (York, 1776), list of subscribers; Darwin used the engraving of the Chinese fern known as the Tartarian Lamb from Hunter's edition as an illustration in the *Loves of the Plants*, 37–9.

12 Derby Philosophical Society, *Rules and Catalogue of the Derby Philosophical Society*, with supplements of 1795 and 1798 (Derby 1793–8), 3, 14, 15, 18, 19, 20, 22, 29.

13 S. Shaw, *The History and Antiquities of Staffordshire*, 2 vols (London, 1802), vol. 1, 68; E. Whittle, *The Historic Gardens of Wales* (London, 1992), 46–8; M. Craven, *John Whitehurst: Innovator, Scientist, Geologist and Clockmaker*, 2nd edn (Croyden, 2015), 69–91; C. Hartwell, N. Pevsner and E. Williamson, *The Buildings of England: Derbyshire* (New Haven, 2016), 51, 186, 221, 352, 403, 476, 487, 509, 558–9, 629.

[14] W. Gilpin, *Essays on the Picturesque as Compared with the Sublime and the Beautiful*, 3 vols (London, 1810), vol. 1, 229–96; Daniels, 'Iconography of woodland', 43–82; S. Daniels, *Joseph Wright* (London, 1999), 68–74; J. Bonehill and S. Daniels eds, *Paul Sandby: Picturing Britain* (London, 2009), 226–33; Elliott, Watkins and Daniels, *British Arboretum*, 25–6; Watkins, *Trees, Woods and Forests*, 102–18; C. Watkins, *Trees in Art* (London, 2018), 27–36.

[15] A. Young, *The Farmer's Tour through the East of England*, 4 vols (London, 1771), vol. 1, 202–3; Hartwell, Pevsner and Williamson, *Derbyshire*, 471–7.

[16] Young, *Farmer's Tour*, 203–4.

[17] Rooke, *Descriptions and Sketches*; Daniels, *Humphry Repton*, 154–66; Watkins, *Trees, Woods and Forest*, 140–52; Sloan, *Landscape Studies of Hayman Rooke*, 64–82.

[18] Young, *Farmer's Tour*, 212–13; J. Pilkington, *A View of the Present State of Derbyshire*, 2 vols (Derby, 1789), vol. 2, 437–40; Farey, *General View*, vol. 2, 219, 223, 237–8, 244–78; Edwards, *The Peak District*, 76–9; J. Pearson, *Stags and Serpents: the Story of the House of Cavendish and the Dukes of Devonshire* (London, 1983), 79–80, 91–115; H. Colvin, *Calke Abbey, Derbyshire: a Hidden House Revealed* (London, 1986), 90–1, 94–5, 118–21, 125; J. Barnatt and T. Williamson, *Chatsworth: A Landscape History* (Macclesfield, 2005), 101–25; J. Barnatt and N. Bannister, *The Archaeology of a Great Estate: Chatsworth and Beyond* (Oxford, 2009), 67–96; Hartwell, Pevsner and Williamson, *Derbyshire*, 245–9, 220–1.

[19] S. Mason, *The Hardwareman's Daughter: Matthew Boulton and his 'Dear Girl'* (Chichester, 2005), 23–5.

[20] R. S. Fitton, *The Arkwrights: Spinners of Fortune* (Manchester, 1989), 273–4.

[21] Derby Philosophical Society, *Rules and Catalogue of the Library belonging to the Derby Philosophical Society* (Derby, 1815), 10, 22; H. Repton, *Observations on the Theory and Practice of Landscape Gardening* (London, 1803); W. Forsyth, *A Treatise on the Culture and Management of Fruit Trees* (London, 1802); W. Pontey, *The Forest Pruner; or Timber Owner's Assistant* (London, 1805); W. Pontey, *The Profitable Planter: a Treatise on the Theory and Practice of Planting Forest Trees* (Huddersfield, 1808); Farey, *General View*, vol. 2, 219–340 (239), vol. 3, 655; R. S. Fitton and A. P. Wadsworth, *The Strutts and the Arkwrights, 1758–1830: A Study of the Early Factory System* (Manchester, 1958), 179; Historic England: Kingston Park Pleasure Gardens, Kingston on Soar: National Heritage List for England, https://historicengland.org.uk/listing/the-list/list-entry/1001716.

[22] Darwin, *Phytologia*, 367–8; Evelyn, *Sylva*, 3rd edn (1801), vol. 1, 97–104.

[23] Darwin, *Phytologia*, 475–6.

[24] Evelyn, *Sylva*, 3rd edn (1801), vol. 1, 3.

[25] J. C. Loudon, *Observations on the Formation and Management of Useful and Ornamental Plantations*, 2 vols (Edinburgh, 1804); J. C. Loudon, *Arboretum et Fruticetum Britannicum*, 2nd edn, 8 vols (London, 1844), vol. 1, 1–2, 221–2, 226–7, vol. 3, 717–949; J. C. Loudon, *An Encyclopaedia of Agriculture*, 5th edn (London, 1844), 633–64; J. and J. Penoyre, *Houses in the Landscape: A Regional Study of Vernacular Building Styles in England and Wales* (London, 1978); P. Borsay, *The English Urban Renaissance: Culture and Society in the Provincial Town, 1660–1770* (Oxford, 1991), 54–9; T. Williamson, *Polite Landscapes: Gardens and Scenery in Eighteenth-century England* (Stroud, 1995), 124–30; Rackham, *Trees and Woodland*, 91–9; Elliott, Watkins and Daniels, *British Arboretum*, 11–23; Williamson, 'The management of trees and woods', 221–5; Rackham, *Woodlands* (London, 2012), 219–45, 345–53.

[26] E. Wade, *A Proposal for Improving and Adorning the Island of Great Britain for the Maintenance of our Navy and Shipping* (London, 1755); S. Johnson, 'Further thoughts on agriculture', *Universal Visitor* (1756), quoted in J. Sitter, 'Sustainability Johnson', in A. W. Lee ed., *New Essays on Samuel Johnson: Revaluation* (Lanham, 2018), 111–30, at 120–1; Hudson and Luckhurst, *Royal Society of Arts*, 86–9; Elliott, Watkins and Daniels, *British Arboretum*, 221–35; Williamson, 'The management of trees and woods', 221–35.

27 [E. Darwin, T. Bentley and J. Wedgwood] *A View of the Advantages of Inland Navigations with a Plan of a Navigable Canal* (London, 1765), 27; R. Fisher, *Heart of Oak: The British Bulwark* (London, 1763), vii–ix, 80.

28 H. L. Duhamel du Monceau, *Trait des Arbres ar Arbustes qui se Cultivent en France*, 2 vols (Paris, 1755); *Traite des Arbres Fruitiers*, 2 vols (Paris, 1768); J. L. Caradonna, 'Conservationism *avant la lettre*? Public essay competitions on forestry and deforestation in eighteenth-century France' and M. Martin, 'Bourbon renewal at Rambouillet', in L. Auricchio, E. Heckendorn Cook and G. Pacini eds, *Invaluable Trees: Cultures of Nature, 1660–1830* (Oxford, 2012), 39–54, 151–70.

29 Darwin, *Phytologia*, 468–9.

30 Darwin, *Phytologia*, 469–70.

31 Darwin, *Phytologia*, 470.

32 Darwin, *Phytologia*, 476.

33 Darwin, *Phytologia*, 471.

34 Darwin, *Phytologia*, 479–80; W. Pitt, *General View of the Agriculture of the County of Stafford*, 2 vols (London, 1796), 98–9.

35 J. Kennedy, *Treatise upon Planting, Gardening and the Management of the Hot House*, 2nd edn, 2 vols (London, 1777), vol. 1, vii, 1–33.

36 Darwin, *Phytologia*, 472; Sloan, *Landscape Studies of Hayman Rooke*, 60–1.

37 Darwin, *Phytologia*, 472–3.

38 Darwin, *Phytologia*, 473.

39 Darwin, *Phytologia*, 473–4; Lichfield Botanical Society, *System of Vegetables*, vol. 1, 186; Lichfield Botanical Society, *Families of Plants*, vol. 1, 140.

40 Darwin, *Phytologia*, 474; Lichfield Botanical Society, *Families of Plants*, vol. 2, 641.

41 Darwin, *Phytologia*, 474–5; I. Cameron, *Riders of the Storm: The Story of the Royal National Lifeboat Institution* (London, 2002).

42 R. Grove, *Green Imperialism: Colonial Expansion, Tropical island Edens and the Origins of Environmentalism, 1600–1860* (Cambridge, 1995); R. Drayton, *Nature's Government: Science, Imperial Britain and the 'Improvement' of the World* (New Haven, 2000); M. Guenther, 'Tapping nature's bounty: science and sugar maples in the age of improvement', in L. Auricchio, E. Heckendorn Cook and G. Paccini eds, *Invaluable Trees: Cultures of Nature, 1660–1830* (Oxford, 2012), 135–49.

43 Darwin, *Phytologia*, 477.

44 Darwin, *Phytologia*, 503.

45 Darwin, *Phytologia*, 447–8.

46 Darwin, *Phytologia*, 448–9, 503.

47 Bowerbank, *Speaking for Nature*, 177.

48 Darwin, *Loves of the Plants*, 'interlude'.

49 Darwin, *Zoonomia*, vol. 1, vii, quoted in Porter, 'Erasmus Darwin: doctor of evolution?' 62. However, Darwin also cautioned that when, through 'licentious activity', analogy 'links together objects, otherwise discordant, by some fanciful similitude', 'philosophy and truth' would 'recoil from its combination'.

50 A. Marshall, 'Erasmus Darwin contra David Hume', *British Journal for Eighteenth-Century Studies*, 20 (2007), 89–111.

51 E. Darwin, letter to W. Burrow, 11 December 1750, in King-Hele ed., *Collected Letters*, 17–19, quotation and attribution from King-Hele.

52 Darwin, 'Address to the Philosophical Society'.

53 Darwin, *Temple of Nature*, 121–2.

54 Darwin, *Temple of Nature*, 122.

55 Darwin, *Loves of the Plants*; *Economy of Vegetation*; *Zoonomia*; Lichfield Botanical Society, *System of Vegetables*, vol. 1, 3–5; Daniels, 'Iconography of woodland', 43–82; Elliott, Watkins and Daniels, *British Arboretum*, 27–48.

56 E. Darwin, letters to J. Cradock, 21 November 1775 and to E. Pole, 1775? in King-Hele ed., *Collected Letters*, 137–40; Bowerbank, *Speaking for Nature*, 178–81.

57 Shaw, *Staffordshire*, 68–70; Bowerbank, *Speaking for Nature*, 177.

58 Darwin, *Temple of Nature*, 105; R. Payne Knight, *The Landscape, a Didactic Poem in Three Books Addressed to Uvedale Price Esq.* (London, 1794).

59 G. White, *The Natural History and Antiquities of Selbourne* (1789), edited by E. T. Bennett and J. A. Harting (London, 1876), 26.

60 Pitt, *General View*, 102–8; Shaw, *Staffordshire*, vol. 1 , 60–70; Brown, *The Forests of England*, 117; R. W. V. Elliott, 'Staffordshire and Cheshire Landscapes in Sir Gawain and the Green Knight', *North Staffordshire Journal of Field Studies*, 17 (1977), 20–49; R. V. W. Elliott, 'Woods and forests in the "Gawain" country', *Neuphilologische Mitteilungen*, 80 (1979), 48–64; M. W. Greenslade and A. J. Kettle, 'Forests', in *A History of the County of Stafford, Vol. 2* (London, 1967), 335–7, 338–48; 'Needwood Forest', in W. Page and N. J. Tringham, *A History of the County of Stafford, Vol. 10: Tutbury and Needwood Forest* (London, 2007); Williamson, 'The management of trees and woods', 228–9; N. Tringham, 'Needwood Forest Surveys of 1649–50 and attempted enclosure', *Transactions of the Staffordshire Archaeological and Historical Society*, 94 (2010), 28–70; N. Tringham, 'The 1655 petition against the enclosure of Needwood Forest', *Transactions of the Staffordshire Archaeological and Historical Society*, 96 (2013), 72–96.

61 H. Green, *Memoir of Amos Green, Esq. Written by his Late Widow* (York, 1823); S. Whyman, 'John Baskerville, William Hutton and their social networks', in C. Archer-Parre and M. Dick eds, *John Baskerville: Art and Industry of the Enlightenment* (Liverpool, 2017), 87–112.

62 Pitt, *General View*, 102–8; Shaw, *Staffordshire*, vol. 1, 60–70; E. Darwin, letter to M. Boulton, 9 November 1800, in King-Hele ed., *Collected Letters*, 554; Green, *Memoir of Amos Green*, 194, 222–3, 225–6; E. Nixon ed., *A Brief Memoir of the Life of John Gisborne, to which are added Extracts from His Diary* (London and Derby, 1852), 28–9.

63 F. N. C. Mundy, *Needwood Forest* (Lichfield, 1776), 41, 45–52; F. N. C. Mundy, *Needwood Forest: Written in the Year 1776: Never Published* (Derby, 1811); A. Seward, letter to E. Jackson, 3 August, 1792 in *The Letters of Anna Seward*, 6 vols, (Edinburgh, 1811), vol. 3, 154–6; manuscript copies of *Needwood Forest* and the *Fall of Needwood* in Mundy's handwriting are preserved in the Mundy Papers, Derby Local Studies Library, along with correspondence with Seward, Boothby and other literary individuals, including the Ladies of Llangollen concerning the poems; A. Seward, 'On Mundy's Needwood Forest', in W. Scott ed., *The Poetical Works of Anna Seward*, 3 vols (Edinburgh, 1810), vol. 3, 394–7; thanks to Teresa Barnard for informing me of Darwin's and Seward's contributions to Mundy's *Needwood Forest* and for showing me with a copy of her article on Georgian literary female authorship in which this case is discussed.

64 Shaw, *Staffordshire*, vol. 1, 67, 69, 70; W. Pitt, *A Topographical History of Staffordshire*, 2nd edn (Newcastle-under-Lyme, 1817), 166; Green, *Memoir of Amos Green*, 190–91, 225.

65 Mundy, *Needwood Forest*, 45–6; Darwin, *Phytologia*, 480–1.

66 Mundy, *Needwood Forest*, 45–52.

67 Elliott, *Derby Philosophers*; J. C. Colquhoun, *William Wilberforce and his Friends* (London, 1867), 200–1.

68 Shaw, *Staffordshire*, vol. 1, p. 67; Colquhoun, *William Wilberforce*, 199–200.

69 Colquhoun, *William Wilberforce*, 199.

70 Colquhoun, *William Wilberforce*, 206, 211; 46, B. Nicholson, *Joseph Wright of Derby, Painter of Light*, 2 vols (New York, 1967), vol. 1, p. 134; Daniels, *Joseph Wright*; J. Wallis ed., *Joseph Wright of Derby, 1734–1797* (Derby, 1997), 46, 103–8.

71 T. Gisborne, *Walks in a Forest*, 4th edn (London, 1799); T. Gisborne, *The Principles of Moral Philosophy* (London, 1786); T. Gisborne, *The Duties of the Higher and Middle Classes*, 2 vols (London, 1794); T. Gisborne, *The Duties of the Female Sex* (London, 1797); T. Gisborne, *Sermons*, 3 vols (London, 1813); T. Gisborne, *The Testimony of Natural Theology to Christianity* (London, 1818); T. Gisborne, *Considerations on the Modern Theory of Geology* (London, 1837); Nicholson, *Joseph Wright*, vol. 1, 133–7; M. Millhauser, 'The scriptural geologists: an episode in the history of opinion', *Osiris*, 11 (1954), 65–86; T. Mortenson, 'British scriptural geologists in the first half of the nineteenth century: part 6: Thomas Gisborne (1758–1846)', *Technical Journal*, 14 (2000), 75–80; W. Page and N. J. Tringham, *A History of the County of Stafford, Vol. 10: Tutbury and Needwood Forest* (London, 2007).

72 J. Gisborne, *The Vales of Wever: A Loco-descriptive Poem* (London, 1797), 42; Nixon ed., *Brief Memoir of the Life of John Gisborne*, 13–14, 18.

73 Green, *Memoir of Amos Green*, 186–8, 190, 225.

CHAPTER 9

TREES IN THE ECONOMY
OF NATURE

Parallels between trees, humans and other animals reinforced Darwin's progressive Enlightenment belief that improvements in science and medicine would ensure much longer and more comfortable lives virtually free from disease. As a doctor fascinated by similarities between bestial and vegetable bodies, Darwin took a strong interest in tree lifecycles as the most anthropomorphic of plants, inspired by the potential for new remedies and opportunities to apply medical approaches to their study. At the same time, as his pleasure in 'unchastised nature' demonstrates, he was also encouraged by – and helped to foster – the new romantic aesthetic of tree representations in art which celebrated individuality and even character and quirkiness, rather than seeing trees merely as undifferentiated ranks in plantations.

Framing other features such as rivers and bold rocky outcrops, trees were often represented within depictions of sublime scenes on increasingly popular tourist itineraries such as the Derwent valley gorge at Matlock, Derbyshire, or the Manifold valley, in the Peak (District). On touring Derbyshire in 1772, William Gilpin (1724–1804) was impressed by the beauty of the dales, particularly the Dove and Derwent valleys, and he described the former as 'a most romantic and delightful scene, in which the ideas of sublimity and beauty are blended in a high degree'. The striking qualities of Matlock High Tor were enhanced by its silvan decoration and the rich, varied colours and surfaces of the rock. Encouraged by authors such as Gilpin and Uvedale Price, the Picturesque movement celebrated striking and unusual features, including trees of venerable antiquity with gnarled and twisted trunks, which increasingly became objects of fascination in their own right. Artists such as Paul Sandby (1731–1809) and Joseph Wright (1734–1797) made strenuous efforts to move away from stylistic representation and to draw and paint trees in a more naturalistic manner, which shaped Darwin's depictions of trees in his poetry and natural philosophy. While they removed or moved some trees, landscape gardeners such as Lancelot Brown, William Emes and Humphry Repton brought veteran specimens and old coppice and plantations into their designs and appreciated the differences between tree kinds, positioning outliers and examples seen as special or 'exotic' at key points in parks and garden.[1]

After examining Darwin's analyses of tree anatomy and physiology, this chapter explores his use of tree products as treatments and how medicine could help support the preservation of trees because of their role in combatting disease. It argues that these studies of trees, combined with Darwin's poetical portrayals, shaped his belief in their centrality in the economy of nature as living worlds and rich ecosystems in microcosm.

TREE PHYSIOLOGY AND ANATOMY

As demonstrated in the previous chapter, the pivotal status of trees in demonstrating Darwin's animal–vegetable analogies was encouraged by his medical practice, psychophysiology and arboricultural poetry. With their longevity, size and anthropomorphic qualities, they more readily illustrated the closeness of animal and vegetable bodies than did smaller, relatively ephemeral plants. Darwin believed that the 'pith of a young bud' so resembled the animal 'brain and spinal marrow' in respect to its central situation that it 'probably gave out nerves' to each living bud fibre, thereby furnishing 'the power of motion, as well as of sensation' to each part of the 'vegetable system'. 'New tree buds' were 'each' individual entities 'generated by the caudex of the leaf', and therefore must 'possess a sensorium' of their own. Inspired by the experiments in Stephen Hales's *Vegetable Statics* (1727), Darwin closely investigated the seasonal motions of sap, regarding it as akin to blood in animals and referring to cuticles on the bark, which was 'furnished with lymphatics to absorb moisture from the air', as resembling 'the external skin of animals' and the 'lympathics' that opened upon it, the bark of the root with its 'lymphatics'resembling the stomach's 'mucous membrane' and 'lacteal' vessels.[2] The proximity of bestial and plant worlds and the nutritional benefits afforded by tree products were evident in other ways, including the 'nutritious mucilage' provided by holly trees to deer in Needwood Forest during the winter months if branches had been removed, while the bark also contained a 'resinous material' that could be obtained by boiling and washing away the other parts. This resembled 'elastic resin' from South America and the 'fossil elastic bitumen found near Matlock in Derbyshire, both in its elasticity and inflammability', suggesting that holly might be more systematically cultivated for this and other products.[3]

Like the operation of vital ethereal fluids such as the spirits of vegetation and animation, for Darwin, the operation of tree bodies again demonstrated how much animal and vegetable physiology had in common. These similarities included the fact that buds were the 'viviparous offspring of vegetables' requiring 'placental vessels for their nourishment' until they formed 'lungs, or leaves', so they could transform the 'common juices of the earth into nutriment'. These vessels existed 'in bulbs and in seeds', supplying young plants with 'a sweet juice' until they started growing leaves, as the horse-chestnut (*Aesculus hippocastum*) demonstrated, with its 'large buds' and flowers visible to 'the naked eye' before they fell in September. Informed by the experiments of Marcello Malpighi (1628–1694), Nehemiah Grew (1641–1712) and Hales on the ascent of 'sap-juice', Darwin conducted his own trials at Radburn from October 1781 on his favourite Mimosa plants to demonstrate the 'vascular connection' of vegetable buds with the leaves 'in whose bosoms they are formed' (Figure 69).[4] On the end of a 'young bud' of a Mimosa he placed a small drop of acid of vitriol (sulphuric acid) with a pen and 'after a few seconds', the leaf 'in whose axilla [the

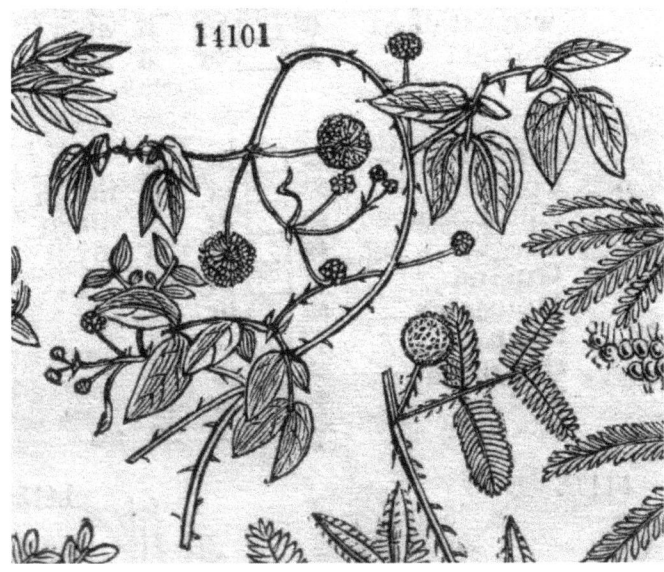

Figure 69.   Mimosa, sensitive plant (14101), from J. C. Loudon, *Encyclopaedia of Plants*, new edn, edited by J. Loudon (1855), 855.

junction of the leaf and stem] it dwelt, closed and opened no more', though the droplet was so small that it just 'injured' the top of the bud. This appeared to demonstrate that leaves and buds had 'connecting vessels', though they had 'arisen at different times and from different parts of the medulla, or pith' and, as the leaf was there first, it must be the 'parent' of the bud.

The similarities that Darwin believed existed between the vegetable frame and the 'animal economy' in terms of physiology, arterial systems, muscles and organs of propagation, reproduction and secretion encouraged him to undertake further investigations of different plant parts to demonstrate the existence of absorbent vessels in all areas besides roots. Continuing his botanical experiments at Radburn Hall, in the summer of 1781 Darwin placed 'the foot-stalks' of large fig leaves an inch down in a 'decoction of madder (*rubia tinctorum*)' and others in a 'decoction of logwood (*Haematoxylum campechense*)' with 'sprigs' cut from an oxtongue (*Picris*) plant. The latter was selected because its 'blood' was white. '[A]fter some hours', on the ensuing day, on taking each of the fig leaf stalks and oxtongue sprigs and cutting off a quarter of an inch of the stalk from the bottom, an 'internal circle of red points' appeared, visible with the magnifying glass, which were the 'ends of absorbent vessels coloured red with the decoction'; likewise, a 'milky juice' quickly bled from an 'external ring' of what Darwin called 'arteries', demonstrating the operation of both 'absorbent' and 'arterial' systems in the alburnum, or sap wood, the latter to convey the nutritious vegetable 'fluid' or 'blood' to various 'glands' for 'growth, nutrition' and the purposes of secretion. In contrast to Grew and Malpighi, who he believed incorrectly interpreted the 'absorbent vessels' as 'bronchi' or air vessels (akin to the airways that lead from the trachea into the lungs in humans, and branch into successively smaller structures, or alveoli, enabling the exchange of oxygen and carbon dioxide), Darwin considered that they operated to imbibe fluids and not gases, with the bubbles emerging upon them being cut providing misleading evidence. When removed from leaves, the vessels proved to be 'spirallines' and not interrupted with valves, as were animal 'lymphatics' (Figure 70). The 'retrograde motion' of the 'absorbent vessels' in plants was demonstrated by removing a 'forked branch' from a tree and immersing one side in water, which would 'prevent the other from withering', or by 'planting a willow branch with the wrong end upward'.[5] The fluids from the soil, 'atmosphere' or plant 'cells and interfaces' that were imbibed by the 'vegetable absorbent vessels' were conveyed to the main stalk of each leaf, at which the 'absorbents' combined into 'branches' and formed numerous 'pulmonary arteries' before being 'dispersed' to the extremities – which was demonstrated by cutting slices from the supporting stalk of a horse chestnut leaf in early autumn before the leaves were shed. Such observations revealed that there was a 'complete circulation' within leaves, with 'pulmonary' veins receiving blood from each 'artery' extremity on the upper sides, which rejoined again in the main stalk, producing numerous 'arteries, or aortas' that dispersed the 'blood' over fresh bark, 'elongating its vessels' and 'producing its secretions'.[6]

Turning to the involvement of leaves in 'vegetable respiration', Darwin noted that Hales had removed leaves from apple tree branches and discovered that the fruit exhaled as much as the leaves, making him doubt that their office was primarily as 'perspiratory organs'. Nor did he consider them 'excretory organs'. Rather, the 'analogy' between leaves, lungs or gills seemed to 'embrace so many circumstances' that he was convinced they undertook similar functions. The 'great surface' area of many tree leaves compared with the trunk and branches suggested that they were an organ well adapted to expose 'vegetable juices' to the air, even if, as he believed, it was only the upper surface of the leaves that performed this operation.[7] To demonstrate the analogy with animal circulation and respiration, in June 1781 at Radburn he placed a stalk with leaves and a seed-head of the sun spurge (*Euphorbia helioscopia*) for 'several days' in a decoction of madder (*Rubia tinctorum*) so the lower stem and two 'undermost leaves' were immersed within. After washing these in clear water, the coloured madder passed along the leaf's 'middle rib' and was 'beautifully visible' on both upper and lower surfaces, with numerous red branches passing from it to the ends of the leaves on the upper side which were invisible on the other side unless light was passed through. On the underside, a system of branching vessels carrying a pale milky fluid was seen coming from the leaf ends and covering the whole underside before converging into two large veins, one on each side of the red artery in the middle rib of the leaf, and along with it passing to the 'foot-stalk or petiole'. Examining cut leaf sections with a magnifying glass showed the 'milky blood' oozing out of the returning veins on each side of the red artery in the middle rib, but 'none of the red fluid from the artery' emerged. An oxtongue leaf subjected to similar analysis demonstrated the effect more clearly, and Darwin concluded from this and other experiments, such as covering upper leaf surfaces with oil, that these leaf surfaces were the 'immediate organ of respiration', a

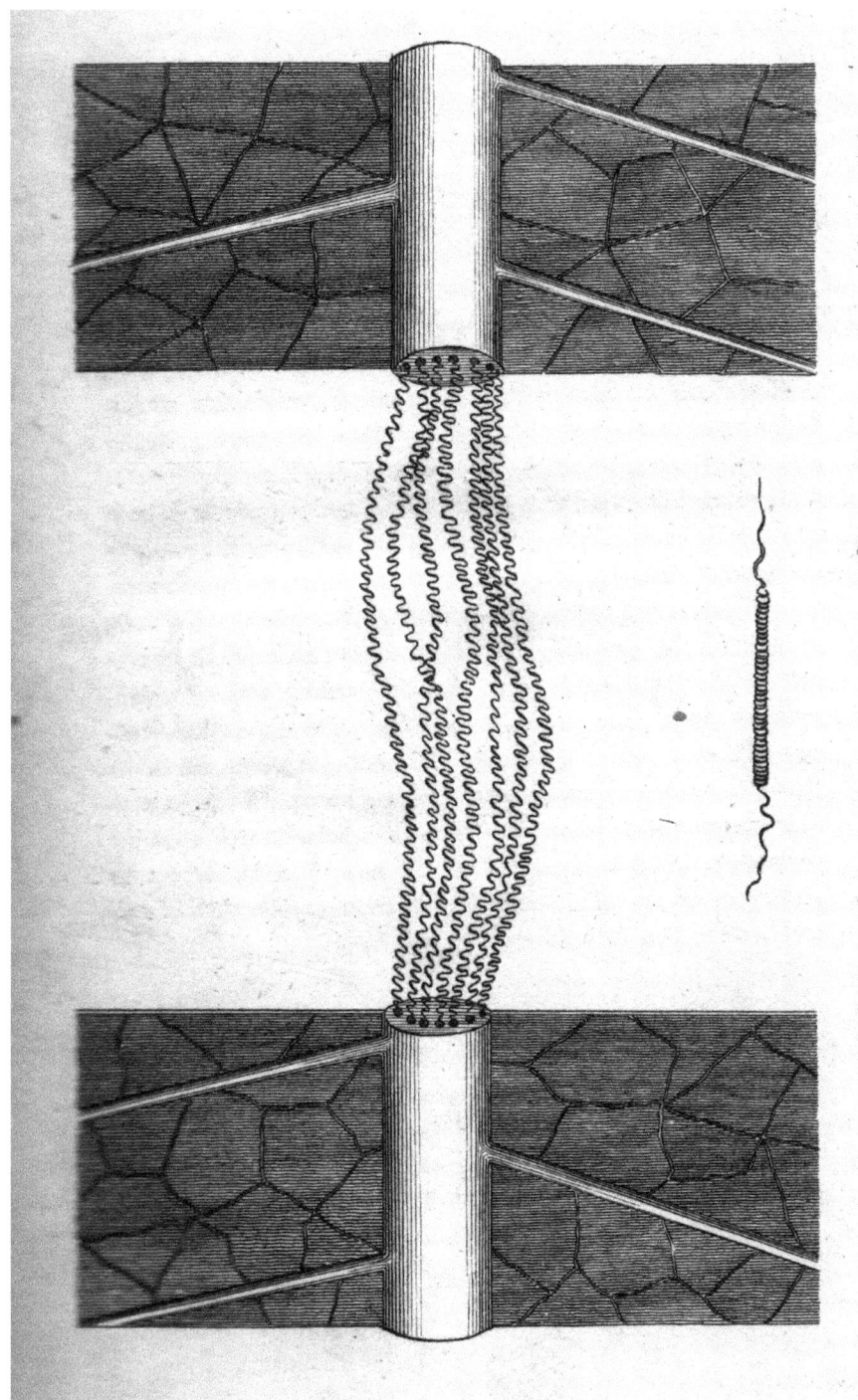

Figure 70. Spiral vessels of vine leaf, from E. Darwin, *Phytologia* (1800), after N. Grew, *The Anatomy of Vegetables Begun* (1672).

kind of vegetable lung that passed out 'phlogistic material to the atmosphere' and absorbed 'oxygene or vital air'. Careful observations of different flowers, including the Christmas rose (*Helleborus niger*), suggested that a similar 'pulmonary system' surrounded fructification involving the corolla (petals) through which the 'vegetable blood' passed to be exposed to 'light and air'. Furthermore, utilising the results of experiments by Carl Willhelm Scheele (1742–1786), Jan Ingenhousz (1730–1799) and his Lunar friend Joseph Priestley, Darwin concluded that the 'great use of light to vegetation' was that by 'disengaging vital air' from perspiration it facilitated a 'union with their blood, exposed beneath the thin surface of their leaves', as 'pure air' was probably 'more readily absorbed'.[8]

MEDICAL USES OF TREE PRODUCTS

As emphasised in chapter 2, organic products were crucial to eighteenth-century medicine, and various tree parts were commonly employed for treatments. With the global circulation of so many trees and tree materials for gardening, food, industry and other purposes, so the range available for medicine increased markedly, but, as Steven King has argued and analysis of recipes in commonplace books demonstrate, the medicine of the 'long' eighteenth century continued to rely strongly upon domestic herbal remedies, although with greater use of imported vegetable materials.[9] Of these, the uses of tree-derived substances specifically has hardly been investigated apart from King's work, despite the fact that the healing power of woodland plants and the employment of materials such as elder bark was ingrained in folklore as well as domestic medical practice. Medical authorities such as Darwin believed that it was possible to take much greater advantage of the special characteristics of tree components such as bark, roots, leaves, heartwood or sapwood, oils, resins and other exudations, seeds and fruits. This was also reflected in the interests of the Derby Philosophical Society's medical members, who at Darwin's instigation acquired studies of vegetable medicines for their library, including works focusing specifically upon the virtues of tree substances. Stuart Harris has argued that the 'economy of vegetation' in Darwin's works represents 'a global economy ... whose geographical extremes might be represented by two "boundary" trees, *Camellia sinensis* and *Cinchona*: extending thus from China to Peru', and, certainly, the 'qualities of plants' and their special value in improving human health were recurring themes in Darwin's poetry.[10] However, what was crucial for Darwin in understanding the current and potential medical uses of tree substances was what lay behind the particular virtues each tree had. Underpinning this, in his view, were the characteristics that trees evolved in order to survive in particular conditions and compete with other animate beings. Medical practitioners and the sick could take advantage of these characteristics to discover new treatments.

Observations concerning the environment and localities in which trees grew shed light on which illnesses each product might be most useful for. Darwin argued that bark, for instance, was made of 'congeries' of the lengthy 'caudexes' of each bud, which were composed of 'absorbent vessels' that imbibed 'nutriment from the earth' and 'arteries and veins' that supplied this nutriment to the 'growing vegetable'. The vegetable juices that trees secreted and the 'acrid, stringent, or narcotic' liquids they utilised to 'defend' themselves from insect 'depredation' were frequently of use for humans, as were the 'various mucilaginous, oily, or saccharine' materials that 'glands' in the bark and buds produced to nourish 'embryon buds'. The fact that there was the 'strongest analogy' between barks and roots in most respects except for where cuticles were 'adapted' for 'moist earth' and a 'dry atmosphere', respectively, was important from a medical perspective because it meant that 'active' vegetable juices with distinctive properties were present in both – which was why barks and roots were usually the most efficacious elements.[11]

As detailed previously, Darwin believed that the adaptation of particular plants to their environment gave them characteristics or encouraged the production of substances that were especially medically beneficial. This also meant that the character and quality of soils or environment might be determined by the 'growth and colour' of vegetables that covered it, which was useful for prospective buyers of land, who could make a quick judgement about its value, although it required 'an experienced eye' to fully appreciate, being not 'easily described in words'. Where plants grew 'wild on soils' this therefore 'in some measure' revealed how likely it was that they would be able to nurture healthy and vigorous growth. For instance, the foxglove (*Digitalis*) and the 'arenarea' – probably sandwort (*Arenaria montana*) – generally thrived on 'sandy soils', while brooklime (*Veronica beccabunga*) and some kinds of cress 'belonged' to watery places and other plants thrived in mountainous conditions. Darwin called for a more comprehensive 'catalogue' of vegetables and their 'natural conditions' based upon where they grew 'spontaneously', which would help determine the most fertile kinds of soil and conditions for optimum growth, just as flowering times revealed seasonal temperatures 'in each climate'.[12]

It was probable that the 'interior barks' of many trees and shrubs, and especially those protected by 'thorns or prickles' intended to dissuade animals, such as hawthorn (*Crataegus*), gooseberry (*Ribes uva-crispa*), gorse (*Ulex*), elm (*Ulmus*) and holly (*Ilex*) – contained 'much mucilaginous or nutritious matter'.[13] Likewise, the many barks that contained 'bitter, resinous, aromatic or acrid materials', which supplied apothecaries' shops, including Peruvian bark, the West Indian and South American shrub cascarilla or willow-leaved croton (*Croton eluteria*) and cinnamon (*Cinnamomum*), were 'designed by nature' to shield vegetables from animal or insect assaults. This was why some woods, when fashioned into furniture, such as cedar, cyprus and mahogany, were resistant to worm attack; though expensive, these were increasingly used instead of walnut during the Georgian period partly because there were shortages of walnut, but also because as much cedar, cyprus and mahogany came from the West Indies, it was regarded as more robust because of the climate where it had been grown. Likewise, animals avoided eating many trees and plants because the juices they produced were 'poisonous' or disgusting to them, with bitter 'flavours to the nose or palate'. Similarly, humans were also unable to eat these plants during periods of food shortages or when in extreme circumstances, such as those 'unfortunate shipwrecked travellers' forced to traverse 'hundreds of miles' through 'uninhabited' tracts without finding enough edible plants to sate their hunger.[14]

The same adaptations for vegetable defence and survival were also useful for dyeing and tanning, in which the 'pores of animal skin' were 'impregnated' by 'vegetable particles'. For example, solutions from the bark, leaves or galls of oak, ash and alder were useful for tanning animal hides, providing durability and defence against 'putrefaction'. Other plants with particular characteristics formed by their environment produced potentially useful products for domestic usage, industry and culinary purposes, including the pasture herb and Board of Agriculture favourite sainfoin (*Onobrychis viciifolia*), French honeysuckle (*Hedysarum*), the 'broad thick leaves' of 'American nightshade' (*Phytolacca*) or even some mosses or lichens; their remnants or by-products then perhaps provided heat as fuel for fires or fermentation, as some gardeners used autumn leaf fall in hothouses. The strength and tenacity of bark fibres were similarly exploited for 'apparel, paper, cordage' and numerous other 'mechanical purposes'; hemp, linen, cannabis, flax, linum and papyrus bark or leaves were 'the first used for paper', while mulberry tree bark was 'still' employed making cloth in Tahiti and on other Pacific Islands.[15]

As emphasised in chapter 2, for Darwin the 'essential oils' that constituted the 'odorous exhalations' of 'aromatic or balsamic' bushes and plants, such as gale or sweet-gale (*Myrica*), the bitter resin from the balsam poplar (*Populous balsamifera*) and the balm tree or balsam of gilead (*Commiphora gileadensis*), had, once again, largely originated from their need to defend themselves against insect and animal predators.[16] Other derivatives of trees and shrubs with medicinal properties that made them ideal for the chemist's cabinet, according to Darwin, included guaiacum gum – oil or resin from the tree *Guaiacum officinale* (Figure 71) – bitter ash or bitter wood (*Quassia amara*), logwood or 'Campech[e] wood' and sassafras (*Sassafras officinale*) – a small tree or bush of North and Central America (Figure 72), and essential oils or balsams such as '*Oleum rhodii*' or 'Oil of Rhodium' ('Rose-root', *Rhodiola rosea*), which was produced by distillation from the wood or root.[17] Such 'bitter' or 'aromatic' flavours that might be 'agreeable' to the human tongue equally owed their origin to the need for defence or preservation, meaning that these plants were taken into 'extensive use' for domestic or healing purposes. Examples included tea (*Camellia sinensis*), ash (*Fraxinus excelsior*), sage (*Salvia*), mint (*Mentha*) and lemon balm (*Melissa officinalis*), which were all beneficial 'in health and sickness', the 'very pleasant aromatic flavour' of sage being 'esteemed salubrious' 'from high antiquity to present times'. These were especially 'nutritive' when combined with cream and sugar, 'certainly' contributing to the health of the population by decreasing reliance upon 'fermented or spirituous liquors' and promoting 'morality' and sociability by helping to bring together both sexes in pleasant sociability.[18]

It was not only the environment in which trees and shrubs grew that ensured the medical efficacy of some of their derivatives, but the fact that, as we have emphasised, for Darwin they had such fundamental structural and vital similarities with animals. Given that each tree or plant was an 'inferior or less perfect' animal in many respects, the many resemblances between both helped make parts of the former an effective treatment because each had 'fibres' that were excitable into different forms of motion by 'irritations of external objects'. The resinous and odourus 'secretions' of trees and plants, such as those from fruit and types of gum, resin, wax and honey, so many of

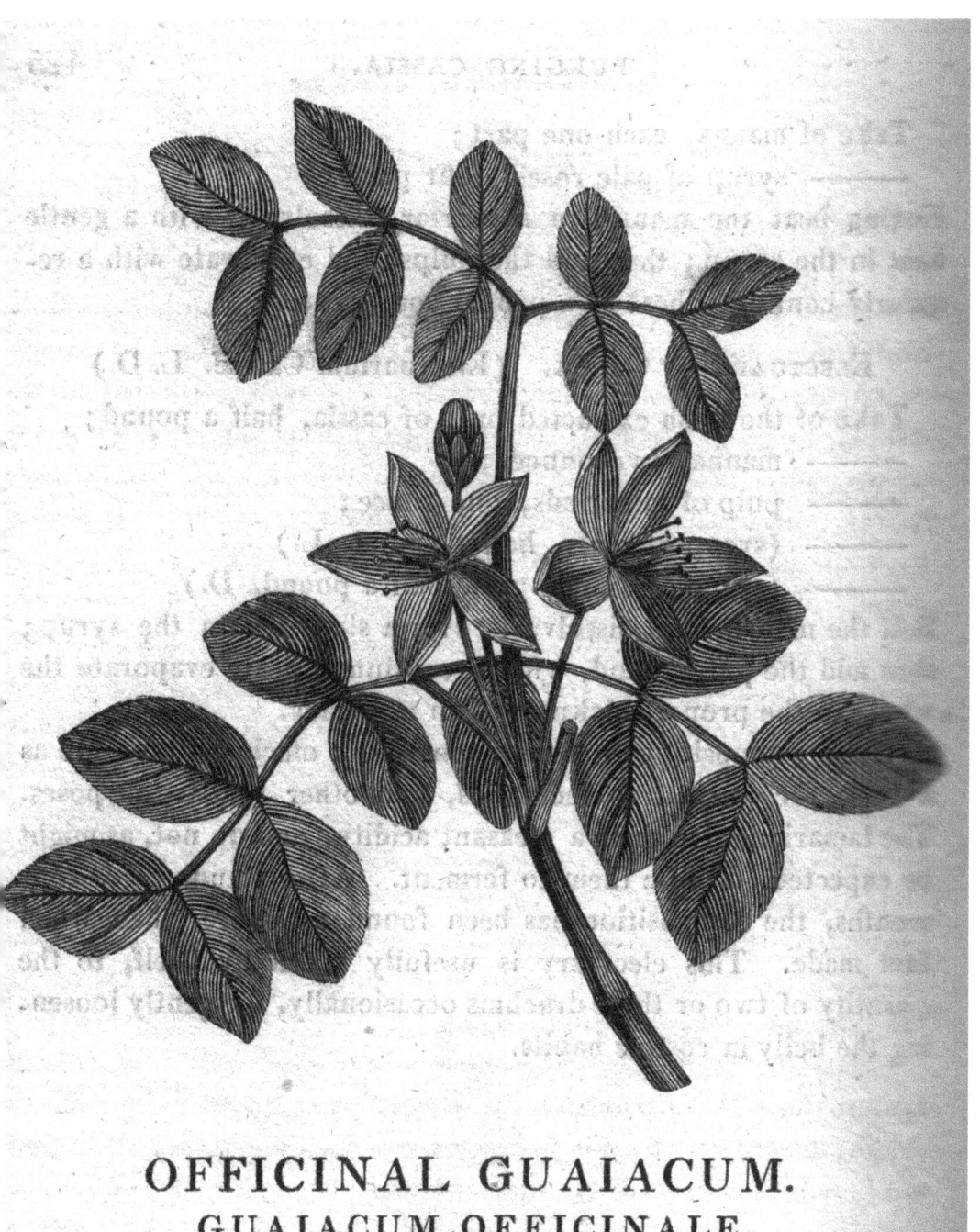

OFFICINAL GUAIACUM.
GUAIACUM OFFICINALE.

71. *Guaiacum officinale* tree, from R. J. Thornton, *A Family Herbal*, 2nd edn (1814), 426.

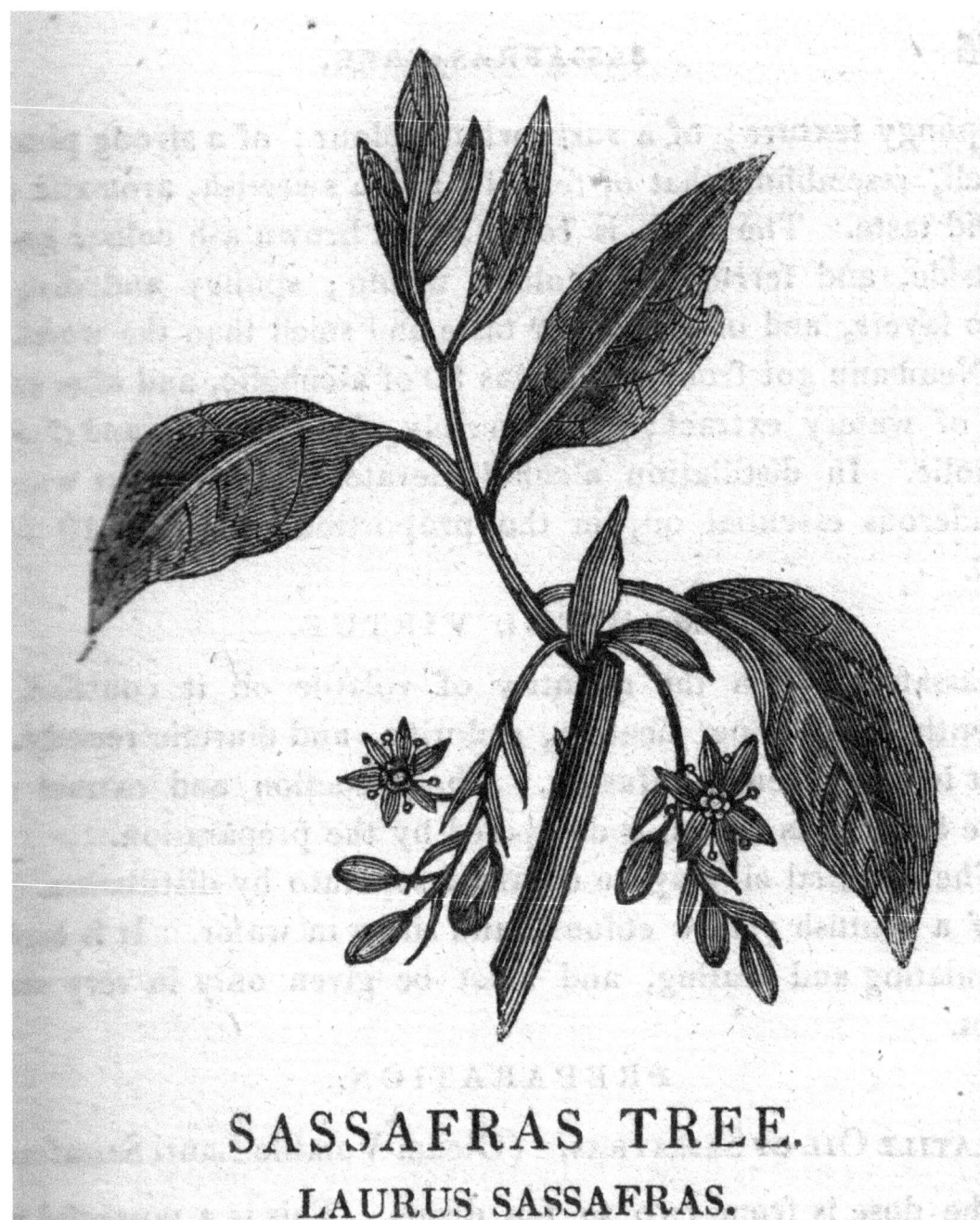

SASSAFRAS TREE.

LAURUS SASSAFRAS.

Figure 72. Sassafras, *Sassafras officinale*, from R. J. Thornton, *A Family Herbal*, 2nd edn (1814), 375.

which were employed in medical practice, appeared to be formed like those substances created by animal glands, which helped to explain, for example, why Peruvian bark provided such a 'powerful stimulus' to the 'spirit of animation' or 'sensorial power' of animals through their fibres or muscles.[19] Among 'reverentia', or substances that restored the 'natural order of the inverted irritative motions' for conditions such as hysteria and vomiting, Darwin employed castor oil, essential oils of cinnamon, nutmeg, charcoal, oil of amber, logwood, resin and turpentine, for example.[20]

For 'sorbentia', which were conditions that increased 'irritative motions', such as ulcers and venereal diseases, many of the recommended substances were also derived from trees. These included such staples of the Georgian household and medical practice as oil of vitriol, lemon, *Aloe vera*, blackthorn (*Prunus spinosa*), crab apples, pear (*Pyrus*), quince (*Cydonia oblonga*), Peruvian bark, the leaves of the wormwood (*Artemisia maritima*), orange peel, cinnamon, nutmeg, mace, tobacco, oak bark, the rich resins of oak galls (*Gallae quercinae*) and logwood. Another example was the extract known as 'Japan earth' or *Terra japonica* (*Catechu mimosa* [Thornton], *Acacia catechu* [Lindley]), which was a product of acacia trees used in India and other Asian countries, which, according to Robert Thornton, was prepared from the central coloured section of the wood by a process of 'decoction, evaporation and exsiccation' in sunshine. Logwood was a species of flowering tree native to Central America and Mexico which had been imported since the seventeenth century, while cinnamon was obtained from the inner bark of the cinnamon tree, native to south-east Asia, and turpentine was produced by distilling resin, especially from coniferous trees (Figure 73). While these tree derivatives were taken internally, others were applied externally over wounds as plasters or dressings, and Peruvian bark could be administered on linen mixed with white lead and glue (often itself derived from tree materials) to form a plaster for 'scrophulous ulcers'.[21]

One example of the encouragement that Darwin's approach to the investigation of vegetable virtues provided to others is the work of Hull physician John Alderson (1757–1829) on the poison sumach (*Toxicodenron vernix*; Darwin and Alderson call it *Rhus toxicodendron* after Joseph Pitton de Tournefort, but this name was later applied to poison ivy, and they mean the poison sumach or 'oak'), a 'deciduous shrub of moderate worth' originally brought to Britain during the seventeenth century from North America, where it was used by Californian Amerindians for medicinal purposes. The sumach was common in North American swamps and meadows and had a trunk from ten to thirty feet high with pale greyish bark. According to John Lindley the 'juice' or even 'exhalation' in the air surrounding the tree 'impregnated with the volatile principle' was 'to many persons a serious poison' causing 'dangerous swellings'. Alderson, who was physician to the Hull Infirmary and later founder of the Hull Botanic Garden and first president of the Hull Literary and Philosophical Society, published a pamphlet on the sumach (1794) that sought to demonstrate its 'efficacy in the cure of paralysis', providing an account of how it was administered. He argued that a serious 'nervous debility' was increasingly pervading all ranks of society, criticised the use of wine, spirits and laudanum and argued that, though not hitherto not in the materia medica, the sumach could help 'enervated' constitutions. Darwin's method of talking to plant

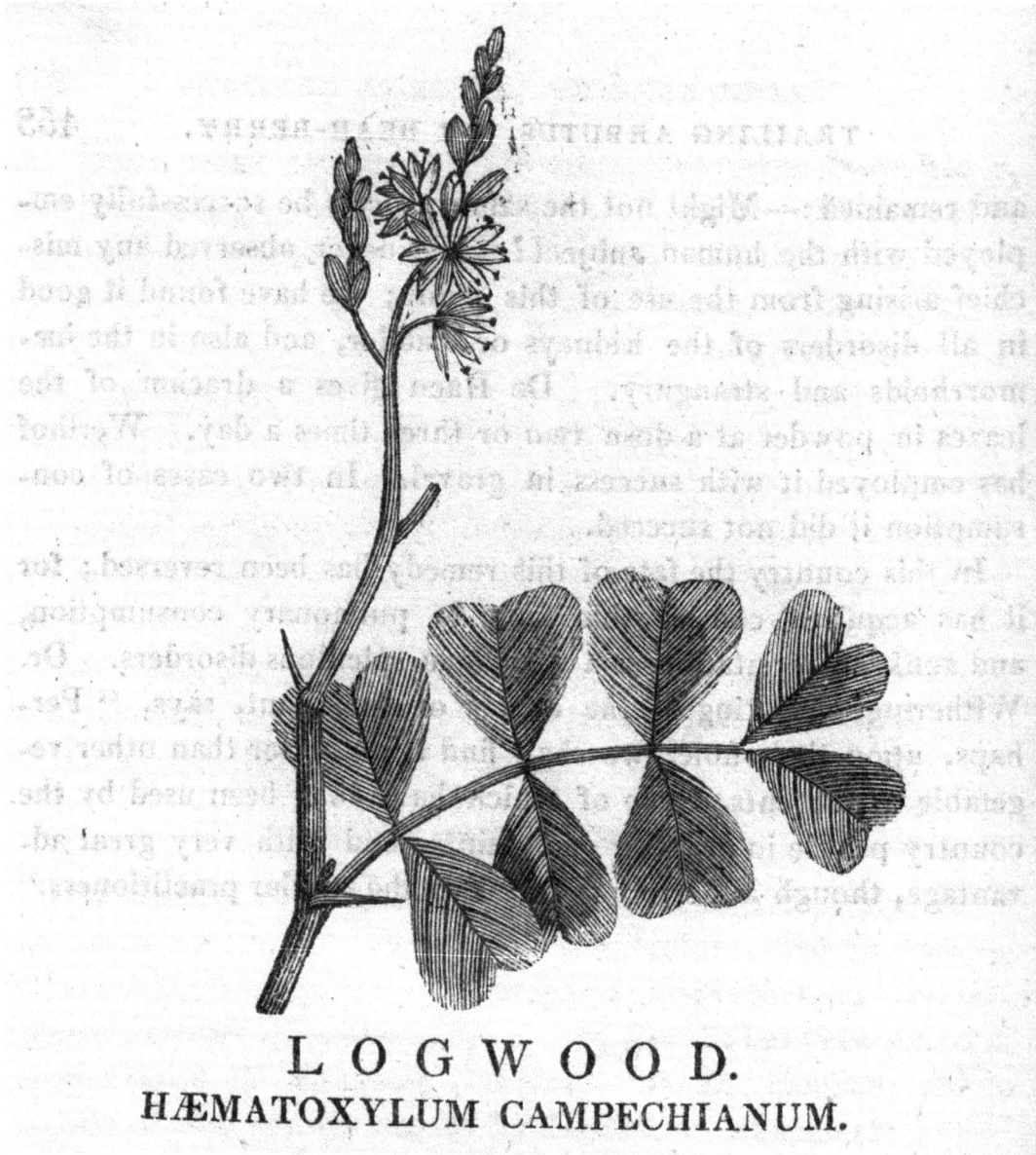

L O G W O O D.
HÆMATOXYLUM CAMPECHIANUM.

Figure 73.   Logwood or 'Campech[e] wood', from R. J. Thornton, *A Family Herbal*, 2nd edn (1814), 454.

enthusiasts, gardeners, farmers, nurserymen and others in the search for potentially medically efficaceous vegetables rather than relying upon well-used herbals is evident in the approach taken by Alderson, who first encountered the plant at the nurseries of Messrs John Philipson and Robert Scales at Cottingham, Yorkshire, assisted by the naturalist Adrian Hardy Haworth (1768–1833) and later James Edward Smith, who compared specimens from the herbarium with his own and provided botanical advice, having been told about the 'wonderful effects of the poison oak'. Alderson initially administered the plant as an infusion distilled in water, which had little effect, but then found that grains of powdered leaves infused in boiling water and given as a bolus successfully treated different forms of muscular paralysis, including those of the limbs, hands and facial muscles.

Alderson's account of his 'trials' to determine the best strength and means of administration of sumach for 'nervous disorders and paralytic affections' followed the methods pursued by William Withering and Darwin in their investigations of foxglove summarised in chapter 2, beginning with Linnaean botanical information concerning plant characteristics, then giving some notice of the typical locations in which it grew before providing a series of case studies of its application and supporting testimonials from favourable fellow medical practitioners. It is striking that Alderson repeatedly cited *Zoonomia* and the Lichfield botanical society's Linnaean translations in support of his arguments, and also that his book was sold by Darwin's publisher Joseph Johnson. Alderson's study of the poison sumach was clearly of great interest to Darwin, who immediately ordered it for the Derby Philosophical Society library (and may have recommended it to the publisher), citing it in the third edition of *Zoonomia* as a treatment for palsy or hemiplegia while cautiously emphasising that it was difficult to know which medicines worked best because muscular movements had to be relearned 'as in infancy, by frequent efforts'.[22]

Of the many tree products utilised medically by Darwin and Georgian practitioners more generally the most common was Peruvian bark (*Cinchona officinalis*) which he usually called 'the bark' because of its ubiquity (Figure 74), although in reality various kinds of 'bark' were in use because of difficulties of supplies from South America and the negotiations involved, which encouraged British efforts to seek out alternatives in their colonies. Probably for these reasons, Darwin ordered the botanist Aylmer Bourke Lambert's *Description of the Genus Cinchona* (1797) for the Philosophical Society library, and Peruvian bark was celebrated in *The Loves of the Plants*:

> Where Andes hides his cloud-wreathed crest in snow,
> And roots his base on burning sands below,
> Cinchona, fairest of Peruvian maids,
> To health's bright Goddess in the breezy glades,
> On Quito's temperate plain an altar rear'd,
> Trill'd the loud hymn, the solemn prayer preferr'd.

In a note, Darwin claimed that discovery of the medicinal powers of *Cinchona* had occurred after some of the trees were felled and fell into a lake at Loxa in Peru (the city of Loja, which, along with Quito, is now in Ecuador) during a serious 'epidemic fever'; the woodsmen, having drunk of the impregnated waters, were found to be cured, which revealed the 'virtues of this famous drug', sacred to Hygeia.[23] In his analysis of *Cinchona*, which was dedicated to the Linnaean Society and Banks as Royal Society president and prefaced by a translation of a dissertation on the subject by the Danish–Norwegian

COMMON PERUVIAN BARK TREE.
CINCHONA OFFICINALIS.

Figure 74.   Peruvian bark tree, from R. J. Thornton, *A Family Herbal*, 2nd edn (1814), 114.

botanist Martin Henrichsen Vahl (1749–1804), Lambert (1761–1842) was concerned to investigate which species the Peruvian and other barks of 'similar quality' were taken from, as the ubiquity of its medical use had created some confusion about this, which Darwin and his medical Derby Society colleagues probably hoped that the book would help resolve. According to Vahl, there was still great ignorance about the origins of several medicines in 'every day use' in the practice of physic, including both plants from 'distant regions' and British 'natives', and about their bodily impacts, and there were many 'frauds' perpetrated by those who mixed Peruvian bark with 'ingredients of similar colour and taste', but there was reason to hope that similar plants would have similar 'virtues'. A vice-president of the Linnaean Society with estates in Jamaica and Ireland whose most influential work was the lavishly produced *Genus Pinus* (1803–24), Lambert included engraved figures by Austrian artist Ferdinand Bauer (1760–1826) of the different species of *Cinchona*, such as *C. pubescens*, *C. macrocarpa* and *C. angustifolia*, and views of flowers, seeds, fruits, leaves and branches with botanical descriptions of each supported by correspondence from botanists, medical practitioners and travellers.[24]

Darwin emphasised that repeated doses of *Cinchona* as a stimulus would restore the 'natural quantity of sensorial power' by irritation, aided by the 'sensorial power of association' becoming an 'acquired habit'. So, when administered at 'intervals for the cure of intermittent fevers' and sixty grains were provided every three hours for one day 'preceding the expected paroxysm' (a stage or stages of the disease associated with particular symptoms), this would 'stimulate the defective part of the system into action' and 'prevent the torpor or quiescence of the fibres', which made up the 'cold fit', revitalising the exhausted 'sensorial power' or 'spirit of animation'. Repetition of this in smaller quantities with about sixty grains of Peruvian bark given morning and evening for a fortnight would prevent the condition through direct impact of the 'strong stimulus' and sensorial association combining with that of irritation because the 'acquired habit' formed by repetition 'assists the power of the stimulus'.[25] Darwin's earliest recorded patient, the Nottingham shoemaker stabbed with a conical knife during a drunken argument with a fellow shoemaker in 1756, was treated with the bark. The man had been bled twice and a small dressing applied to the wound, which had caused a small hole about two inches deep. Darwin administered one dram of *Cinchona*, followed by further drams of the bark, one of which was mixed with sweet almond oil.[26] Unfortunately the patient died, which hastened Darwin's departure from Nottingham but did not reduce his faith in *Cinchona*, which remained a staple of his medicine for the next forty-five years. For instance, after Josiah Wedgwood's wife Sally (1734–1815) suffered a miscarriage in 1772, he prescribed it with steel filings (iron) with wine, sweet meats, spring water and other substances to make it more bearable. In 1794 Darwin prescribed '60 grains of powder or (what is better) of good extract of bark, twice a day for a fortnight, or 3 weeks, either as a bolus, or powder, or diffused in two ounces of water made a little warmish' to Tom Wedgwood (1771–1805) for general physical debility. He had sufficient faith in the benefits of *Cinchona* to recommend that children might be cured of ague by placing it 'quilted between two shirts, or strewn in their beds', and he tried to administer it to his son Edward (1782–1829) when he was ill with the condition in 1784, although the boy resisted and his mother Elizabeth had to remain with him.[27]

TREE DISEASES AND THEIR TREATMENTS

Insights into how trees might be protected from disease and the treatments that could be used to heal them were obtainable, according to Darwin, from applying knowledge gained working with humans and animals. As with certain field crops and plants, there were sometimes propensities towards particular illnesses in particular trees and, in line with Darwin's neo-Hippocratic medical system, restoring the equilibrium of their bodies was often more useful than worrying about problems caused by external putrefaction and miasmas. He was most interested in those tree ailments that could be observed and studied more easily, and in *Phytologia* devoted considerable space to analysing those diseases that he believed had 'internal' causes, and how these might be combated or cured. As Darwin's contemporary the York physician Alexander Hunter emphasised, despite the tendency of the Linnaean system to lump them together there were 'obvious and striking differences' between trees and other plants, especially in terms of 'size' and 'duration', which had been asserted by almost all naturalists and philosophers from antiquity to the eighteenth century, including Andrea Cesalpino (1519–1603), John Ray (1627–1705), John Evelyn and Joseph Pitton de Tournefort (1656–1708), although, like Darwin, Hunter emphasised how varied even the same species could be depending upon 'climate, soil and management'.[28]

As we saw in the previous chapter, with their fixity in the landscape and lengthy lifespans, trees offered more opportunity for a tactile sensory relationship with humans over time, and Darwin observed, touched and tasted their bodies and secretions at different hours of the day and in varied conditions, just as he utilised his own body for other scientific experiments (such as using his tongue to detect the presence of electric charge). To maximise bark growth, he advocated encouraging as many leaf-buds as possible at the expense of flower buds, which ought to be pinched off 'as soon as they appear' because the flower buds acquired 'no new caudexes' and died as soon as they had 'ripened their seed'. Leaf buds, on the other hand, were 'viviparous offspring' or numerous small plantlets attached to – and nurtured by – parent trees that acquired 'new caudexes extending down' earthwards, thus extending bark thickness, which would increase the volume of timber produced because it 'gradually changed into wood'. For Darwin the caudexes of the buds were not just the stem or trunk but specifically the 'connecting vessels' between the plumes (leaf buds) and radicles (roots) which formed a combined resilient 'intertexture' of bark fibres to support the tree. Ideally, bark needed to be preserved by 'rubbing off' 'parasitic vegetable[s]' such as moss with a 'hardish brush', which probably also facilitated increased 'motion to the vegetable circulation' and promoted 'the ascent' of juices, while washing the trunks of wall-grown trees, such as fruit trees, with a 'water-engine' also facilitated greater bud production and perhaps prevented the 'canker'. Equally, where necessary, trees ought to be protected from too much lichen, fungi, mistletoe, ivy, clematis, woodbine virgin's bower and other 'parasite vegetables' and 'climbers'.[29]

Some wounded trees were particularly vulnerable to insect assault, especially those whose wood contained 'less acrimony', which were sometimes 'bored into and eaten' by large worms or maggots 'as thick as a goose quill', which he had seen happen to a pear tree; its entire 'internal wood' was consumed, leaving nothing to hold it up, after which it was easily brought down by wind.[30] Some human efforts to protect trees or enable them to heal their own wounds quickly backfired because materials that could injure the trees when they were absorbed were employed. Darwin saw peach and nectarine tree branches in leaf destroyed when sprayed by arsenic solution and spirit of turpentine – as was commonly practised – to kill off insects, while 'several young elm trees' were inadvertently exterminated because their boles were covered by quick lime mixed with cow dung to deter horses.[31] He generally favoured using means of countering animal or insect problems other than indiscriminate applications of such chemicals, like encouraging natural predators to control insect populations.

Many older trees were assaulted by canker or what Darwin called *gangrena vegetabilis* – a 'phagedenic ulcer of the bark' – which spread around the trunk and branches and was particularly prone to attacking old apple and pear trees (Figure 75), and also honey-dew (*suffusio mellita*) – sweet clammy drops that glittered on tree foliage during hot

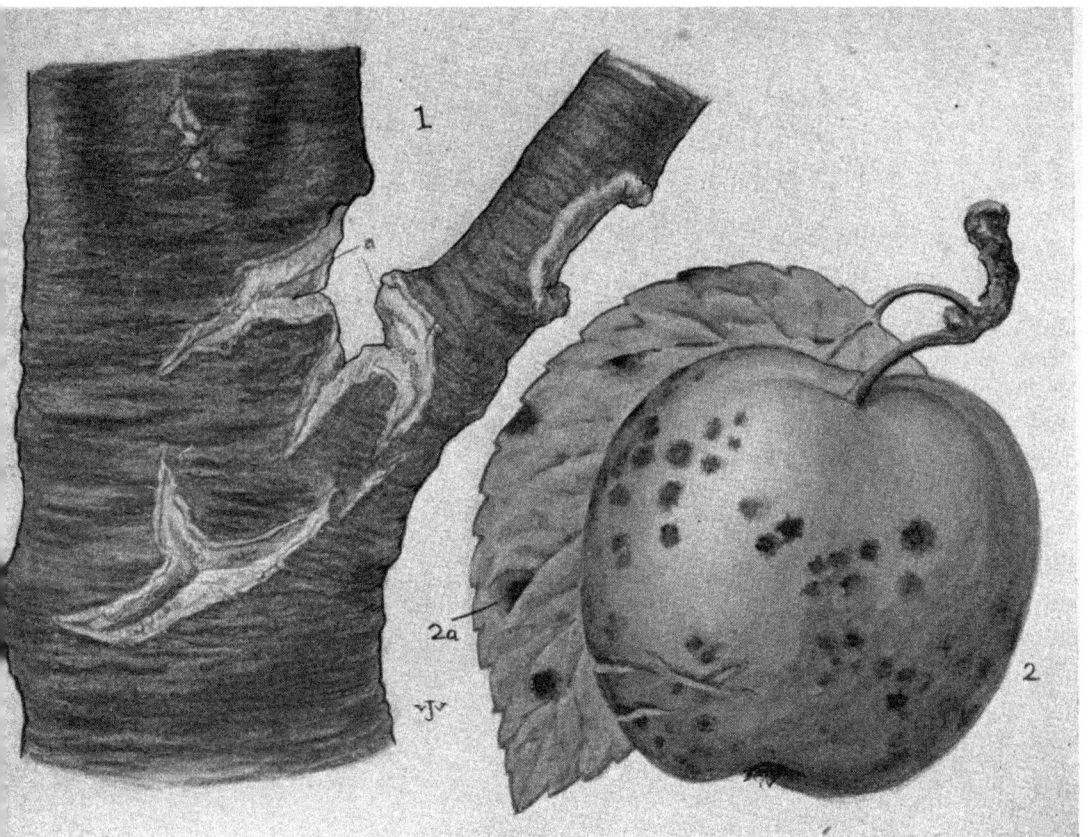

Figure 75.   Apple tree attacked by canker and scabs on apple also caused by a fungus, from A. D. Webster, *Tree Wounds and Diseases* (1916).

weather – and miliary sweat (*exsudatio miliaris*), which caused problems for sap flow.[32] Tree canker was sometimes caused by 'external violence', as when spades damaged bark, and might be remedied by applying and binding a section of the living bark of 'less valuable' trees to cover the wounded part. It was also possible to pare the edges of gangrenous bark ulcers with a knife, admitting air and preventing insect depredations and moisture retention, thereby combating the 'putrefaction' of 'stagnant juices' and halting the cankerous spread. The technique involved cutting away the 'dead lips' of the tree 'wound' without injuring healthy bark, and Darwin recommended covering the 'naked alburnum or sap-wood on a dry day' with thick white paint to prevent rot from rain or dew and insect attacks. However, because the paint consisted of a quarter ounce of 'sublimate of mercury, *hydrargyrus muriatus*' mixed with a pound of white lead to ward off insects, giving it a 'poisonous quality', it was crucial to ensure that it did not touch the wound's edges and cause further debilitation to the tree.[33]

When tree wounds were deep enough to reveal the underlying sapwood, Darwin believed that the upper parts tended to heal because they received help from the 'nutritive juices secreted by the vegetable blood'. He recommended that 'plasters', 'elastic bandage' or 'shred of flannel' be employed for binding wounds, using other pieces of the same tree, which was 'strictly analogous' to uniting 'inflamed or wounded parts' in animal bodies, such as in the case of cleft lip or inserting living teeth to replace those lost. This approach and the language employed by Darwin to describe it recalls his use of plasters in medicine and underscores the extent to which his approach to arboricultural ailments paralleled the treatments he employed for human illnesses. Not only did trees and shrubs provide an array of medicinal products, but ideas and methods from wider medical practice were applied quite closely for healing them. In his Derby orchard Darwin experimented with a method of trying to save the branches of two reliable fruiters from the canker by enclosing the diseased part and some inches above it in a garden pot of earth previously divided and supported by stakes and tied together round a branch. The idea was that this part might 'strike' new roots into the earth within the plant pot and that the newly revitalised branch might then be separated off some months later and planted afresh as a new tree with all the productive qualities of the old one. His medical experience and notion that bark was the most fundamental tree organ meant that he was keen that this should be protected as much as possible. Care should therefore be taken, when tying up transplanted fruit trees to support early growth, to prevent untoward pressure from the twine, wooden props or 'bandage'. When supporting fruit trees in his Derwentside orchard he therefore 'fastened one prop by a strong nail' to each tree and found that this provided support with much less injury than using the normal methods of supporting by 'three props and adapted cordage'.[34]

Honey-dew was a 'saccharine juice' secreted by the tree through what Darwin called the 'retrograde motions of the cutaneous lymphatic vessels', which were associated with either the 'common sap vessels' or the 'umbilical vessels'. In normal circumstances it was used by the tree to provide nutriment or encourage the growth of leaf buds when in their 'embryon state' but when this went wrong it was akin to attacks of '*diabetes mellitus*' – in other words, this was the tree version of diabetes, which in humans, of course, resulted in an inability to absorb blood sugar with a range of common symptoms, including sweating and, as Darwin knew from cases he had examined, the expelling of large quantities of sugar in urine. The 'saccharine and nutritious quality' of honey-dew was likewise evident from the taste and number of insects it attracted (which helped make it so damaging) and it was similar to the sap secreted in autumn by birch and maple trees. Although Pierre Augustin Boissier de Sauvages (1710–1795) had argued that there were two kinds of honey-dew, one from the tree and the other secreted by aphids, Darwin was doubtful because it seemed very unlikely that an insect would apparently needlessly secrete such a 'nutritive' substance, and there were no similar examples of this in the natural economy.[35] Instead he thought that honey-dew was only a 'morbid exudation' or symptom of tree disease that might perhaps pass through insects almost unchanged rather than originating in them, which appeared to be confirmed by the presence of black powder on the upper surfaces of leaves in place of the honey-dew and the fact that aphids were so common on peach and nectarine trees, which secreted no honey-dew.

Darwin's observations of trees growing around the Full Street area provided further information on the circumstances in which honey-dew was exuded. On the morning of 18 June 1798, for example, he saw a 'remarkable' amount of the substance dripping from leaves among a row of hazel trees (*Corylus avellana*) growing near a pond, which reminded him of observations made by the French naturalists Henri-Louis Duhamel du Monceau, in his *Physique des Arbres* (1758), and Jacques Renéaume de la Tache (1725–1796), who had seen it coming from maples (*Acers*).[36] Darwin saw the upper surfaces of the leaves, which were glistening in the bright morning sunshine, were covered with a 'viscid juice' which he tasted and found to be 'as sweet as diluted honey'. During the day so much dripped down that it moistened the gravel path beneath each tree, with more forming as the temperature rose and continuing over the next few days in fine weather, while those leaves obscured from the sun's rays seemed to have less or none of the sparkling dew. He concluded that, rather than being secreted by the aphids, this honey-dew was caused by the 'inverted action' of 'external lymphatics' hastened by the 'debility' caused by constant high temperatures, with the 'moisture' present in the local environment from pond and river being another factor. It resulted from a second period of midsummer sap flow produced by the trees to provide 'vegetable nutriment' for their new buds, and was, according to Darwin, either made by the aphids or may have passed through them.[37]

Observations from medical practice also informed Darwin's interpretation of 'miliary sweat' in *Phytologia*, which appeared to be produced by 'too great and continued heat', as occurred when vines in hothouses were kept at too high a temperature or allowed insufficient ventilation. The secretion produced by miliary sweat was not sweet-tasting like the honey-dew dripping from the Derby hazel trees, but consisted 'of mucilage', which remained upon the plant in 'small round globules' like the seeds of millet (which was the origin of the name of the condition). Darwin likened it to the 'very similar' appearance of small 'hard round globules on the skin', which he examined in *Zoonomia* and had seen upon various patients, including a dying woman suffering from 'miliary fever' (or miliaria) who had 'round pellucid globules' smaller than pin heads upon her neck and bosom, which easily 'rubbed off with the finger' – and he believed these to be probably caused by the surfeit of heat and 'exclusion of air', just as occurred in the vegetable version. Animal or plant bodies stuck in overheated and under-ventilated hothouses or bedrooms, because 'perspirable matter' contained more 'mucilaginous than natural' water and the watery part was exhaled, formed a 'mucilaginous' element upon leaves or skin, a process that Darwin thought was akin to the formation of stalactites on the roofs of Derbyshire caverns from calcareous solutions 'simply by the evaporation of the water' over time.[38]

Unusual gum secretion and sap flow in trees and plants also demonstrated that they were wounded or diseased from 'internal' causes almost as surely as bleeding betokened a wounded animal. In the case of sap flow, this occurred when the sapwood – which nourished the 'young buds' – was wounded in spring or when, for example, vines in hothouses were pruned too late in the season, upon which they might 'bleed to death'. This seasonal vegetable phenomenon could, however, be used to advantage by farmers to more easily destroy weeds such as thistles by cutting them down in early spring. Sponges placed upon tree wounds and held in place by bandages was a means of offering protection and helping them to heal.[39] Wounds could also be bound with leather, India rubber or even a piece of bark transplanted from similar trees of 'inferior value' to 'prevent further effusion'.[40] Problems caused by gum secretion in deciduous trees and resin from pine trees were similar; this was a 'nutritious fluid' for new buds that exuded from the 'wounded alburnum', but it contained no 'saccharine quality' and would naturally harden in dry weather or could be treated in the same ways as sap flow.

Darwin believed that the prevalence of human, animal and vegetable diseases and propensities towards illness were partially determined by hereditary factors, which sexual reproduction helped to prevent. As the Herefordshire horticulturist Thomas Andrew Knight showed, tree canker was prevalent in those that had been propagated by 'ingrafting' for centuries, and Darwin maintained that it was therefore a 'disease of old age' akin to the limb 'mortification' suffered by elderly people owing to the growing 'inirritability' of parts of their bodily system.[41] His contention that tree buds were a 'lateral progeny' rather than continuations of old trees encouraged him to conclude that canker was probably 'an hereditary disease' because the resemblance of engrafted trees to 'their parents' made them more susceptible to ailments that slowly or cumulatively developed through the 'influence of soil or climate', without the 'probability of improvement' and renewal represented by 'the progeny of sexual generation'.[42]

The decline of trees produced from grafting compared with the fate of those arising from sexual reproduction paralleled the animal propensity for disease and debility when too much close interbreeding occurred. As he expressed it poetically in *The Temple of Nature*: 'the feeble births acquired diseases chase, till death extinguish the degenerate race', so, as grafted trees grew, they spread 'their fair blossoms' through the air. But as 'canker' tainted the vegetable blood' and destroyed the wood, so through successive generations 'from perennial roots, the wire or bulb with lessen'd vigour shoots' until curling leaves and 'barren flowers' revealed a 'waning lineage, verging to decay'.

Darwin noted how Knight had observed that apple and pear trees long propagated from grafts became so debilitated that they were scarcely worth cultivating. Similarly, he suspected that potato diseases resulting in curled leaves and those in strawberry plants manifest by barren flowers had both probably occurred because the plants had been propagated by roots or 'solitary reproduction' for too long, rather than through 'seeds or sexual reproduction', leading to hereditary debility.[43]

TREES IN THE ECONOMY OF NATURE

To Darwin there seemed potentially few limits to tree growth given the right conditions, driven by all the potential agricultural and horticultural improvements that might be made in the future, provided that the perils of over-zealous enclosure and harvesting could be avoided. This belief encouraged him to make incredible predictions that one day individual trees could expand over thousands of years to cover vast territories. The manner in which internal wood was produced by the 'induration of the sap wood' and discoveries of massive trees such as some African '*Linneus adansonia*' – presumably the baobab, *Adansonia digitata*, which produced the 'Ethiopian sour-gourd' fruit (Figure 76) – with trunks twenty-five feet thick suggested that there might be 'no natural boundary' to some examples.[44] Even if individual trunks were a mile apart they 'might be enlarged' until they joined together and covered the entire globe with 'ligneous mountains' formed by 'successive generations of vegetable buds', just as some oceans were 'crowded with calcareous rocks' created by 'successive generations of coralline insects'. It was only the inexorable and inevitable processes of 'internal decay' that had prevented giant trees from covering much of the globe prior to human interventions.

Decay as much as the processes of living was of fascination to Darwin, partly because each was equally essential for organic life and the non-organic world, and he made many comparisons concerning how these operated in animals and vegetables. Like tenacious animals or humans battling against adversity or old age, some trees clung to life even if their internal heartwood was seriously decayed or even entirely lost – which was one of the fascinating aspects of ancient oaks such as those in Needwood and Sherwood forests. In such venerable examples, the 'internal wood' became 'gradually detached from the alburnum' as

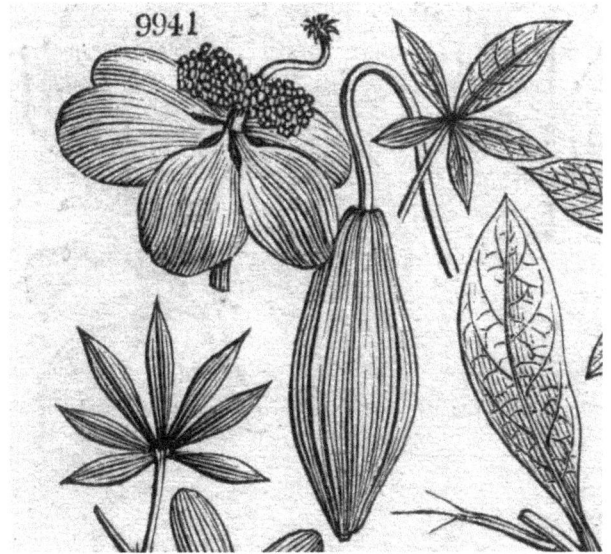

Figure 76.   Baobab, *Adansonia digitata*, flower and fruit (9941), from J. C. Loudon, *Encyclopaedia of Plants*, new edn, edited by J. Loudon (1855), 593.

decay overtook their frame, which was evident in many venerable willows and oaks – yet these determined fighters resisted death and even carried on growing externally, even though they must be absorbing 'putrid matter', and were only felled by gales tearing down their hollow trunks once their internal support had vanished. Examples of ancient hollow oaks of 'uncommon dimensions' included the 'remarkable' Greendale oak at Welbeck Park in Sherwood Forest, Nottinghamshire (Figure 77), which was just about large enough to drive a coach through, and, as we've seen, the celebrated Swilcar oak in Needwood Forest, the subject of Darwin's poem.[45] Like animal flesh, tree 'vascular' systems underwent processes of 'chemical decomposition' through time, losing through 'fermentation and putrefaction' their 'carbonic and phosphoric acids', which hastened their dissolution into dust, as was evident from rotten collapsing trunks, which lost their carbon and had the weird quality of luminosity as a result of the 'escape or production of phosphoric acid'. Further separation of the component organic parts produced decaying morasses that eventually fossilised to become 'eternal monuments of departed vegetable life', just as the remains of sea creatures provided a parallel in limestone rocks.[46]

The 'uncommon dimensions' of Darwin's favourite ancient oaks demonstrated how large trees could become even in the British climate and, as we have emphasised, he suggested that highland areas and mountains such as the Pennines might be usefully and productively planted with coniferous trees for timber production.[47] His observations of natural history and trees in gardens and plantations encouraged him to temper picturesque planting with attention to their habits and their relationship with climate. At one level, geology, topography and climate helped to determine where particularly species of tree would survive or should be planted. For example, he observed in *The Economy of Vegetation* that deciduous exotic trees, plants and fruit trees were much more susceptible to cold weather damage and required moving or protection using conservative walls, covers or glass. Drier or 'more resinous plants', on the other hand, including 'pines, yews, laurels and other evergreens', were less susceptible to damage from cold, while trees planted in valleys were more liable to injury from spring frosts than those on high ground because their 'succulent shoots' came out earlier. The fact that the 'common heat' rising from the earth in the British climate was four degrees allowed 'tender trees' that could be bent down and were not susceptible to attack from pests, such as figs, to be 'secured from the frost' by turning them to the ground and covering them with straw or fern.[48]

However, Darwin also acknowledged that trees also played a role as active entities in shaping climate and topography, through, for example, photosynthesis (with leaves operating akin to 'lungs', as will be examined shortly) and in channelling water to the ground. In his note on springs in *The Economy of Vegetation* (1791) for example, Darwin contended that trees helped to trap moisture and facilitated the passage of water from the atmosphere to the ground. He emphasised that the prevalent 'condensation of moisture' upon mountain tops partly occurred because of the 'dashing of moving clouds against them', which could be clearly observed in misty conditions where single trees on otherwise bare summits obstructed the mist in its passage. These had a far

Figure 77. South-east view of the Greendale Oak, Welbeck Park, drawn by S. H. Grimm and engraved by T. Nivares (1776), from J. Evelyn, *Silva: or a Discourse of Forest Trees*, edited by A. Hunter, 3rd edn, 2 vols (1801), vol. 2.

greater amount of water falling from their leaves than dropped at the same time from trees near the valley floor nearby.

Between the publication of *The Loves of the Plants* (1789) and the first edition of *The Economy of Vegetation* (1791) Darwin read White's *Natural History of Selborne* (1789), and seems to have been immediately impressed with aspects of the work – although, indirectly, he also made some criticisms. White's approach to the natural history of his immediate vicinity, as we have emphasised, in many ways parallels that of Darwin. White argued that the many woods and hills around Selbourne helped to shape the local climate, which was sheltered from 'strong westerly winds' by The Hangar, a large wooded chalk hill, the 'soft' air being moistened 'from the effluvia of so many trees' – which were yet 'perfectly healthy and free from agues'. White believed that the Selbourne woodlands combined with the 'dense clayey soil' and tree shade moderated temperatures, while air currents and wind with 'vast effluvia' emanating from the trees tempered and moderated the 'heats'.[49] Darwin noted that White had provided a description of a large tree from which, because of its location, a plentiful stream would flow during 'moving' mists, which was so vigorous as to fill normally dry cart ruts in a nearby lane. White had 'ingeniously' suggested that 'trees planted about ponds of stagnant water' in this way might therefore help to supply reservoirs. The 'spherules' of water from which mist or clouds were formed were kept separated by such a small 'power' that merely agitating them against tree leaves or the 'greater attraction of a flat moist surface' would cause condensation or precipitation.[50]

As well as their role in shaping local weather conditions, in his *Natural History of Selborne* White frequently emphasised how important trees and woodlands were as habitats for innumerable other plants and animals, placing weight, as did Thomas Gisborne in *Walks in a Forest*, upon their mutual inter-dependency within the natural economy. White closely observed the extent to which birds were dependent upon trees and the impact of unusual weather patterns or of lopping or felling upon this. Around his house was a large rookery and the birds spent much of the day on 'nest trees' during mild weather, retreating to roost in 'deep woods' every evening in winter and revisiting 'their nest trees' the following morning.[51] White described how a 'large excrescence about the middle of the stem' of one 'shapely and tall' oak at the centre of Losel's wood served as a nest for ravens to the extent that it became known as the 'Raven Tree'. The ravens were protected from climbing youths by the difficulty of reaching the nest, but when the tree was cut down the birds were killed.[52] Another example were 'willow wrens', who tended to inhabit the canopy of high beech woods and White enquired about the kinds of tree containing most heron's nests and whether these consisted of entire groves, woods or merely a few trees.[53] He witnessed the extent to which birds relied upon insects and their larvae – that were themselves supported by trees – for food, including during the winter months when sustenance elsewhere was scarce.[54] When trees were affected by unusual weather, seasonal changes or human activity the lives of the birds were likewise impacted. When, for example, severe or late frosts delayed the spring, this prevented access to the 'produce' of 'tender and curious trees' that was 'conducive' to the survival of birds. Similarly, White believed that the wood pigeons that 'abounded' around Selbourne until the 1750s had 'much decreased in number' after the beech woods were 'greatly thinned'. Long hot and dry summers, such as those of 1781 and 1783, caused severe problems for the fruit on White's fruit trees and also led to many being eaten by the myriads of wasps that appeared.[55]

While equally aware of the importance of trees and woodlands as habitats, Darwin emphasised the significance of rivalry as much as the mutual interdependency of living beings within his deistic natural economy. Although the 'artificial' Linnaean system, which contemporaries such as James Edward Smith likened to an alphabet, might encourage a static, passive view of creation, tree observations, medical practice and his favourite animal–plant analogies encouraged Darwin to believe that, no matter how cultivated, the earth was bursting with vitality and teaming with competing life forms. In marked contrast to his friend Gisborne's presentation of woodland nature in *Walks in a Forest*, Darwin recognised that vitality entailed fierce competition. Plants fought, cannibalised or attacked each other, preying upon each other in a 'vegetable war' where 'herb, shrub, and tree' battled for 'light and air' with insects and other creatures as roots struggled for 'moisture and for soil' below. The hollies surrounding the veteran oaks of Needwood were not merely tributary bands but competing beings. Ivies clasped and slowly strangled the 'tall elm' while 'insect hordes with restless tooth devour the unfolded bud, and pierce the ravell'd flower'. The competitiveness and prodigality of nature was fully evident as 'each pregnant oak a thousand acorns forms, profusely scatter'd by autumnal storms' while 'countless aphides, prolific tribe, with greedy trunks the honey'd sap imbibe' swarming on leaves as 'pendant nations' tenanting every twig (Figure 78). Vegetables were, in succession, 'an inferior order of animals fixed to the soil' to be preyed on by 'locomotive animals' who also attacked each other, forming a world that was 'one great slaughter-house'.[56]

However, as Darwin believed the 'digested food of vegetables' to be largely sugar, from which came 'their mucilage, starch, and oil', and as animals relied upon such 'vegetable productions', so what he regarded as the sugar-creating manufactories within 'vegetable vessels' were the essential living engine 'for all organised beings'.[57] The fixity of plants in the earth with 'leaves, innumerable, waving in the air' was required for water's decomposition and reconstitution as 'saccharine matter', a process that would have been more difficult to execute in 'cumbrous', moving animal bodies. In a fascinating and evocative image, Darwin argued that humans or quadrupeds could never have carried on their heads or backs 'a forest of leaves' or had 'long branching lacteal or absorbent vessels terminating in the earth'. Animals therefore relied upon vegetables creating the nutritious substances that they required to survive and employed specialist bodily organs to prepare them 'further for the purposes of higher animation and greater sensibility.' Similarly, the 'apparatus of green leaves and long roots' was 'found' too cumbersome for the 'more animated and sensitive parts of vegetable flowers' such as the anthers and stigmas. Though they shared many characteristics in common with animals, according to Darwin, plants therefore had to be be 'separate' kinds of 'beings' endowed with the 'passion and power of reproduction'. As we have seen, he maintained that the male flower parts of anthers and filaments and female parts of stigma and style were 'separate and distinct animated beings' that may originally have been insects and that, after expanding

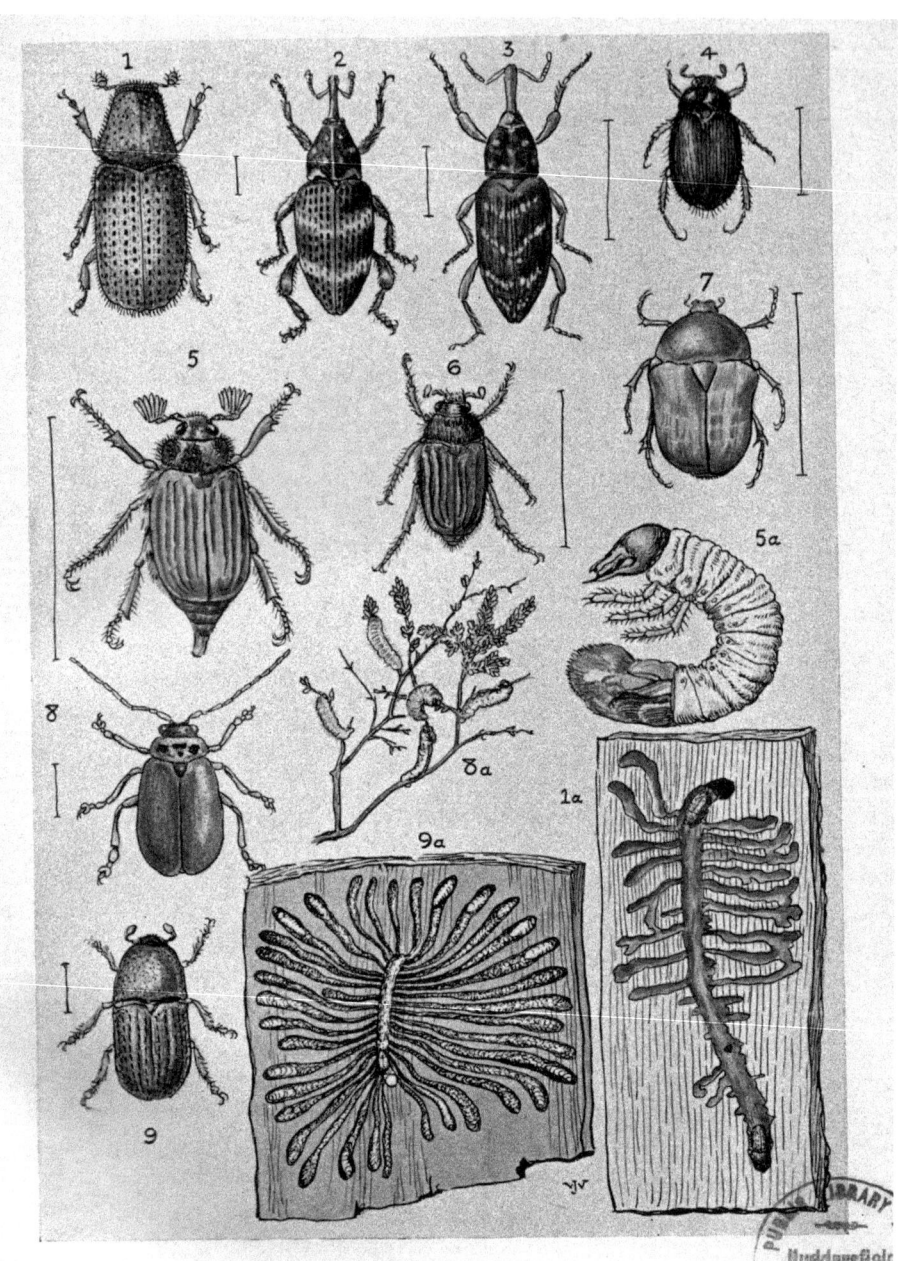

1, 1a. PINE BEETLE AND GALLERIES.
2, 3. PINE WEEVILS.
4. GARDEN CHAFER.
5, 5a. COCKCHAFER AND GRUB.
6. SUMMER CHAFER.

7. GREEN ROSE CHAFER.
8, 8a. HEATHER BEETLE AND GRUBS.
9, 9a. ELM BARK BEETLE AND
GALLERIES.

Figure 78.   Beetles and grubs that attack trees, from A. D. Webster, *Tree Wounds and Diseases* (1916).

the petals or flower 'lungs', they were nourished by honey (nectar). This was a food 'ready prepared' by long roots and green leaves for the 'placentation or support' of the plant's 'seeds, bulbs and buds', which were also separate nascent beings who took it in through 'their absorbent mouths'.[58]

Tree lifecycles demonstrated that competition actually ensured mutual interdependency, which provided Darwin with a 'glimpse' of the fundamental 'economy of nature' and prompted him to exclaim that 'God dwells here'. 'Births and deaths' contended with 'equal strife' while 'every pore of nature' teemed with animation, yet reproduction strove to 'vanquish death' and ensure the survival of happiness as 'young renascent nature conquers time'. Darwin argued that large trees were much more than single beings but progenitors of future forests through the spread of 'a thousand acorns', while harbouring thousands of life forms within trunks and branches, such as the 'countless aphids' that 'tenant every twig'. The potential for vitality was tenacious and universal, as large ancient trees clearly demonstrated. True, life forms were checked by 'war, pestilence, disease and dearth', but this prevented the planet from being overcome by 'superfluous myriads', and from this cycle, as the rapidly dying insects 'swarming in the noontide bower' demonstrated, life arose anew and 'vanquish'd death'.[59]

CONCLUSION

We have seen in this chapter how trees were central to Darwin's understanding of botany and of plant and animal physiology, and supported his evolutionary ideas in various ways. Although an enthusiastic Linnaean, whose *Loves of the Plants* did much to publicise the benefits of that classification system, Darwin recognised that it was less useful for arboriculture than for other plants, as the difficulties he experienced identifying fossil plant specimens demonstrated. While he continued classifying trees according to abstract Linnaean divisions, inspiration for Darwin's arboriculture came from elsewhere. During the 1770s, responding to threats of disafforestation in the name of enclosure and estate improvement and with a renewed vigour for life fostered by love, he created the Lichfield botanic garden and returned to poetry. Significantly, the subject of the first series of poems he had composed since being a student was the importance of trees in the economy of nature and human history, culture and society. Poetic arboricultural personifications, the efficacy of tree products in medicine, encounters with midlands estate trees and plantations, experimental observations of trees in his gardens and a practical knowledge of British botany encouraged Darwin to draw close analogies between trees and animals. As venerable witnesses of human history, exemplars of the tenacity, prodigality and interdependency of life, miniature worlds and monarchs of the forest, trees such as the mighty Swilcar oak exemplified Darwin's vision of the natural economy. Like all life, they faced inevitable destruction as a result of ageing, as well as elemental forces such as winds and lightning, yet from the 'shattered antlers' which 'strew the glade' as tree 'fibres' crumbled into 'mouldering dust', new 'rising forests' would appear, sheltered by 'old branches and decaying trunk', and 'bloom forever in the immortal lay'. Just as geological evidence from Peak limestone revealed how the death of numerous forms of oceanic life created new worlds, so the successive lives and deaths of trees also revealed the constant rebirth of multiple worlds. The individual experience of one venerable oak tree encapsulated in microcosm Darwin's grand overarching cyclical cosmos, where past life forms, as 'monuments of past felicity', provided the nutrients for the renewal of life as earths and stars were born and died through 'millions of ages'.[60]

NOTES

[1]   W. Gilpin, *Essays on the Picturesque as Compared with the Sublime and the Beautiful*, 3 vols (London, 1810), vol. 1, 229–96; S. Daniels, 'The political iconography of woodland in later Georgian England', in D. Cosgrove and S. Daniels eds, *The Iconography of Landscape* (Cambridge, 1988), 43–82; S. Daniels, *Joseph Wright* (London, 1999), 68–74; J. Bonehill and S. Daniels eds, *Paul Sandby: Picturing Britain* (London, 2009), 226–33; P. A. Elliott, C. Watkins and S. Daniels, *The British Arboretum: Trees, Science and Culture in the Nineteenth Century* (London, 2011), 25–6; C. Watkins, *Trees, Woods and Forests: A Social and Culture History* (London, 2014), 102–18; C. Watkins, *Trees in Art* (London, 2018), 27–36.

[2]   E. Darwin, *Phytologia; or the Philosophy of Agriculture and Gardening* (London, 1800), 449–54; S. Hales, *Vegetable Statics* (London, 1727); J. Walker, 'Experiments on the motion of sap in trees', *Transactions of the Royal Society of Edinburgh*, 1 (1788), no. 2, 237–46; J. Von Sachs, *History of Botany*, translated by H. E. F. Garnsey and revised by I. B. Balfour (Cambridge, 1890), 476–82; J. Reynolds Green, *A History of Botany in the United Kingdom* (London, 1914), 198–207; A. E. Clark-Kennedy, *Stephen Hales, DD, FRS: An Eighteenth-century Biography* (Cambridge, 1929), 59–75.

[3]   Darwin, *Phytologia*, 449–54.

[4]   E. Darwin, *The Economy of Vegetation*, 4th edn (London, 1799), 'additional note xxxv: vegetable placentation', 437–40; Hales, *Vegetable Statics*; Clark-Kennedy, *Stephen Hales*, 63–75.

[5]   Darwin, *Economy of Vegetation*, 'additional note xxxvi: vegetable circulation', 441–8.

[6]   Darwin, *Economy of Vegetation*, 'additional note xxxvi: vegetable circulation', 441–8.

[7]   Darwin, *Economy of Vegetation*, 'additional note xxxiv: vegetable perspiration', 448–57.

[8]   Darwin, *Economy of Vegetation*, 'additional note xxxiv: vegetable perspiration', 450–4.

[9]   S. King, 'The healing tree' in L. Auricchio, E. Heckendorn Cook and G. Pacini eds, *Invaluable Trees: Cultures of Nature, 1660–1830* (Oxford, 2012), 237–50, at 237–9.

[10] S. Harris, 'Myth and medicine in Erasmus Darwin's epic poetry', in C. U. M. Smith and R. Arnott eds, *The Genius of Erasmus Darwin* (Aldershot, 2005), 321–35, at 325.

[11] Darwin, *Phytologia*, 429.

[12] Darwin, *Phytologia*, 550–1; Lichfield Botanical Society, *A System of Vegetables … translated from the thirteenth edition of the Systema Vegetabilium of the late Professor Linneus*, 2 vols (Lichfield, 1783), vol. 1, 345, 59; Lichfield Botanical Society, *The Families of Plants … Translated from the Last Edition of the 'Genera Plantarum'*, 2 vols (Lichfield, 1787), vol. 1, 307, 15.

[13] Darwin, *Phytologia*, 448–9.

[14] Darwin, *Phytologia*, 450; R. Fastnedge, *English Furniture Styles, 1500–1830* (Harmondsworth, 1970), 106–9.

[15] Darwin, *Phytologia*, 450–1, 466–8; Lichfield Botanical Society, *Families of Plants*, vol. 1, 318, vol. 2, 508; R. J. Thornton, *A Family Herbal: or Familiar Account of the Medical Properties of British and Foreign Plants* (London, 1814), 805–7; J. Lindley, *Flora Medica* (London, 1838), 178–9.

16   Darwin, *Phytologia*, 466; Lichfield Society, *Families of Plants*, vol. 2, 695; Thornton, *Family Herbal*, 372–4; Lindley, *Flora Medica*, 176.

17   Darwin, *Phytologia*, 472; Lichfield Botanical Society, *System of Vegetables*, vol. 2, 752; Lichfield Botanical Society, *Families of Plants*, vol. 2, 696.

18   Darwin, *Phytologia*, 465.

19   E. Darwin, *Zoonomia; or the Laws of Organic Life*, 3rd edn, 4 vols (London, 1801), vol. 1, 108–11, 135–6; vol. 2, 479–84.

20   Darwin, *Zoonomia*, vol. 2, 529–36.

21   Darwin, *Zoonomia*, vol. 2, 472–518; for *Cathechu mimosa*, Thornton, *Family Herbal*, 866–70; Lindley, *Flora Medica*, 268–9.

22   J. Alderson, *The Essay on the Rhus Toxicodendron: Pubescent Poison Oak, or Sumach*, 2nd edn (Hull, 1796), v, 1–2, 4, 6, 8, 12–13, 15–18, 25, 28–66; Lichfield Society, *Families of Plants*, vol. 1, 197; E. Darwin, 'An account of the successful use of foxglove in some dropsies and pulmonary consumption, *Medical Transactions*, 3 (1785), 255–308; W. Withering, *Account of the Foxglove and some of its Medical Uses* (Birmingham, 1785); C. Darwin, *Experiments Establishing a Criterion between Mucaginous and Purulent Matter* (Lichfield, 1780), 103–15; Darwin, *Zoonomia*, vol. 4, 129–30; Lindley, *Flora Medica*, 284–5; P. A. Elliott, *Enlightenment, Modernity and Science: Geographies of Scientific Culture and Improvement in Georgian England* (London, 2010), 159–63.

23   E. Darwin, *The Loves of the Plants*, 4th edn (London, 1799), 103–4; Derby Philosophical Society, *Rules and Catalogue of the Derby Philosophical Society*, with supplements of 1795 and 1798 (Derby 1793–8), 35. Quinine, one of the active substances in *Cinchona* was used to treat malaria.

24   A. B. Lambert, *A Description of the Genus Cinchona* (London, 1797), vii–ix, 1, 6–7, 9–57; W. Blunt, *The Art of Botanical Illustration* (London, 1950), 195–202; Elliott, Watkins and Daniels, *British Arboretum*, 34, 70–81.

25   Darwin, *Zoonomia*, vol. 1, 108–11, 135–6; vol. 2, 479–84.

26   E. Darwin, letter to A. Reimarus, 20 September 1756, in D. King-Hele ed., *The Collected Letters of Erasmus Darwin* (Cambridge, 2007), 36–9.

27   Darwin, letters to J. Wedgwood, 30 September 1772, 11 March 1784 and T. Wedgwood, 10 August 1794, in King-Hele ed., *Collected Letters*, 36–9, 119–23, 229 and 445–7; Darwin, *Zoonomia*, vol. 2, 479–84.

28   J. Evelyn, *Sylva: or a Discourse of Forest-Trees* with notes by A. Hunter, 3rd edn, 2 vols (York, 1801), vol. 1, 4.

29   Darwin, *Phytologia*, 452; J. Evelyn, *Sylva, or a Discourse of Forest Trees*, 4th edn (1706) with introduction by J. Nisbet, 2 vols (London, 1908), vol. 1, 314–35.

30   Darwin, *Phytologia*, 453.

31   Darwin, *Phytologia*, 453.

32   Darwin, *Economy of Vegetation*, 229; Darwin, *Phytologia*, 293.

33   Darwin, *Phytologia*, 293–4, 447–54.

34   Darwin, *Phytologia*, 449–54.

35   T. Wildman, *A Treatise on the Management of Bees* (London, 1768), 45–7.

36   Darwin, *Phytologia*, 295–6; H. L. Duhamel du Monceau, *La physique des arbres, où il est traité de l'anatomie des plantes et de l'économie végétale: pour servir d'introduction au traité complet des bois et forests ... Partie 2*, 2 vols (Paris, 1758), vol. 1, 150.

37   Darwin, *Phytologia*, 294–6.

38   Darwin, *Phytologia*, 297–8; Darwin, *Zoonomia*, vol. 3, 400–2.

39   Darwin, *Phytologia*, 298–9.

40   Darwin, *Phytologia*, 299–300.

41   Darwin, *Economy of Vegetation*, 229; Darwin, *Phytologia*, 293; T. A. Knight, *A Treatise on the Culture of the Apple & Pear and on the Manufacture of Cider and Perry* (Ludlow, 1797), 9–19.

42   Darwin, *Phytologia*, 293–4, 447–54.

[43] Darwin, *Phytologia*, 449–54; E. Darwin, *The Temple of Nature: or the Origins of Society* (London, 1803), 56–7; Porter, 'Erasmus Darwin'.

[44] Darwin, *Phytologia*, 479; Lichfield Botanical Society, *System of Vegetables*, vol. 2, 508; Lichfield Botanical Society, *Families of Plants*, vol. 2, 469.

[45] Darwin, *Phytologia*, 477–8, 479.

[46] Darwin, *Phytologia*, 478–9.

[47] Darwin, *Phytologia*, 447–9, 469–71, 479–80.

[48] Darwin, *Economy of Vegetation*, note, 50–1. Having witnessed the severe damage caused to the 'laurustines, bays, laurels and arbutuses' in his garden by repeated snow melting and night-time freezing, Gilbert White urged that planters who wanted to 'escape the cruel mortification of losing in a few days the labour and hopes of years' needed to cover all their most delicate trees and shrubs during this period or at least to dislodge the snow. Like Darwin, he also suggested that planting under walls provided another solution, especially for exotics. By the spring thaw, White found that his Portugal laurel and American junipers had survived well, but the ilexes were 'much injured, the cypresses were half destroyed, the arbutuses lingered on, but never recovered' and the bays, laurustines and laurels 'were killed to the ground', while the wild hollies on the warmer aspects had lost all their leaves (White, *Natural History and Antiquities of Selbourne*, 288–90).

[49] White, *Natural History and Antiquities of Selbourne*, 301.

[50] Darwin, *Economy of Vegetation*, Additional Note XXVI on Springs, 387–90; White, *Natural History and Antiquities of Selbourne*, 204–6.

[51] White, *Natural History and Antiquities of Selbourne*, 166.

[52] White, *Natural History and Antiquities of Selbourne*, 6.

[53] White, *Natural History and Antiquities of Selbourne*, 55, 64.

[54] White, *Natural History and Antiquities of Selbourne*, 105–6.

[55] White, *Natural History and Antiquities of Selbourne*, 34, 142, 299–300.

[56] Darwin, *Temple of Nature*, 132–4; T. Gisborne, *Walks in a Forest, or Poems Descriptive of the Scenery and Incidents Characteristic of a Forest*, 3rd edn (London, 1797).

[57] Darwin, *Temple of Nature*, 132–4.

[58] Darwin, *Economy of Vegetation*, additional note xxxix; Darwin, *Temple of Nature*, 107.

[59] Darwin, *Temple of Nature*, 156–9, 166.

[60] Darwin, *Phytologia*, 480–1.

CONCLUSION

Beginning and ending in two botanic gardens at Lichfield and Kew, as Erasmus Darwin did in *The Economy of Vegetation*, we will conclude by highlighting some of the main themes that have emerged in our analysis of his approaches to gardening, botany, horticulture, tree cultures and farming, especially as expressed in *Phytologia*. These include the role of critical personal observations, medical practice and family members, patients and friends in nurturing his ideas and the usage he made of his body as an experimental tool to investigate potential novel foodstuffs. Secondly, we will examine some of the short-term and longer-term impacts that his contributions to these endeavours had, including the stimulus his arguments concerning agriculture and the agency of animals and plants provided to writers and scientists such as his grandson Charles Darwin and the chemist Humphry Davy. Finally, we will take a stroll with Darwin through the royal botanic gardens beside the Thames, encountering George III and Queen Charlotte, exploring some of the international dimensions of his medico-botany and presenting an offering to Hygeia, Greek goddess of health, in her sacred grove.

Darwin's *The Loves of the Plants* (1789) and *The Economy of Vegetation* (1791) captured the imagination of late Georgian society partly because of the combination of poetry and illuminating scientific notes. Of these, the long essays on botany attracted much attention. As his grandson Charles Darwin remarked, their author's success 'was great and immediate'; his grandfather made much money from the publication and there were various British and foreign editions. Contemporaries such as Horace Walpole hailed Darwin's 'most beautifully and enchantingly imagined' creation, while Richard Lovell Edgeworth claimed that sections 'seized hold of his imagination' to such a degree that his 'blood thrilled back through his veins'. The young poets Samuel Taylor Coleridge, William Wordsworth and Percy Bysshe Shelley were initially excited and inspired, even if they later turned against Darwin's style, while the older generation of poets, including William Cowper and William Hayley, were equally enthused.[1]

The presentation of gardening, botany and horticulture in the epic poems and *Phytologia* had a major impact on how these endeavours were portrayed in literature and also helped to make picturesque botanical gardens more fashionable. Darwin's gardens undoubtedly were, as his friend Anna Seward contended, places that united 'Linnean science' with landscape 'charm', but they were also sites for observation and experimentation in plant physiology and anatomy that enabled him to make recommendations concerning agricultural improvements. One manifestation of this was the succession of subscription and semi-public botanical gardens developed between 1780 and 1840 that increasingly combined systematic and practical physic collections with attention to landscape beauty and design, which the *Botanic Garden* helped to make fashionable.[2] Not least of these was the Derby Arboretum (1840), one of the earliest and most influential British public parks, which was donated and planned by the wealthy textile manufacturer Joseph Strutt, one of Darwin's protégés and patients, and designed by the Scottish landscape gardener John Claudius Loudon with strong support from Derby Philosophical Society members, including the physician and president Richard Forester – another of Darwin's younger friends, who was a trustee. Originally projected as a public botanic garden, the Arboretum featured a systematic collection of trees and shrubs in a picturesque landscaped setting intended, like Darwin's poems, to be the portal to a captivating world of natural history and scientific knowledge. The Darwinian inspiration was underlined by calls for the 'late excellent old physician' and author of the *Botanic Garden* to be honoured with a statue in the park to add to the various busts of him recently displayed at the Derby Mechanics' Institute exhibition (1839).[3]

Examining Darwin's approach to gardening, botany, horticulture and farming, the similarities between his methods and those of his grandson Charles are strongly evident, and, as Desmond King-Hele, Mea Allan, Howard Gruber and Peter Ayres have convincingly demonstrated, there are similarities between how different members of the Darwin family approached these endeavours from the mid-eighteenth to the early twentieth century.[4] But there is also a danger of slipping into whiggishness by picking those elements of Darwinian farming and gardening that seem to anticipate later science, especially with regard to Erasmus Darwin. The tendency to contrast his botanising as a hobbyist and 'theoriser' with the more precise and systematic approach of his grandson Charles and great grandson Francis wrenches their predecessor from his historical mileau and underplays how much careful observation and experiment he too undertook in places such as gardens, woodland, farms and fields.[5] And concentrating upon the poetry, even if that includes *The Temple of Nature*, and downplaying his medical practice also means that many dimensions of Darwin's work are missed or underestimated, especially if his voluminous correspondence (as revealed by Desmond King-Hele) and notebooks are included.

One dimension of the real-world observations and experiences encouraged by his medical practice, as we have seen throughout the book, is the degree to which Darwin carefully witnessed and recorded practices in gardening, farming and horticulture as closely as he witnessed human bodies and behaviours. While, with its neo-classical language, divine beings and imagery, some of his poetry utilises common Georgian pastoral tropes and Augustan literary conventions, it is clear from all his compositions and correspondence that Darwin had in his final decades what might be defined as a romantic ecological approach to agriculture and the natural economy. The numerous animals and plants inhabiting Darwin's works are active beings with agency, shaping their own lives, rather than merely reactive entities or passive literary creations, and constantly respond to the challenges of their environments. They have an insatiable appetite for life, whether individually or collectively through the generations, and are driven by intricately refined internal bodily processes fuelled by the nourishment they absorb from the world around.[6] As we have seen, for Darwin the very qualities that some plants developed to survive in harsh environments or protect themselves against predators in warring nature, such as aromatic, bitter, narcotic or acrid juices, provided them with their medicinal virtues. His analyses of animal and vegetable diseases likewise demonstrate the extent to which he employed careful observations and experiments and where possible sought information from friends, family, patients and others – whom he encouraged to undertake their own studies and experiments. Many of his informants were farmers and most were dependent upon the land in one form or another, as were the Darwin family themselves for part of their income. As has been emphasised and Peter Worsley has shown, the Darwin family had landholdings in Derbyshire, Nottinghamshire and Lincolnshire, which increased during the eighteenth and nineteenth centuries.[7]

Darwin walked around fields and farms and travelled across the midland countryside by horse and carriage along the newly turnpiked but still sometimes poor-quality roads in order to treat thousands of patients during a career of almost half a century, observing seasonal and longer-term alterations. As we have seen, he witnessed everything from changing livestock and economic uses of plants in fields – such as the camomile farms in the Chesterfield area – to swarms of May-chafers along miles of hedges near Pleasley. He traversed the fields surrounding Lichfield with his son Charles, pointing out plants and insects and helping him to produce leaf prints for his notebook. He conversed with farmers, estate managers, gardeners, fellow doctors and surgeons and others with professional interests in agriculture and the health of plants and animals. He discussed the efficay of different drill plough designs with the renowned Thomas William Coke of Norfolk and animal breeding with farmers and landowners such as Sir Harry Harpur at Calke and Hugo Meynell (1735–1808) of Quorn Hall, celebrated master of the Quorn Hounds and one of the most influential hunting squires of the second half of the century. Meynell helped make fashionable a new kind of foxhunting centred upon north Leicestershire, breeding and using foxhounds more effectively, spreading the sport across a much wider terrain and attracting visits from aspiring huntsmen from across the British Isles and Ireland who came to observe his methods. As his close attention to everything from bee battles to aphidious exudations under the magnifying glass demonstrates, Darwin was a keen observer of even the most apparently minute and mundane creatures. He also encouraged family, friends and patients to provide information and undertake observations for him.[8]

Like his grandson, Darwin often found the most apparently lowly animals and plants, which were usually seen as economically or culturally insignificant, particularly engaging. Both men were fascinated with earthworms and parasitic worms (helminths) alike, and were mightily impressed by their tenacity for life. Equally, as we have seen, cryptogamia such as moss, lichen and fungi attracted Darwin's attention, though awkwardly accommodated within the Linnaean system, and he believed them to connect animals and vegetables, with some beginning their lives within animal bodies.

Later in life, Darwin saw himself as a 'farmer', keeping livestock such as pigs and seeking to care for them as effectively as possible. He carefully observed animal behaviour, almost as closely as he marked human conduct, and one result was that one of his most successful medical interventions was non-intervention, or knowing when not to actively intervene, to allow nature to take its course and bodies to restore their own equilibrium. Many of Darwin's human patients doubtless survived as a result of his employment of a light touch in the medical arts, as did animal patients, such as some of the south Derbyshire cattle suffering from distemper, who were placed on vegetable-rich diets and left alone to recover on his instructions.

Another dimension of his close observation and engagement with gardening and horticulture is the eagerness which he employed the very tactile experience of tasting and smelling substances to determine their chemical qualities, which was clearly inspired by his keen medical sense. Somewhat recklessly, it must be said, he tasted a whole range of organic materials, including tree and plant secretions and different types of vegetable product, using his mouth as a scientific instrument to judge the composition of these at some risk to his own body. His medical interest in and concern for the health of the general population, stimulated by concerns over food supplies during the revolutionary wars, led him to recommend that different tree and plant substances might have major economic benefits or provide important new foodstuffs. Emphasising that animals and other human cultures consumed foods novel to the British palate, Darwin therefore suggested that insects might provide a 'delicious' food and a major source of nourishment in the future.[9]

Gardens, farms and orchards were places for Darwin to observe and experiment, providing both many examples of the natural economy in operation, albeit managed by human intervention, and inspiration for his writing. We have seen how much experience of these places was a Darwinian family affair, providing food and medical supplies as well as beautiful places to rest, retreat, sojourn and play. Gardens were arenas for entertainment for wider family members and friends and were shaped by these wider networks, with parcels of seeds and plants as well as books and information in correspondence passing backwards and forwards through the post. Robert Waring Darwin senior, author of the *Principia Botanica* (1787), was often consulted for his expertise, while his nephew Robert Waring junior, as we know from his son Charles, took a 'great pleasure' in gardening and had a strong 'love of plants', planting 'ornamental trees and shrubs' in his Shrewsbury gardens (Figure 79).[10] Family members all contributed to the development of Darwin's Derby hothouse and Erasmus Darwin junior considered building a house upon the eight-acre Lichfield botanic garden site as a place to retire from business, underlining the special status it had for him and the family as a place of beauty, solace and regeneration.[11]

But it was not only family networks that supported and determined Darwin's approaches to gardening, horticulture and botany. This book has also underlined how significant some of Darwin's female friends and patients were (as well as the goddesses of health and botany!) in shaping and encouraging these endeavours and in engaging in critical dialogue and discussion with him about plants and related subjects. As Martin Priestman has emphasised, 'many of the period's leading women writers

ure 79.   Mezzotint of Robert W. Darwin, after painted portrait from K. Pearson, *The Life, Labours and* *ers of Francis Galton*, vol. 1 (1914).

showed some kind of response to Darwin's poeticised science, and to the botany of [*Loves of the Plants*] in particular'. For women such as the very wealthy Mary Cavendish Bentinck, duchess of Portland (1715–1785), and Mary Delaney (1700–1788) and their networks, gardening, collecting and botanical study provided a means of enjoyment, expression and meaning, and, although Seward thought he had misrepresented her methods, Darwin praised Delaney in *The Loves of the Plants* for her 'wonderful', 'accurate and elegant' Linnaean 'hortus siccus' or 'paper garden'. Though usually not formally members of literary and scientific societies – even though they did attend some meetings of these – some women delighted in scientific pursuits, with botany and horticulture being favoured because of their association with gardening, decorative flower painting and the domestic sphere.[12]

Of most importance for Darwin were Anna Seward and his second wife Elizabeth, but other women, such as the botanist, artist and writer Maria Elizabeth Jacson (1755–1829) of Somersal Hall, Derbyshire, William Strutt's wife Barbara (1761–1804) and Jane Borough (1759–1838) were part of Darwin's gardening and botanical networks. The fact that Seward's name has appeared so often in these pages underlines how central their relationship was to both their perceptions of and experiences with gardens, trees, plants and the natural world. Darwin's work encouraged Jacson and she sent him plant drawings and a copy of her *Botanical Dialogues* (1797), to which Darwin and Brooke Boothby subscribed. Similarly, Darwin despatched copies of the Lichfield Linnaean translations to Jane Borough, whose family park and gardens at Castlefields lay on the banks of the Derwent in Derby, and exchanged plant information and specimens with Barbara Strutt after her family had taken over St Helen's House, with its extensive park, gardens and hothouses, from the Gisbornes.[13] Seward's role in encouraging and celebrating the Lichfield botanic garden and in stimulating the composition of *The Botanic Garden* is very clear and has been demonstrated here and by Teresa Barnard, Sylvia Bowerbank and others. The active interest taken by Elizabeth Darwin and Seward in forming and celebrating gardens and plants accords well with Stephen Bending's striking demonstration of the manifold ways in which eighteenth-century elite women experienced and shaped these endeavours.[14]

Elizabeth's role is less well documented because correspondence between husband and wife living together was, of course, not usually necessary, but it is evident that, like her half sister Lady Caroline Curzon, she took a keen interest in improving the gardens (Figure 80). Where these matters are referred to in the letters it is clear that Elizabeth's views were respected by her husband and that, as this book has demonstrated, she encouraged and determined the direction of many of his plant experiments and helped motivate the striking turn he took towards botany, gardening, horticulture and farming from the 1770s onwards. As Darwin's 1775 love letter in the guise of a wood nymph to his future wife demonstrates, she was already directing sylvicultural operations. Equally, Elizabeth tended and improved The Priory gardens for another quarter century after her husband's death in 1802, until reunited with him in Breadsall church (Figure 81).[15]

Another of Darwin's female friends with determined botanical interests was Millicent French (c. 1746–1789), whom he commemorated in one of the most heartfelt

Figure 80. Silhouettes of Elizabeth and Erasmus Darwin mounted on opal glass, Derby, 1800, from K. Pearson, *The Life, Labours and Letters of Francis Galton*, vol. 1 (1914).

Figure 81. Pen and ink sketch of the garden front of Breadsall Priory, from K. Pearson, *The Life, Labours and Letters of Francis Galton*, vol. 1 (1914).

parts of *The Economy of Vegetation*, describing the devastation of her husband Richard French at her loss and his longing for their reunion at the resurrection of the dead at the end of time. Darwin was so concerned to get the tone right that he wrote to Seward asking for advice, enclosing a draft of the epitaph and emphasising 'how difficult' it was 'to write an epitaph', which was a task he had never really attempted before. The significance he attached to the epitaph and the memory of Millicent French is evident from his inclusion of it in *The Economy of Vegetation*. Here Darwin described how the Derwent flowed through Derby, overshadowed by the tower of All Saint's Church: 'Waved o'er his fringed brink a deeper gloom, And bow'd his alders o'er MILCENA'S tomb'.[16] Walking beside the river she

> Explored his twinkling swarms, that swim or fly,
> And mark'd his florets with botanic eye.-
> 'Sweet bud of spring! How frail thy transient bloom,
> Fine film,' she cried, 'of Nature's fairest loom!
> Soon Beauty fades upon its damask throne!'
> - Unconscious of the worm, that mined her own!-[17]

In a note below, Darwin emphasised that Millicent was 'a lady who to many other elegant accomplishments added a proficiency in botany and natural history' and his epitaph, made reference to how she led their (seven) children around the fields (including Richard [Forester] French [1771–1843], a future president of the Derby Philosophical Society), carefully observing the flowers and creatures beside the river, marking the Derwent's 'florets with botanic eye' and 'printing ... his spangled plain'. Darwin's allusion to Millicent's close observations of the changing patterns of light and colour showed her drawing and painting flowers, animals and plants, rather than just passively observing them. Darwin's description is also reminiscent of his accounts of the botanising expeditions he undertook around Lichfield with his son Charles, when they talked about plants, drew sketches and produced the book of nature prints.

Millicent French was sister of Darwin's close friend the poet Francis Noel Clarke Mundy, whose passion for botany and natural history is demonstrated in his Needwood Forest poems. Her husband Richard was a founder member of the Philosophical Society and, with her brother Francis, was witness to the marriage between Darwin and Elizabeth Pole at Radbourne church in 1781. Richard and Millicent French were also friends of Joseph Wright and the poet William Hayley, who described her as having 'an interesting tenderness of character like the painting of Corregio', while Wright painted their portraits (although only that of Richard seems to survive); they also owned and may have commissioned a landscape painting of Chee Tor with rocks, river and vegetation below, probably undertaken during the 1780s.[18] In that decade Richard selected various books for Darwin to order for the Philosophical Society library, including *Voyage to the Cape of Good Hope* (1785) by Anders Sparrman (1748–1820), volumes of the *Letters of the Bath Society* on agriculture and gardening and the *Stirpes Novae* (1784–5) of Charles-Louis L'Heritier de Brutelle (1746–1800). Others chosen by French included William Marshall's *Rural Economy of Norfolk* (1788), the

Lichfield botanical society's *System of Vegetables* (1783) and the *Essay Towards a System of Mineralogy* (1788) by the Swedish mineralogist and chemist Axel Frederic Cronstedt (1722–1765), which was translated by the Swedish chemist Gustav von Engeström (1738–1813) and enlarged by the Portuguese natural philosopher Jean Hyacinthe de Magellan (1723–1790). Although borrowed by Richard, given her interests in natural history and botany it seems likely that these and works on similar subjects were read by – and perhaps borrowed for – Millicent.[19]

L'Héritier's *Stirpes Novae* provided detailed botanical information alongside striking engravings and has been described as 'one of the most delightful flower books' of the period and 'an imposing piece of eighteenth-century book making' with 'spacious descriptions', taxonomic information and 'magnificently produced' plates. L'Héritier was a wealthy botanist who employed the Belgian–French artist Pierre-Joseph Redouté (1759–1840), draughtsman to the cabinet of the French queen Marie-Antoinette and the 'most celebrated' and 'popular' flower painter of his day, the French artist Louis-Barthélémy Fréret (1755–1831) and the British naturalist James Sowerby (1757–1822) to provide paintings that were then engraved to illustrate his book. Although never completed, his book detailed and sought to classify new and unusual plants according to the Linnaean system, and L'Héritier stayed in London in 1786 and 1787, along with Redouté for some of the time, to observe plants, obtaining help from Joseph Banks and Darwin's erstwhile Lunar friends William Withering and Jonathan Stokes, among others.[20]

While the fashionable pleasures of gardening were attractive to men and women alike and appeared far removed from political concerns, changes in poetical taste and the association between Darwin's philosophical ideas, political beliefs and reforming friends caused greater hostility during the revolutionary period, impacting upon the reception of his wider studies of horticulture and farming, especially *Phytologia*. This is exemplified by the ridicule Darwin was subjected to in contemporary cartoons along with his friend Joseph Priestley, while satirical poems such as the 'Loves of the Triangles' cleverly parodied the Darwinian style and focused on his amorous, animated 'Jacobin plants'.[21] A poem entitled *The Golden Age* (1796), ostensibly written by Darwin to his friend the physician Thomas Beddoes, satirised the close analogies between humans, other animals and vegetables in Darwin's works and the notion that plants differed 'from animals alone in name'. While his use of plant personification was acceptable in poetry, given the multiple classical and post-classical precedents, it was unacceptable to draw overly close analogies between humans, other animals and plants, especially at a time of great political uncertainty.[22] The writer of the *Golden Age* asked his readers to:

> See the dull clown survey with stupid stare
> Where leaves once grew, now periwigs of hair!
> While fluids, which a wondrous change betray,
> Ooze from the vernal bud, the summer spray,
> Differing from animals alone in name,
> (As botanists already half exclaim).

Darwin's portrayal of vegetables was held ludicrous because it was claimed he believed them to be 'susceptible of joy and woe', feeling everything that humans experience and knowing 'whate'er we know!' They were equally 'inclin'd to watch or sleep' and, like humans, glowed with 'warm desire' and caught 'from beauty's glance celestial fire!' And, if once wandering through 'shady' grove:

> Musing on absent lovers you would rove,
> And there with tempting step all heedless brush
> Too near some wanton metamorphos'd bush,
> Or only hear perchance the western breeze
> Steal murmuring through the animated trees,
> Beware, beware, lest to your cost you find
> The bushes dangerous, dangerous too the wind,
> Lest ah! Too late with shame and grief you feel
> What your fictitious pads would ill conceal![23]

The anonymous writer of *The Millenium* (1801), another satirical poem, claimed that in Darwin's work:

> Already plants may hate, desire, and love,
> Think, reason, Judge!- with microscopic eye
> See the keen sage, womb, stomach, brains despy!
> Profanely curious mark their loose amours,
> And find for Auckland vegetable whores!

The poem went on to satirise this idea, suggesting that it implied that vegetables therefore deserved full legal and state protection. The writer noted that this was 'the serious creed of Dr. Darwin as a natural philosopher, as well as his system as a creative and vivacious poet' supported by the 'late bulky quarto' *Phytologia*, which was 'replete with entertainment' that combined 'about an equality of fanciful and unfounded opinions and of novel and highly valuable facts'. They claimed to have heard that in response to Darwin's arguments concerning parallels between vegetable and animal physiologies and behaviours a 'great multitude ... particularly among the ladies' (thereby poking fun at Darwin's encouragement of women's gardening and botanical contributions) had become 'converts', so much so that in 'more than one instance' the 'fair proselyte[s]' had 'extended it to so extreme an attitude' as to give their geraniums and other plants 'an airing' in their carriages 'wherever they appeared sickly and seemed to require exercise'.[24]

With regard to Darwin's work on vegetable physiology and anatomy, even sympathetic contemporary philosophers were bewildered and sometimes infuriated by the constant elision between the animal and vegetable worlds. Joseph Banks, the president of the Royal Society, knew Darwin's work very well and encouraged his Linnaean translations and the experiments on plant physiology and anatomy. However, his candid view of him as expressed to others was mixed. Banks observed to Thomas

Andrew Knight that Darwin combined 'truth and falsehood, ingenuity and perversity of opinion, exactly in the manner we mix the ingredients of punch'. However, the regard in which Darwin's work in this field was held during the early nineteenth century is evident from the degree to which authorities such as James Edward Smith, the Linnean Society president and most senior figure in British botany after Banks, cited him with approval in, for instance, his *Introduction to Physiological and Systematical Botany* (1802), where he emphasised that *Phytologia* was 'a store of ingenious philosophy'. Sir John Sinclair, president of the Board of Agriculture, who helped to inspire the composition of *Phytologia*, stated that it was 'highly gratifying to me, to have prevailed on so able a writer as the celebrated Dr. Darwin ... to draw up a work on practical agriculture' and judged the book to be 'a most valuable performance', though 'of too philosophical a description' to qualify as practical agriculture.[25]

Darwin's analyses of vegetable physiology and anatomy and his efforts to apply these to horticulture were also influential during the nineteenth century. However, they were initially overshadowed by Humphry Davy's *Elements of Agricultural Chemistry* (1813) – published for the Board of Agriculture and based upon a series of Royal Institution lectures on 'Chemistry and Vegetable Physiology' from 1803. Davy began his career as an assistant to Darwin's friend the physician Thomas Beddoes at the Bristol Pneumatic Institution – which Darwin and the Derby Philosophers supported financially – before making his name as a chemist and popular lecturer at the new Royal Institution in London from 1801, serving as president of the Royal Society between 1820 and 1827.[26] Already, in his first book of 1800, when examining the effects of nitrous oxide upon vegetation, Davy argued that an experimental chemical approach would shed light upon the operation of manures and proclaimed that the 'chemistry of vegetation', though so obviously 'connected with agriculture' – the 'art on which we depend for our subsistence' – had been 'but little investigated', despite all the recent advances by Joseph Priestley, the Dutch chemist and naturalist Jan Ingenhousz and others, and his next study, *Analysis of Soils* (1805), began this process.[27]

Davy highlighted the work of Darwin as one of the 'enlightened' philosophers who had previously examined the 'physiology of vegetation', along with Nehemiah Grew, Marcello Malpighi, the Swiss naturalist Jean Sennebier and Knight, and emphasised that Darwin had used his knowledge of the 'living powers' of plants, the 'stimulus of air' upon their leaves and 'moisture upon their roots' to explain why they always grew in an upward direction where possible. He was, however, largely critical of Darwin's approach to plant physiology, probably encouraged by how much friends he had gained while in Bristol, such as Samuel Taylor Coleridge, Robert Southey and William Wordsworth, had ostensibly turned against Darwin's work (even if they continued to frequently use his ideas), as well as, perhaps, the satirical 'Loves of the Triangles' and his own approach to writing poetry. There were close parallels between Davy's introductory discourse, which sought to cast the science of chemistry as fundamental to all other sciences, and the manifesto for poetry's universal concern, laid out by Wordsworth and Coleridge in their preface to the *Lyrical Ballads* (1800), for whom Davy had corrected the proofs. In particular, Davy did not approve of the extent to which Darwin drew analogies between vegetable and animal systems, arguing that, as far as he was concerned, the former were only 'living systems' in a more limited sense operating by 'common physical agents' (as Davy thought Stephen Hales had shown so well) to convert 'the elements of common matter into organised structures', whereas the latter had 'animation' and 'voluntary locomotion'. Although he did not name Darwin's 'spirit of vegetation' directly, this was clearly Davy's target in his argument that there was nothing 'above common matter' or 'immaterial' in the 'vegetable economy'.

Paralleling the critique of Darwin's Augustine-style poetry made by Coleridge and Wordsworth in the *Elements of Agricultural Chemistry*, Davy argued that, while vegetable personification might be acceptable in 'poetic form' using the language of 'Dryads to our trees, and Sylphs to our flowers', it was not relevant to 'vegetable physiology nor indeed were ideas of 'animation' and 'irritability'. As if to balance these criticisms, however, he did credit Darwin for first asserting the 'beautiful principle' of the operation of sap flow and the 'deposition' of 'nutritive matter' to assist the 'opening of buds' in spring in the vegetable economy, which was experimentally verified by Knight. Davy was also aware of Darwin's arguments in *Phytologia* concerning soil composition and the means in which plants gained some nutrition from the soil, but he doubted, for chemical reasons, Darwin's belief that paring and burning land was effective because 'torrefaction' enabled clay to imbibe 'nutritive principles from the atmosphere', which were later absorbed by plants. Davy thought that paring and burning probably worked by breaking up clays and destroying 'inert' and 'useless' 'vegetable matter'.[28]

Although Davy's *Agricultural Chemistry* was more immediately influential than Darwin's *Phytologia* during the first half of the nineteenth century, in some respects, and in the longer run, especially through its enunciation of photosynthesis, plant physiology and nutrition – including the role of nitrogen and phosphorous, and attention to the evolutionary dimensions of botany, Darwin's book had greater impact. The Scottish landscape gardener John Claudius Loudon – a student of Andrew Coventry, holder of the first chair in agriculture at Edinburgh University and probably the most influential gardening authority of the first half of the nineteenth century – utilised Darwin's work on gardens, horticulture and agriculture in his voluminous publications, referencing *Phytologia* alongside Davy's *Agricultural Chemistry* in his *Encyclopaedia of Gardening*, noting the poem on the Swilcar oak in the *Arboretum et Fruiticetum Britannicum* (1838) and using it in his *Encyclopaedia of Agriculture* (1843). According to Loudon, while as a poet Darwin was deemed 'rather gaudy and fanciful', and apt to 'indulge in hypothesis', he possessed 'the great quality of being totally exempt from every kind of prejudice'.[29]

Charles Darwin also maintained quite a high opinion of his grandfather's *Phytologia* and sought to correct the fact that the German biologist Ernst Ludwig Krause (1839–1903) made little mention of it in his *Scientific Works of Erasmus Darwin* (1879) by turning his 'preliminary notice' into a fuller memoir. The serious attention that Charles Darwin devoted to plants was of course much informed by his experiences on HMS *Beagle*, voyaging around the world, but his close observations of – and experiments with – garden plants also contributed. Walking on a Sussex heath in 1860, Charles Darwin was amazed at how many insects were caught by the leaves of the common sun-dew (*Dorosera rotundifolia*) and was enticed into a lengthy study of insectivorous plants, the means in which they caught and digested their prey and the evolutionary processes they had undergone to provide them with this opportunity for carnivorous nourishment (Figure 82). His fascination with climbing and insectivorous plants, along with the powers of vegetable movement, is evident in his arguments concerning 'distributed' intelligence and his contention that the tips of radicles or roots were 'endowed' with special 'sensitiveness' and the 'power of directing the movements of the adjoining parts'. According to Charles Darwin, the radicle tip acted like the 'brain of one of the lower animals', 'receiving impressions from the sense-organs and directing the several movements' – views very redolent of those for which his grandfather had been ridiculed. Terms such as 'consciousness', 'cognition' and 'mind' are now often applied to plants, and it has been reaffirmed that they learn 'by association'.[30]

Phytologia and Darwin's work were subsequently praised by Edward John Russell (1872–1965), one of the most distinguished agricultural scientists in Britain during the twentieth century and director of the Rothamsted Experimental Station at Harpenden, Hertfordshire – one of the oldest agricultural research institutions in the world. Rothamsted was established by the Victorian entrepreneur

Figure 82. Common or round-leaved sundew (*Dorosera rotundifolia*), from E. Step, *Wayside and Woodland Blossoms*, new edition (1905).

and scientist John Bennet Lawes (1814–1900) in 1843 and made 'striking advances' using the 'application of modern scientific disciplines to agriculture'.[31] Russell emphasised how Darwin too had sought to apply the 'chemical and botanical knowledge of his day' to the 'problems of plant physiology' to benefit horticulture and agriculture. Although he judged Darwin's verse of 'no special distinction' and believed that he held 'unusual views' on the extent to which plants had similar sensation and volition to animals, he had made important contributions to 'agricultural science', especially in relation to plant nutrition, demonstrating how carbonic acid was taken up by the roots from soil and that sugar and mucilage might also be 'useful manures'. Likewise, Darwin's highlighting of the significance of the decomposition of water and of the role of nitrogen, ammonia and phosphorous in plant nutrition – a 'remarkable piece of intuition' – was a significant contribution. His suggestions concerning the application of bone ash, decaying animal and vegetable material, phosphate of lime and other materials as manures were equally important. Finally, Darwin's recommendations concerning soil analysis as a means of improving fertility and the use of chemical tests to find the 'main properties' of 'primary' calcareous, argillaceous and siliceous earths was also noteworthy.[32]

In Desmond King-Hele's view, *Phytologia* is 'the best of Darwin's prose works' and 'free from the fallacies that mar *Zoonomia*', being 'solidly grounded in the good earth' with numerous 'new ideas and some major discoveries'. Of these, his most 'important contribution' related to defining for the first time probably the 'most all-embracing of biological processes' after that of evolution in his description of photosynthesis, which went 'much further' than previous investigations. Likewise, his analysis of 'essential plant nutrients' was 'impressive', as was his work defining what would now be described as the 'carbon and nitrogen cycles in nature'. His 'suggestions for pesticides', such as the 'biological control of aphids' using the syrphid, or hover fly, were equally 'ingenious'.[33]

Finally, Darwin's arguments concerning animal agency, behaviour and intelligence – although these too attracted some contemporary hostility in the context of the French Revolution and ensuing wars – had an immediate and a longer-term impact upon natural history. The anonymous writer of *The Millenium* (1801) ridiculed Darwin's emphasis upon the wonders of insect behaviour, asserting that he would even condemn the 'hard-hearted wretch who exults upon having cleared his head of vermin' and who ought to be reminded that he had 'destroyed by so cruel and unjustful an action' 'thousands of free republics of lice'. The author claimed that it afforded 'a most sublime consolation' to a person of 'universal sympathy' who is 'covered with the itch' that so much 'organised matter' 'stolen' from their 'own system' might be transmuted into 'the forms of millions of microscopic animals – the sum total of whose happiness [was] perhaps greater than his could ever be, by whose destruction they have gained their existence!'[34] However, after the first edition of *Zoonomia* appeared in 1794, the *Analytical Review* proclaimed that Darwin's dissertation on the 'instinctive actions of animals' had a 'degree of ingenuity to which we imagine nothing comparable will be discovered in the multitude' of preceding works on the subject. The editor called it 'the most masterly' part of 'the whole work' and carried extensive quotations, including Darwin's descriptions of Indian birds, wasp ingenuity and arguments in favour of animal contracts and against perceptions of them as mere machines.[35]

The agricultural writer John Lawrence (1753–1839), who maintained that the 'rights of beasts' should be formally acknowledged and legally protected, seems to have drawn extensively upon Darwin's arguments concerning animal sagacity and behaviour. Drawing parallels with contemporary radicals and reformers, who emphasised the centrality of human rights, Lawrence argued that rights for animals arose 'spontaneously, from the conscience, or sense of moral obligation in man', who was:

> indispensably bound to bestow upon animals, in return for the benefit he derives from their services, good and sufficient nourishment, comfortable shelter, and merciful treatment; to commit no wanton outrage upon their feelings, while alive, and to put them to the speediest and least painful death, when it shall be necessary to deprive them of life.

He therefore proposed that 'the Rights of Beasts be formally acknowledged by the state' and that laws be designed on these principles to 'guard and protect them from acts of flagrant and wanton cruelty, whether committed by their owners or others'.[36] Lawrence had read Darwin's 'celebrated Zoonomia', which he referred to in his *Treatise on Horses* (1796/98) in relation to the health benefits to be derived from riding.[37] When seeking to explain the reasons for human cruelty to animals, Lawrence, like Darwin, partly blamed defects in 'early tuition' rather than 'a natural want of sensibility in the human heart' and the tendency to look upon animals as 'mere machines; animated yet without souls; endowed with feelings, but utterly devoid of rights'.[38] Darwin's work on animal instinct in *Zoonomia* was likewise cited by Henry Brougham in his *Dialogues on Instinct* (1844). Brougham repeated Darwin's account of observing the industrious wasp to show that it 'shows no little sagacity', as well as his descriptions of the crows in the north of Ireland smashing mussels on rocks and his observation that cuckoos would build their own nests if they could not steal others. The example of the monkey in the Exeter Exchange using stone tools also featured, although Brougham made his speaker argue that this might have been accomplished by training rather than learning, 'as apes have often times been taught human habits'.[39]

But probably the most important long-term impact of Darwin's studies of animal physiology and behaviour was the stimulus it provided for his grandson's globally influential scientific studies. In *The Descent of Man* (1871) Charles Darwin argued that there were many similarities between the mental systems of humans and animals that betrayed their common evolutionary origins. Likewise, although there are only a small number of direct references to *Zoonomia* in his *Expression of the Emotions in Man and Animals* (1872) and the former's psychology is not rated highly in his 'Life of Erasmus Darwin', there is much evidence that it had a strong impact upon his whole programme of research, not least the behavioural and physiological parallels and analogies drawn from across the human and animal worlds and the context and manner in which emotions were physically expressed.[40] The way in which Charles Darwin closely observed the behaviour and development of domestic animals and his own children to understand both the relationship between instinctual and socially acquired behaviours and assumptions concerning animal emotions and intelligence all drew upon his grandfather's work. Charles Darwin's observations on infant behaviour and the stages of his children's development, which were recorded in a diary begun in 1839 after the birth of his first son William and utilised in his *Expressions of the Emotions* and in an article on infant development in *Mind* (1877), included notes upon their first signs of pleasure, anger and fear, how they behaved during breastfeeding and parallel behaviours from across the animal and vegetable worlds.[41]

Having ranged across the natural history of the world and the wonders of arts, sciences and culture, the *Botanic Garden* concluded fittingly just as it had started, in a botanic garden, and it is worth returning to those special places where we began to conclude this book. In the final peroration of *The Economy of Nature*, presenting what is, in some ways, the culmination of the whole epic poem, Darwin returned to plant science as the agent of good health, underlining how much his interest in this was determined by his medical studies and practice. Journeying to 'Imperial Kew' and sitting on the south bank of the 'glittering' Thames 'enthron'd in vegetable pride', he brought together medicine and the landscape in a hymn of praise to Hygeia, the goddess of health, for the life-giving properties of the vegetable world that furnished so many of his favourite treatments, which Seward believed to be one of the most 'eminently beautiful' poetical depictions in the work. Here:

> Obedient sails from realms unfurrow'd bring
> For her the unnam'd progeny of spring.

In the gardens, nymphs carefully nurtured the young seedlings and bulbs to maturity, supporting 'the weak stem' while training 'the erring tendril'. They fanned:

> in glass-built panes the stranger flowers
> With milder gales, and steep with warmer showers.
> Delighted Thames through tropic umbrage glides,
> And flowers antarctic, bending o'er his tides;
> Drinks the new tints, the sweets unknown inhales,
> And calls the sons of science to his vales.

Enlarged and much developed under Princess Augusta of Saxe-Gotha-Altenburg (1719–1772), dowager Princess of Wales, and John Stuart, third earl of Bute (1713–1792), with assistance from John Hill and William Aiton (1731–1793), and superintended by Joseph Banks from around 1773, while remaining the 'private property' of the king, Kew grew into the largest botanical garden in the empire, combining various houses and gardens acquired at different times. A living herbarium and centre for study and plant exchange that brought the 'sons of science [and some daughters!] to its vales' with its physic and systematic collections and hothouses, under Banks it operated as part of an international network of botanical gardens. Like the earlier Chelsea Physic Garden, Kew's Thames location enabled the river to serve as a grand highway for trans-oceanic exchanges of plants and botanical knowledge, while landscaping and planting beautified the gardens with avenues, plantations, hothouses and structures such as Sir William Chambers' famous pagoda (1761). By the 1780s Kew had a major plant collection and large herbarium, as the catalogues published by Hill (1768) and Aiton (1789) demonstrate.[42]

With its global acquisitions, 'Imperial Kew' operated under Banks as a centre for the exchange of plants and vegetable knowledge from the British colonies and further afield. The analyses of Darwin's medical and botanical investigations throughout this book have demonstrated the strength of international Enlightenment inter-connections between trade, manufactures, medico-botanical practice and the sciences. Darwin was an eager believer in Enlightenment internationalism and the ideals of the republic of letters, and enthusiastically sought out and embraced scientific ideas, practices and plants from international sources, including botanists from France, Sweden, the German states, Italy and North America, even during periods of war, an attitude also evident among other members of the Derby Philosophical Society, who acquired many international works for their library and accounts of global exploration and travel.[43] The supply of plants, vegetable substances and information ultimately from sources such as the Amerindians of North America, indigenous peoples of Brazil and West Indian plantations through the competing maritime trading and colonial activities of the Portuguese, Spanish, Dutch, French, English and other nations was certainly tainted by the pernicious dimensions of European ambition, as the miseries of the Atlantic slave trade triangle between Africa, Europe and the Americas demonstrate. The colonies and plantations provided crops such as tobacco, logwood, cotton and sugar, as well as markets for European products, while the difficulties experienced by Europeans and transplanted peoples facing the climatic conditions and diseases of the 'Torrid Zone' and their efforts to mitigate these provided new sources of medical knowledge. Darwin was able to take advantage of information and materials drawn from the colonies, such as medico-botanical knowledge and plant substances. He was, however, strongly opposed to the more aggressive dimensions of colonialism, including international warfare and the slave trade.[44]

The national campaign against slavery led by William Wilberforce (1759–1833), Thomas Clarkson (1760–1846), Granville Sharpe (1735–1813) and others, including former slaves such as Olaudah Equiano (1745–1797), focused upon organising meetings and efforts to pass legislation and get parliament to debate the issue, circulating accounts of the trade's brutality such as Equiano's autobiographical *Interesting Narrative* (1789) and mounting a very effective West Indian sugar boycott campaign. Darwin and friends such as Josiah Wedgwood, Francis Noel Clark Mundy and Thomas Gisborne helped lead the midland agitation. In 1788 a petition supported by the Derby mayor and corporation and many inhabitants was presented to the House of Commons by the borough MPs, which stated that the slave trade was 'the greatest cruelty and oppression' and an 'inhuman commerce ... universally' to be 'held in detestation'. At the same time, William Shakespeare's play 'Othello the Moor of Venice' was performed for the benefit of actor Thomas Grist (c. 1750–1808) at the theatre in Derby, along with a farce and a recitation by Grist of lines from 'The Task' (1785) by William Cowper (1731–1800) on the 'subject of the AFRICAN SLAVE TRADE'. Another petition for the abolition of the 'horrible and detestable' trade was prepared by the county in 1792 by Thomas Gisborne.[45]

There was, however, a powerful national lobby of merchants and investors fighting back, and local supporters who opposed the petition, included Bache Heathcote (1759–1826) of Littleover near Derby, who maintained that ending slavery would destroy the British economy and simply provide opportunities for international rivals and that it could be continued with greater regulation. Henry Redhead Yorke (1772–1813) was a Creole of dual heritage from a plantation-owning, gentry family from the West Indian island of Barbuda, whose mother, Sarah Bullock, was a dual heritage slave. He was educated as a lawyer at Cambridge University and brought up at Little Eaton, Derbyshire, and published a pamphlet dedicated to Heathcote arguing that the trade's evils had been exaggerated and that it was too vital to national prosperity to be suddenly curtailed. A strong advocate of political reforms, like Darwin, Redhead Yorke was a supporter of the Derby Society for Political Information and accompanied the physician William Brookes Johnson in 1792 as one of two delegates who presented the society's petition to the French National Assembly in Paris. In the booklet, which he later renounced, Redhead Yorke claimed that slaves on British-owned islands were treated with 'enlightened reason and humanity', that they were 'happier' than many European freeman and that they tended to bring problems of ill health upon themselves, exacerbated by a harsh climate, while brutalities might be prevented by new government regulations such as limiting the size of ships transporting them from Africa.[46]

In *The Loves of the Plants*, while detailing the movement of the black seeds of cassia (wild senna, *Senna hebecarpa*) whose 'tawny children' were conveyed by currents across the Atlantic from North America to northern Europe, Darwin argued forcefully that if parliament knew that 'OPPRESSION' was occurring yet failed to act it 'SHARES THE CRIME':

> E'en now, in Afric's groves with hideous yell
> Fierce SLAVERY stalks, and slips the dogs of hell;
> From vale to vale the gathering cries rebound,
> And sable nations tremble at the sound!

In *The Economy of Vegetation* he doubtless intentionally shocked some of his readers by likening the British-led slave business to 'murder, rapine, theft' that passed under the name 'trade' and comparing it to the massacres of the Spanish Conquistadors in the Americas undertaken in the name of Roman Catholicism. He also included a reproduction of Wedgwood's medallion of a slave in chains imploring to be free against the 'detestable traffic in human creatures', explaining it in a footnote below.[47] In 1789, on hearing that muzzles and gags for slaves were being made in Birmingham, Darwin urged that 'long whips or wire tails' and other instruments 'of torture of our own manufacture' might be exhibited in parliament during the debates to 'great effect'. He introduced the subject of abolition again when discussing sugar cane in *Phytologia*, imploring the 'Great God of Justice' to 'grant that the crop might 'soon be cultivated only by the hands of freedom', thereby providing 'happiness' for labourers, merchants and consumers of sweet treats.[48]

It is probable that Erasmus and Elizabeth Darwin visited Kew when they met Joseph and Dorothea Banks during their honeymoon stay in London in April and May 1781, and it was after this that Darwin's correspondence with its superintendent began concerning the Lichfield botanical society's Linnaean translation. The Derby Philosophical Society acquired Aiton's *Hortus Kewensis* (1789) for their library as soon as it was published, which Darwin probably used to inform the portrayal of Kew in *The Economy of Vegetation*.[49] The royal botanical gardens were already the subject of at least two poems by George Ritso (1763) and the Irish poet and dramatist Henry Jones (1767), which Darwin may have been familiar with, although his emphasis was primarily upon the union of botany with medicine embodied by the gardens and signified by Hygeia. Jones celebrated the transformation of an unpromising Thames-side site into a royal paradise, a 'cultivated world' of 'sublime and grandeur' and 'blooming majesty benign' in which, guided by the king's 'princely soul' and Bute's 'Genius, sense and taste', 'teeming' exotic 'vegetating life' from across the globe intermingled its 'aromatic fragrances' and flourished for the benefit of 'godlike science' and 'arts sublime'.[50]

Conscious of the king's recent recovery from a crippling bout of insanity that had almost resulted in a regency under George, Prince of Wales (the future George IV), Darwin portrayed the gardens as the place where the 'ROYAL PARTNERS' George and Queen Charlotte could now find healthful solace walking among the 'exotic glades' from Britain and numerous foreign lands. The wonderful collections at Kew and other botanic gardens formed for the benefit of science and medicine were as if Nature's 'admiring eyes' had come to feast on the beauty of its own 'fruits and foliage'. Transporting precious plants from far-flung countries and 'discordant skies' under the beneficence of royal patronage to 'Britain's happier clime' enabled them to reach the perfection enjoyed by the 'sweet blooms', roses and 'towering' oaks of the nation's own 'fair Scions', which graced and guarded its 'golden lands'.[51] As Kew and his medical practice demonstrated, botanical gardens brought together 'rich balms' and 'sweet flowers' from 'Arabia's shades', fruits from the 'banks of Arno, or of Po', fragrant tea leaves from China, spices from 'sultry India', which scented 'the night air round her breezy coasts', roots from 'bleak Siberia' and Peruvian bark, whose 'broad umbrage' waved over 'Andes steeps' – all in honour more than anything else of 'HYGEIA'S shrine'. So, according to Darwin, the goddess waved her 'serpent wand' over 'BRITANNIA'S throne' and marked it for her 'own', shedding over these 'realms' her 'beamy smile' of health and prosperity – aided by the medical arts and his own practice.[52]

Yet, despite delighting in Kew and being offered the chance to live in the capital, the largest cultural, scientific and political centre in the nation, Darwin never did so. After he successfully treated one of the daughters of naval captain Charles Fielding and his wife Sophia (Finch), who stayed in Derby with the Darwin family, word reached the king and queen. This was because Sophia was a woman of the bedchamber of Queen Charlotte, while her mother Lady Charlotte Finch (1725–1813) was governess to the royal children. According to Charles Darwin, who had it from one of the granddaughters of King George and Queen Charlotte, having been told about how brilliant a doctor Darwin was George asked 'why does not Dr. Darwin come to London? He shall be my physician if he comes', which he repeated 'in his usual manner'. Darwin also treated Fanny Coutts (1773–1832) in Derby, daughter of Thomas Coutts (1735–1822), the royal banker, and after the three travelled to Matlock for her health in 1796 the latter remarked to prime minister William Pitt that he had been 'detained in Derbyshire' longer than intended because of her illness and the 'desire I had for the advice of Dr. Darwin, whose genius and skill I hold in high esteem'. Darwin also knew the royal physicians Sir George Baker, whose medical lectures he attended when a student and who contributed an appendix to his paper on the medicinal uses of the foxglove, and later Richard Warren (1731–1797), who, according to Violetta Darwin, consulted her father at Derby as a last resort in 1797, to be candidly told he had no chance of recovery, which helped him put his affairs in order.

Despite the testimonials and offers of lucrative practice that would have surely followed, after considering the offer of being a royal physician, according to Charles

Darwin, Elizabeth and his grandfather agreed that they 'disliked the thoughts of a London life so much that the hint was not acted on'. Darwin's reformist politics and affiliation with the Cavendish family, who were associated with Prince George and Charles James Fox's Whig party, might have presented difficulties too, although he had patients of all political persuasions. It must also be said that, while Darwin wished the royal couple good health in the 'exotic glades' of Kew and had eaten and drunk 'for the king with all our might' during his illness in 1789, as he remarked to Josiah Wedgwood, he also took 'great pleasure' in the 'success of the French against a confederacy of Kings' in 1792, as he told his 'dear old friend' Richard Dixon. He also noted to Dixon, with some undeferential relish, that Thomas Paine thought a 'monkey, or a bear, or a goose, may govern as well' and much more more cheaply than 'any king in Christendom, whether idiot or madman in his royal senses', although for Darwin the former was not as much of an insult as it might have been for many of his contemporaries.[53] Instead of coming to London, with its million inhabitants, he chose to enjoy the intellectual stimulation and comraderie of midlands Enlightenment society, the beauties and productivity of the countryside and the company of his family, friends and numerous patients, forming his own green Lichfield and Derby shrines to the goddess Hygeia. These became rich founts of botanical and medical wisdom and poetical inspiration, thereby helping give him a national fame he might never have obtained if he had disappeared into the great metropolis.

NOTES

[1]   C. Darwin, *Life of Erasmus Darwin*, edited by D. King-Hele (Cambridge, 2003), 31–4; D. King-Hele, *Erasmus Darwin and the Romantic Poets* (Basingstoke, 1986), 62–147, 187–226, 272–81.

[2]   P. A. Elliott, *Enlightenment, Modernity and Science: Geographies of Scientific Culture and Improvement in Georgian England* (London, 2010), 125–66.

[3]   P. Elliott, 'The Derby Arboretum (1840): The first specially designed municipal public park in Britain', *Midland History*, 26 (2001), 144–76; P. A. Elliott, C. Watkins and S. Daniels, *The British Arboretum: Trees, Science and Culture in the Nineteenth Century* (London, 2011), 135–54, 185–209.

[4]   M. Allan, *Darwin and His Flowers: The Key to Natural Selection* (London, 1977); P. Ayres, *The Aliveness of Plants: The Darwins at the Dawn of Plant Science* (London, 2008).

[5]   Ayres, *Aliveness of Plants*, 5.

[6]   J. Bate, *Romantic Ecology: Wordsworth and the Environmental Tradition* (London, 1991); J. Bate, *The Song of the Earth* (London, 2000); L. Ottum and S. T. Reno ed., *Wordsworth and the Green Romantics* (Lebanon NH, 2016), especially S. Reno, 'Rethinking the Romantics' love of nature', 28–58; B. P. Robertson ed., *Romantic Sustainability: Endurance and the Natural World, 1780–1830* (Lanham, 2016).

[7]   P. Worsley, *The Darwin Farms: The Lincolnshire Estates of Charles and Erasmus Darwin and their Family* (Lichfield, 2017).

[8]   E. Darwin, *Phytologia; or the Philosophy of Agriculture and Gardening* (London, 1800), 329, 335–7, plate IX, 432–3; R. Longrigg, *The English Squire and his Sport* (London, 1977), 117–24.

[9]   Darwin, *Phytologia*, 229–330.

[10] R. W. Darwin, *Principia Botanica: or a Concise and Easy Introduction to the Sexual Botany of Linnaeus* (Newark, 1787); F. Darwin ed., *The Life and Letters of Charles Darwin*, 2 vols (New York, 1925), vol. 1, 9.

11 E. Darwin, letter to R. Darwin, 8 August 1799, in D. King-Hele ed., *The Collected Letters of Erasmus Darwin* (Cambridge 2007), 530; Darwin, *Life of Erasmus Darwin*, 13.

12 E. Darwin, *The Loves of the Plants*, 4th edn (London, 1799), 88–9; A. Seward, *Memoirs of the Life of Dr. Darwin* (London, 1804), 314–17; A. B. Shteir, *Cultivating Women, Cultivating Science: Flora's Daughters and Botany in England, 1760–1860* (Baltimore, 1999); J. Uglow, 'But what about the women?: The Lunar Society's attitude to women and science and to the education of girls', in C. U. M. Smith and R. Arnott eds, *The Genius of Erasmus Darwin* (Aldershot, 2005), 163–78; M. Laird and A. Weisberg-Roberts, A. eds, *Mrs. Delany & Her Circle* (New Haven, 2009); P. A. Elliott, *The Derby Philosophers: Science and Culture in British Urban Society, 1700–1850* (Manchester, 2009), 217–34; Elliott, *Enlightenment, Modernity and Science*, 17–47; T. M. Kelley, *Clandestine Marriage: Botany and Romantic Culture* (Baltimore, 2012), 90–125; M. Priestman, *The Poetry of Erasmus Darwin: Enlightened Spaces, Romantic Times* (Farnham, 2013), 250; M. Laird, *A Natural History of English Gardening* (New Haven, 2015), 231–325.

13 M. E. Jacson [originally published anonymously by 'a Lady'], *Botanical Dialogues between Hortensia and her Four Children* (London, 1797); E. Darwin, letters to M. E. Jacson, 24 August 1795, J. Borough, 1 March 1802, and B. Strutt, Summer 1801, 16 March, March/April 1802 in King-Hele ed., *Collected Letters*, 482–3, 576–7, 574, 577, 579.

14 S. Bowerbank, *Speaking for Nature: Women and Ecologies of Early-Modern England* (Baltimore, 2004); T. Barnard, *Anna Seward: A Constructed Life: A Critical Biography* (Aldershot, 2009); S. Bending, *Green Retreats: Women, Gardens and Eighteenth-Century Culture* (Cambridge, 2013).

15 E. Darwin, letter to J. Craddock, 21 November 1775 and letter to E. Pole, late 1775? in King-Hele ed., *Collected Letters*, 137–40; D. King-Hele, *Erasmus Darwin: A Life of Unequalled Achievement* (London, 1999), 357.

16 The tomb and memorial for Millicent may have been originally planned for St Alkmund's Church, Duffield, the parish church of the Mundy family, which contains various memorials to them and seems to fit Darwin's description. However, this was apparently never executed, as Millicent's memorial was placed in Mackworth church near Derby instead, which is not near the Derwent.

17 E. Darwin, *The Economy of Vegetation*, 4th edn (London, 1799), 158–61; marriage between Richard French of All Saint's Parish and Millicent Mundy on 8 February, 1770 recorded in the marriage register of St Peter's Church, Derby, Derbyshire Record Office, D1792/A/PI/3/1, transcription available at http://www.tinstaafl.co.uk/eandwhmi/derbyshire/church%20pages/derby_st_peter.htm [accessed 19 January 2021].

18 B. Nicholson, *Joseph Wright of Derby, Painter of Light*, 2 vols (London, 1968), vol. 1, 97–8, 155, vol. 2, plates 85 and 265, pp. 58, 168; King-Hele, *Erasmus Darwin: A Life of Unequalled Achievement*, 172, Elliott, *Derby Philosophers*, 216–34, 269.

19 Derby Local Studies Library, Manuscript catalogue and charging ledger of the Derby Philosophical Society, 1785–9, BA 106, 9229; Lichfield Botanical Society, *A System of Vegetables … translated from the thirteenth edition of the Systema Vegetabilium of the late Professor Linneus*, 2 vols (Lichfield, 1783); A. Sparrman, *A Voyage to the Cape of Good Hope: Towards the Antarctic Polar Circle, and Round the World*, translated by G. Forster (London, 1785); F. Cronstedt, *Essay Towards a System of Mineralogy*, translated by G. von Engestrom and enlarged by J. H. de Magellan (London, 1788); W. Marshall, *The Rural Economy of Norfolk*, 2 vols (London, 1788).

20 C. L. L'Héritier de Brutelle, *Stirpes Novae, aut minus Cognitae, quas descriptionibus et iconibus illustravit* (Paris, 1784–5); J. Britten and B. B. Woodward, 'L'Héritier's Botanical Works', *The Journal of Botany*, 43 (1905), 266–273; 325–329; W. Blunt, *The Art of Botanical Illustration* (London, 1950), 173–82, 190–2; G. Bucheim, 'A bibliographical account of L'Héritier's *Stirpes Novae*', *Huntia*, 2 (15 October 1965), 29–58; F. A. Stafleu, 'L'Héritier du Brutelle: The man and his work' in C. L. L'Héritier de Brutelle, *Sertum Anglicum: Facsimilie with Critical Studies and a Translation*, edited by G. H. M. Lawrence (Pittsburgh, 1963), xiii–xliii.

[21] Darwin, *Life of Erasmus Darwin*, 31–4; King-Hele, *Erasmus Darwin and the Romantic Poets*, 62–147, 187–226, 272–81; P. Fara, *Erasmus Darwin, Sex, Science and Serendipity* (Oxford 2012), 30–42, 259–80.

[22] Anonymous, *The Golden Age: A Poetical Epistle from Erasmus D---N to Thomas Beddoes, MD* (London, 1794), 7–8.

[23] Anonymous, *The Golden Age*, 7–8.

[24] Anonymous, *The Millenium; a Poem in Three Cantos*, 2 vols (London, 1801), 99–102.

[25] *Life and Works of Sir John Sinclair*, vol. 2, p. 85, quoted in King-Hele ed., *Collected Letters*, 517; Darwin, *Life of Erasmus Darwin*, 40–1.

[26] H. Davy, *Elements of Agricultural Chemistry*, 2nd edn (London, 1814); A. and N. Clow, *The Chemical Revolution: A Contribution to Social Technology* (London, 1952), 466, 493–501; H. Hartley, *Humphry Davy* (Wakefield, 1972), 44, 96–9; King-Hele, *Erasmus Darwin and the Romantic Poets*, 23–5, 91–3, 176–7; S. Wilmot, 'The Business of Improvement': Agriculture and Scientific Culture in Britain, c1700–c1870, Historical Geography Research Series, 24 (1990), 22–3; J. Golinski, *The Experimental Self: Humphry Davy and the Making of Man of Science* (Chicago, 2016), 133–9.

[27] H. Davy, *Researches Chemical and Philosophical Chiefly Concerning Nitrous Oxide* (Bristol, 1800), 563–4; H. Davy, *On the Analysis of Soils as Connected with their Improvement* (London, 1805).

[28] Davy, *Elements of Agricultural Chemistry*, 10, 33, 248–51, 349; S. Coleridge and W. Wordsworth, *Lyrical Ballads*, edited by R. L. Brett and A. R. Jones, 2nd edn (Abingdon, 2005 [1800]), 29, 286–314; Hartley, *Humphry Davy*, 22–5, 41; King-Hele, *Erasmus Darwin and the Romantic Poets*, 62–79, 91–147; Priestman, *The Poetry of Erasmus Darwin*, 224–37.

[29] J. C. Loudon, *Arboretum et Fruticetum Britannicum*, 2nd edn, 8 vols (London, 1844), vol. 3, 1769; J. C. Loudon, *An Encyclopaedia of Gardening*, 2nd edn (London, 1824), 1109, new edition edited by J. Loudon (London, 1865), xxv; J. C. Loudon, *An Encyclopaedia of Agriculture*, 5th edn (London, 1844), 1211; Wilmot, 'The Business of Improvement', 32–3.

[30] Darwin, *Life of Erasmus Darwin*, 40–1; C. Darwin, *Insectivorous Plants*, 2nd edn, revised by F. Darwin (London, 1908), 1–17; K. Thompson, *Darwin's Most Wonderful Plants: Darwin's Botany Today* (London, 2019), quotation, 85; S. Manuso and A. Viola, *Brilliant Green: The Surprising History and Science of Plant Intelligence* (Washington DC, 2015).

[31] For the significance and development of Rothamsted: E. J. Russell, *A History of Agricultural Science in Great Britain, 1620–1954* (London, 1966), 88–107, 143–75, 232–43, 271–7, 289–332.

[32] Russell, *History of Agricultural Science*, 62–3, 67–76, 107–9.

[33] King-Hele, *Erasmus Darwin A Life of Unequalled Achievement*, 333; King-Hele, *Erasmus Darwin and the Romantic Poets*, 23–24.

[34] Anonymous, *The Millenium*, 99–102.

[35] *Analytical Review* (August 1794), 337–44.

[36] J. Lawrence, *A Philosophical and Practical Treatise on Horses and on the Moral Duties of Man towards the Brute Creation*, 2 vols (London, 1796–8), vol. 1, 119–20, 123.

[37] Lawrence, *Treatise on Horses*, 273; E. Darwin, *Zoonomia; or the Laws of Organic Life* 2nd edn, 2 vols (London, 1796), vol. 1, 293–7.

[38] Lawrence, *Treatise on Horses*, 118; compare Darwin, *Zoonomia*, 3rd edn, 4 vols (London, 1801), vol. 1, 187.

[39] H. Brougham, *Dialogues on Instinct* (London; 1844), 101–3, 122.

[40] C. Darwin, *The Descent of Man and Selection in Relation to Sex* (London, 1871); C. Darwin, *The Expression of the Emotions in Man and Animals*, 2nd edn, edited by F. Darwin (London, 1904), 30, 48, 77; Darwin, *Life of Erasmus Darwin*, 36–7.

[41] Darwin, *Expression of the Emotions in Man and Animals*; C. Darwin, 'A biographical sketch of an infant', *Mind*, 2 (1877), 285–94; C. Darwin, 'Observations on his children', in F. H. Burkhardt ed., *The Correspondence of Charles Darwin* (Cambridge, 1988), vol. 4, 410–33; H. E. Gruber, *Darwin on Man: A Psychological Study of Scientific Creativity* (London, 1974), 218–42; S.

Shuttleworth, 'The psychology of childhood in Victorian literature and medicine', in H. Small and T. Tate eds, *Literature, Science, Psychoanalysis, 1830–1970: Essays in Honour of Gillian Beer* (Oxford, 2003), 86–101; S. Shuttleworth, *The Mind of the Child: Child Development in Literature, Science and Medicine, 1840–1900* (Oxford, 2010), 221–32.

[42] J. Hill, *Hortus Kewensis* (London, 1768); Seward, *Erasmus Darwin*, 280–1; Loudon, *Encyclopaedia of Gardening*, 1068; J. Smith, *Records of the Royal Botanic Gardens, Kew* (London, 1880); D. P. Miller, 'Joseph Banks, empire and "centres of calculation" in late Hanoverian London', in D. P. Millar and P. H. Reill eds, *Visions of Empire: Voyages, Botany and Representations of Nature* (Cambridge, 1996), 21–37; R. Desmond, *Kew: The History of the Royal Botanic Gardens* (London, 1998), 1–126; R. Drayton, *Nature's Government: Science, Imperial Britain, and the Improvement of the World* (New Haven, 2000), 152–92; L. H. Brockway, *Science and Colonial Expansion: the Role of the British Royal Botanic Gardens* (New Haven, 2002), 1–8.

[43] Works acquired by the Derby Philosophical Society included: G. Juan and A. Ulloa, *A Voyage to South America*, translated from Spanish with notes by J. Adams, 3rd edn, 2 vols (London, 1772); A. Sparman, *A Voyage to the Cape of Good Hope ... and Round the World*, translated from Swedish, 2 vols (London, 1786); P. J. Marshall and G. Williams, *The Great Map of Mankind: British Perceptions of the World in the Age of Enlightenment* (London; J. M. Dent and Sons, 1982); D. N. Livingstone and C. W. J. Withers eds, *Geography and Enlightenment* (Chicago, 1999); K. Sloan ed., *Enlightenment: Discovering the World in the Eighteenth Century* (London, 2003); S. Seymour, L. Jones and J. Feuer-Cotter, 'The global connections of cotton in the Derwent Valley Mills in the later eighteenth and early nineteenth centuries', in C. Wrigley ed., *The Industrial Revolution: Cromford, the Derwent Valley and the Wider World* (Cromford, 2015), 150–70.

[44] P. Wallis, 'Consumption, retailing and medicine in early-modern London', *Economic History Review*, 61 (2008), 26–53; L. Schiebinger, *Plants and Empire: Colonial Bioprospecting in the Atlantic World* (Cambridge MA, 2017); P. Chakrabarti, *Materials and Medicine: Trade, Conquest and Therapeutics in the Eighteenth Century* (Manchester, 2010); J. Carney and R. N. Rosomoff, *In the Shadow of Slavery: Africa's Botanical Legacy in the Atlantic World* (Berkeley, 2011); P. Wallis, 'Exotic drugs and English medicine: England's drug trade, c1550–c1800', *Social History of Medicine*, 25 (2012), 20–46; L. Schiebinger, *Secret Cures of Slaves: People, Plants and Medicine in the Eighteenth-Century Atlantic World* (Chicago, 2017); S. Seth, *Difference and Disease: Medicine, Race and the Eighteenth-Century British Empire* (Cambridge, 2018).

[45] *Derby Mercury*, 31 January, 14 February 1788, 22 March 1792; W. Cowper, *The Task* (London, 1817 [1785]); the reference is probably to the lines on pp. 32–3.

[46] *Derby Mercury*, 2 October, 17 July 1788, 28 June 1792, 31 August 1797; H. Redhead (Yorke), *A Letter to Bache Heathcote esq., on the Fatal Consequences of Abolishing the Slave Trade, both to England and her American Colonies* (London, 1792), 7–9, 12–13, 49–54; Elliott, *Derby Philosophers*, 91–8; A. Goodrich, 'Radical "Citizens of the World", 1790–85: the early career of Henry Redhead Yorke', *Journal of British Studies*, 53 (2014), 611–35; A. Goodrich, *Henry Redhead Yorke, Colonial Radical: Politics and Identity in the Atlantic World, 1772–1813* (2019). Yorke's original (family) name was Redhead but he changed it to Yorke in 1792.

[47] Darwin, *Loves of the Plants*, 160–5; Lichfield Botanical Society, *Families of Plants ... Translated from the Last Edition of the 'Genera Plantarum'*, 2 vols (Lichfield, 1787), vol. 1, 280; Darwin, *Economy of Vegetation*, 101; Darwin, *Life of Erasmus Darwin*, 77–8, 86–7; King-Hele, *Erasmus Darwin: A Life of Unequalled Achievement*, 231–2; Fara, *Erasmus Darwin*, 164–84.

[48] Darwin, letter to J. Wedgwood, 18 July 1789, in King-Hele ed., *Collected Letters*, 345–6; Darwin, *Phytologia*, 70; King-Hele, *Erasmus Darwin: A Life of Unequalled Achievement*, 232.

[49] W. Aiton, *Hortus Kewensis*, 3 vols (London, 1789); Derby Local Studies Library, manuscript catalogue and charging ledger of the Derby Philosophical Society; E. Darwin, letter to J. Banks, 13 September 1781, in King-Hele ed., *Collected Letters*, 186–8; King-Hele, *Erasmus Darwin: A Life of Unequalled Achievement*, 172–3, 179–82.

[110] G. Ritso, *Kew Gardens: A Poem, humbly inscribed to Her Royal Highness the Princess Dowager of Wales* (London, 1763); H. Jones, *Kew Garden: a Poem in Two Cantos* (London, 1767), 8, 13, 19, 22–3, 32–3, 44.

[111] Darwin, *Economy of Vegetation*, 234–5; I. Macalpine and R. Hunter, *George III and the Mad Business* (London, 1969).

[112] Darwin, *Economy of Vegetation*, 236–8.

[113] Darwin, *Life of Erasmus Darwin*, 37, 69; E. Darwin, letters to R. Dixon and T. Coutts, 25 October 1792 and 5 November 1796, in King-Hele ed., *Collected Letters*, 409–11, 504–5; King-Hele, *Erasmus Darwin: A Life of Unequalled Achievement*, 13, 21, 207.

SELECT BIBLIOGRAPHY

MANUSCRIPT MATERIAL

Cambridge University Library, Department of Manuscripts and Special Collections

Cambridge University Botanical Garden, minute books of the trustees and other papers

Darwin Papers (DAR)

Darwin, C., manuscript nature printed notebook, MS. DAR, AD 101.41 reproduced with notes by A. Secord at: https://cudl.lib.cam.ac.uk/view/MS-ADD-10141

Darwin E., manuscript catalogue of hardy plants (1796), DAR, 227.2.11

Derby Local Studies Library, Derby

Bennet, A., 'Memoranda Miscellania', commonplace book

Derby Philosophical Society, manuscript catalogue and charging ledger, 1785–9 and cash ledger, 1813–1845 (BA 106, 9229–9230)

Erasmus Darwin, commonplace book, microfilm copy of original at Down House Museum

Mundy Papers

Nun's Green box of broadsides and pamphlets (box 27).

Paving and Lighting Commissioners, Minute Book of the Paving and Lighting Commissioners, 1789–1825 (BA 625/8, 16048)

Strutt papers and correspondence on microfilm (D125/-). Originals moved to Derbyshire County Record Office

Derbyshire County Records Office, Matlock

E. Darwin, prescriptions for Sir Henry and Lady Frances Harpur, 1771–1787, Harpur Crewe Family Papers, Derbyshire County Record Office, Matlock, D2375/F/G/1/5/7

Strutt family papers, including deeds, estate and family papers and correspondence (D1564, D2912, D2943M, D3772, D5303)

Birmingham Central Library, Archives and Heritage

Galton papers, Birmingham Central Library, Archives and Heritage (MS3101)

Fitzwilliam Museum, Cambridge
Strutt papers (MS 48 – 1947)

Keele University Library
Wedgwood Manuscripts

Harris Manchester College, Oxford
Martineau papers

Lichfield, Staffordshire County Record Office (now moved to Stafford)
Account book of the Lichfield Conduit Lands Trust, 1741–1856, D126/2/1

Lichfield, Samuel Johnson Birthplace Museum
Anna Seward correspondence (2001.76.6)

Linnean Society Library, London
W. B. Coyte, letter to J. E. Smith, 27 February 1805, GB110/JES/COR/21/87

PERIODICALS

Derby Mercury
Gardener's Magazine
Gentleman's Magazine
Harrison's Derby and Nottingham Journal
Philosophical Transactions

WORKS PRINTED BEFORE c. 1900

Adam, W., *Gem of the Peak*, 2nd edn (Derby, 1840)

Addington, S., *An Inquiry into the Reasons for and Against Inclosing Open-Fields* (Coventry, 1772)

Aikin, C. R., *A Concise View of all the most Important Facts ... respecting the Cow Pox* (London, 1800)

Aikin, J., *A Description of the County from Thirty to Forty Miles round Manchester* (London, 1795)

Aikin, J., *An Essay on the Application of Natural History to Poetry* (Warrington, 1777)

Aikin, L., *Memoir of John Aikin, MD ... with a selection of his miscellaneous pieces, biographical, moral and critical*, 2 vols (London, 1823)

Aiton, W., *Hortus Kewensis*, 3 vols (London, 1789)

Alderson, J., *The Essay on the Rhus Toxicodendron: Pubescent Poison Oak, or Sumach*, 2nd edn (Hull, 1796)

Anderson, J., *Description of a Hot House* (London, 1803)

Anderson, J., *Essays Relating to Agriculture and Rural Affairs* (London, 1797)

Anonymous, *The Golden Age: A Poetical Epistle from Erasmus D---N to Thomas Beddoes, MD* (London, 1794)

Anonymous, *The Improved Culture of Three Principle Grasses, Lucerne, Sainfoin and Burnet* (London, 1775)

Anonymous, *The Millenium; a Poem in Three Cantos*, 2 vols (London, 1800–1)

Bacon, F., *The Works of Francis Bacon*, 10 vols (London, 1824)

Bakewell, R., *Observations on the Influence of Soil and Climate upon Wool* (London, 1808)

Banks, J., *A Short Account of the Causes of the Disease in Corn, Called by Farmer's the Blight, the Mildew and the Rust*, 2nd edn (London, 1806)

Barker, J., *Epidemicks, or General Observations on the Air and Diseases* (Birmingham, 1795)

Barker, J., *An Inquiry into the Nature, Cause, and Cure of the Present Epidemick Fever* (London, 1742)

Baron, J., *Life of Edward Jenner*, 2 vols (Colburn, 1838)

Beddoes, T., *Contributions to Physical and Medical Knowledge Principally from the West of England* (Bristol, 1799)

Bennet, A., 'An account of a doubler of electricity, or a machine by which the least conceivable quantity of positive or negative electricity may be continually doubled, till it becomes perceptible by common electrometer, or visible sparks … communicated by the Rev. Richard Kaye, LLD, FRS', *Philosophical Transactions*, 77 (1787), 288–96

Bennet, A., 'Description of a new electrometer, in a letter from the Rev. Abraham Bennet, MA to the Rev. Joseph Priestley, LLP, FRS', *Philosophical Transactions*, 76 (1786), 26–34

Bennet, A., 'Letter on attraction and repulsion: communicated by Dr. Percival, October 11, 1786', *Manchester Memoirs*, 3 (1788), 116–23

Bennet, A., *New Experiments on Electricity* (Derby and London, 1789)

Bennet, A., 'A new suspension of the magnetic needle, invented for the discovery of minute quantities of magnetic attraction … communicated by the Rev. Sir Richard Kaye, Bart, FRS', *Philosophical Transactions*, 82 (1792), 81–98

Bentham, J., *An Introduction to the Principles of Morals and Legislation* (1789), 2 vols (London, 1823)

Bentley, T., Darwin, E. and Wedgwood, J., [attributed] *A View of the Advantages of Inland Navigations with a Plan of a Navigable Canal* (London, 1765).

Bewick, T. (and R. Beilby), *A General History of Quadrupeds*, 7th edn (Newcastle-upon-Tyne, 1820)

Black, W. G., *Folk Medicine: A Chapter in the History of Culture* (London, 1883)

Boswell, J., *Life of Johnson*, edited by R. W. Chapman (Oxford, 1980 [1799])

Britton, J. and Brayley, E. W., *The Beauties of England and Wales*, vol. 3, Cumberland, Isle of Man and Derbyshire (London, 1802)

Brooke, A., *Miscellaneous Experiments and Remarks on Electricity, the Air-Pump and the Barometer* (London, 1789)

Brougham, H., *Dialogues on Instinct* (London, 1844)

Brown, J. C., *The Forests of England and the Management of Them in Bye-Gone Times* (London, 1883)

Brown, T., *General View of the Agriculture of the County of Derby* (London, 1794)

Browning, J., 'On the effects of electricity on vegetables', *Philosophical Transactions*, 114 (1747), no. 2, 373–5

Bryant, C., *Flora Dietica; or History of Esculent Plants* (London, 1783)

Bryant, C., *A Particular Enquiry into the Causes of that Disease in the Wheat Commonly Called Brand* (Norwich, 1783)

Buffon, G. L. Leclerc, *Buffon's Natural History Abridged*, translated by W. Smellie, 2 vols (London, 1792)

Buffon, G. L. Leclerc, *Histoire Naturelle*, 44 vols (Paris, 1749–1804)

Burke, E., *The Works of Edmund Burke*, 2 vols (London, 1834)

Cavallo, T., *A Complete Treatise on Electricity in Theory and Practice*, 4th edn, 3 vols (London, 1795)

Chambers, J., *A General History of the County of Norfolk*, 2 vols (Norwich, 1829)

Cochrane, A., Earl of Dundonald, *A Treatise [on] Agriculture and Chemistry* (London, 1795)

Coleridge, S. and Wordsworth, W., *Lyrical Ballads*, edited by R. L. Brett and A. R. Jones, 2nd edn (Abingdon, 2005 [1800])

Colquhoun, J. C., *William Wilberforce and his Friends* (London, 1867)

Coyte, W. B., *Hortus Botanicus Gippovicensis, or, A systematical enumeration of the plants cultivated in Dr Coyte's botanic garden at Ipswich* (Ipswich, 1796)

Coyte, W. B., *Index Plantarum* (Ipswich, 1807)

Crump, S., *Inquiry into the Nature and Properties of Opium* (London, 1793)

Culpeper, N., *The Complete Herbal and English Physician Enlarged* (Ware, 1995 [1653])

Curtis, S., *Lectures on Botany as Delivered in the Botanical Garden at Lambeth by the Late William Curtis FLS*, 2 vols (London, 1805)

Curtis, W., *A Catalogue of the British, Medicinal, Culinary and Agricultural Plants cultivated in the London Botanic Garden* (London, 1783)

Curtis, W., *Companion to the Botanical Magazine; or a Familiar Introduction to the Study of Botany* (London, 1788)

Curtis, W., *Flora Londinensis; or Plates and Descriptions of Such Plants as Grow Wild in the Environs of London* (London, 1777)

Darwin, C., 'A biographical sketch of an infant', *Mind*, 2 (1877), 285–94

Darwin, C., *The Descent of Man and Selection in Relation to Sex* (London, 1871)

Darwin, C., *Experiments Establishing a Criterion between Mucaginous and Purulent Matter* (Lichfield, 1780)

Darwin, C., *The Expression of the Emotions in Man and Animals*, 2nd edn, edited by F. Darwin (London, 1904)

Darwin, C., *The Life of Erasmus Darwin*, edited by D. King-Hele (Cambridge, 2003)

Darwin, C., 'Observations on his children', in F. H. Burkhardt ed., *The Correspondence of Charles Darwin* (Cambridge, 1988), vol. 4, 410–33

Darwin, C., *The Power of Movement in Plants* (London, 1880)

Darwin, E., 'An account of the successful use of Foxglove in some dropsies and in the pulmonary consumption', *Medical Transactions*, 3 (1785), 255–86

Darwin, E., *The Economy of Vegetation*, 4th edn (London, 1799)

Darwin, E., 'Elegy on the much lamented death of a most ingenious young gentleman', in S. Harris ed., *The Prince, My Son, A hero: Three Elegies* (Sheffield, 2009)

Darwin, E., 'Life of Charles Darwin', in C. Darwin, *Experiments Establishing a Criterion between Mucaginous and Purulent Matter* (Lichfield, 1780)

Darwin, E., 'Frigorific Experiments on the Mechanical Expansion of Air, Explaining the Cause of the Great Degree of Cold on the Summits of High Mountains, the Sudden Condensation of Aerial Vapour, and of the Perpetual Mutability of Atmospheric Heat', *Philosophical Transactions*, 78 (1788), 43–52

Darwin, E., *The Loves of the Plants*, 4th edn (London, 1799)

Darwin, E., *Phytologia; or the Philosophy of Agriculture and Gardening* (London, 1800)

Darwin, E., *Plan for the Conduct of Female Education* (Derby, 1797)

Darwin, E., *The Prince, My Son, A hero: Three Elegies*, edited by S. Harris (Sheffield, 2009)

Darwin, E., 'Remarks on the opinion of Henry Eeles Esq., concerning the ascent of vapour', *Philosophical Transactions*, 50 (1757), 240–54

Darwin, E., *The Temple of Nature; or the Origins of Society* (London, 1803)

Darwin, E., *Zoonomia; or the Laws of Organic Life*, 2nd edn, 2 vols (London, 1796), 3rd edn, 4 vols (London, 1801)

Darwin, F., ed., *The Life and Letters of Charles Darwin*, 2 vols (New York, 1925)

Darwin, R. W., *Principia Botanica: or a Concise and Easy Introduction to the Sexual Botany of Linnaeus* (Newark, 1787)

Davy, H., *Elements of Agricultural Chemistry in a Course of Lectures for the Board of Agriculture*, 2nd edn (London, 1814)

Davy, H., *On the Analysis of Soils as Connected with their Improvement* (London, 1805)

Davy, H., *Researches Chemical and Philosophical Chiefly Concerning Nitrous Oxide* (Bristol, 1800)

Demainbray, S., 'An application of electricity towards the improvement of vegetation', *The Scots Magazine*, 9 (January 1747), 40, also *Gentleman's Magazine*, 17 (February 1747), 80–1

Derby Philosophical Society, *Rules and Catalogue of the Library of the Derby Philosophical Society*, with supplements of 1795 and 1798 (Derby, 1793–8)

Derby Philosophical Society, *Rules and Catalogue of the Library of the Derby Philosophical Society* (Derby, 1815)

Dickson, R. W., *Practical Agriculture; or Complete System of Modern Husbandry* (London, 1805)

Dimsdale, T., *The Present Method of Inoculation for the Small Pox* (London, 1767)

D'Ormoy, M. L'Abbe, 'Experiments on the influence of electricity on vegetation', *Annals of Agriculture*, 15 (1791), 28–60

Duhamel de Monceau, H. L., *La physique des arbres, où il est traité de l'anatomie des plantes et de l'économie végétale: pour servir d'introduction au traité complet des bois et forests ... Partie 2*, 2 vols (Paris, 1758)

Duhamel du Monceau, H. L., *Trait des Arbres ar Arbustes qui se Cultivent en France*, 2 vols (Paris, 1755)

Duhamel du Monceau, H. L., *Traite des Arbres Fruitiers*, 2 vols (Paris, 1768)

Edgeworth, M., *Letters for Literary Ladies* (1795)

Edgeworth, R. L. and Edgeworth, M., *Memoirs of Richard Lovell Edgeworth, Esq. begun by Himself and Concluded by his Daughter, Maria Edgeworth*, 2 vols (London, 1820)

Edgeworth, R. L. and Edgeworth, M., *Practical Education*, 2 vols (London, 1798)

Eeles, H., 'Letter concerning the cause of the ascent of vapour and exhalation, and those of winds; and of the general phenomena of the weather and barometer', *Philosophical Transactions*, 49 (1755/6), 124–54

Evelyn, J., *Directions for the Gardener and other Horticultural Advice*, edited by M. Campbell-Culver (Oxford, 2009)

Evelyn, J., *Sylva; or a Discourse of Forest Trees* (London, 1664)

Evelyn, J., *Sylva: or a Discourse of Forest Trees*, with notes by A. Hunter, 1st edn, 2 vols (York, 1776), 3rd edn, 2 vols (York, 1801)

Evelyn, J., *Sylva or a Discourse of Forest Trees*, 4th edn (1706) edited by J. Nisbet, 2 vols (London, 1908)

Farey, J., *A General View of the Agriculture and Minerals of Derbyshire*, 3 vols (London, 1811–17)

Faulkner, T., *An Historical and Topographical Description of Chelsea and its Environs* (London, 1810)

Ferguson, J., *Analysis of a Course of Lectures on Mechanics, Pneumatics, Hydrostatics and Astronomy* (London, 1761)

Field, H., *Memoirs of the Botanic Garden at Chelsea belonging to the Society of Apothecaries at London*, continued by R. H. Semple (London, 1878)

Fisher, R., *Heart of Oak: The British Bulwark* (London, 1763)

Floyer, J. and Baynard, E., *The Ancient Psychrolusia Revived, or, an Essay to Prove Cold Bathing Both Safe and Useful* (London, 1702)

Fordyce, G., *Elements of Agriculture and Vegetation* (London, 1771)

Fordyce, J., *Historia Febris Miliaris, et de Hemicrania Dissertatio* (London, 1758)

Fordyce, W., *The Great Importance and Proper Method of Cultivating and Curing Rhubarb in Britain for Medical Uses* (London, 1784)

Forsyth, W., *A Treatise on the Culture and Management of Fruit Trees* (London, 1802)

Fowler, T., *Medical Reports of the Effects of Arsenic in the Cure of Agues, Remitting Fevers and Periodic Headaches* (London, 1786)

Fowler, T., *Medical Reports of the Effects of Tobacco in the Cure of Dropsies and Dysuries*, 2nd edn (Stafford, 1788)

Gilpin, W., *Essays on the Picturesque as Compared with the Sublime and the Beautiful*, 3 vols (London, 1810)

Gilpin, W., *Three Essays on Picturesque Beauty*, 2nd edn (London, 1794)

Gisborne, T., *Considerations on the Modern Theory of Geology* (London, 1837)

Gisborne, T., *The Duties of the Female Sex* (London, 1797)

Gisborne, T., *The Duties of the Higher and Middle Classes*, 2 vols (London, 1794)

Gisborne, T., *An Enquiry into the duties of Men in the Higher and Middle Classes of Society in Great Britain*, 7th edn, 2 vols (London, 1824)

Gisborne, T., 'On the benefits and duties resulting from the institution of societies for the advancement of literature and philosophy', *Manchester Memoirs*, 5 (1798–1802), 70–88

Gisborne, T., *The Principles of Moral Philosophy* (London, 1786)

Gisborne, T., *Sermons*, 3 vols (London, 1813)

Gisborne, T., *The Testimony of Natural Theology to Christianity* (London, 1818)

Gisborne, J., *The Vales of Wever: A Loco-descriptive Poem* (London, 1797)

Gisborne, T., *Walks in a Forest*, 4th edn (London, 1799)

Glover, S., *History and Gazetteer of the Town of Derby*, 2 vols (Derby, 1833)

Goldsmith, O., *A History of the Earth and Animated Nature*, 4 vols (York, 1804)

Good, J. M., *The History of Medicine in so far as it Relates to the Profession of the Apothecary* (London, 1795)

Green, H., *Memoir of Amos Green, Esq. Written by his Late Widow* (York, 1823)

Hales, S., *Vegetable Statics* (London, 1727)

Hanbury, W., *A Complete Body of Planting and Gardening*, 2 vols (London, 1770)

Hanbury, W., *Essay on Planting, and a Scheme for making it Conducive to the Glory of God and the Advantage of Society* (London, 1758)

Hankin, C. C., ed., *Life of Mary Anne Shimmelpenninck*, 2nd edn (London, 1858)

Henly, W., 'Experiments and observations in electricity', *Philosophical Transactions*, 67 (1777), 85–143

Hill, J., *Family Herbal* (London, 1812 [1755])

Home, H., Lord Kames, *Elements of Criticism*, 7th edn, 2 vols (Edinburgh, 1788)

Home, H., Lord Kames, *The Gentleman Farmer* (Edinburgh, 1776)

Hume, D., *Essays, Moral, Political, and Literary*, edited by T. H. Green and T. H. Grose, 2 vols (London, 1889)

Hunter, J., 'An account of the *gymnotus electricus*', *Philosophical Transactions*, 65 (1775), 395–407

Hunter, J., 'Anatomical observations on the Torpedo', *Philosophical Transactions*, 63 (1773), 481–9

Hutton, C., 'Authentic memoirs of the life and writings of the late John Whitehurst FRS', *Universal Magazine* (November 1788), 225–9

Hutton, W., *The History of Derby from the Remote Ages of Antiquity to the Year MDCCXCI*, 2nd edn (London, 1817)

Jackson, J., *History and Antiquities of the Cathedral Church of Lichfield* (Lichfield, 1796)

Jackson, J., *History of the City and Cathedral of Lichfield* (London, 1805)

Jacson, M. E. [originally published anonymously by 'a Lady'], *Botanical Dialogues between Hortensia and her Four Children* (London, 1797)

James, R., *Dr. Robert James's Powder for Fevers* (London, 1748)

James, R., *A Vindication of the Fever Powder* (London, 1776)

Jenner, E., *A Complete Statement of Facts and Observations relative to the Cow-pock*, 2 vols (London, 1800)

Jenner, E., *An Inquiry into the Cause and Effects of the Variolae Vaccinae* (London, 1798)

Jenner, E. and Woodville, W., *A Comparative Statement of Facts and Observations relative to the Cow Pox* (London, 1800)

Jewitt, L., ed., *The Ballads and Songs of Derbyshire* (London, 1867)

Johnson, S., *Dictionary of the English Language* [1755], edited by J. Lynch (London, 2004)

Johnson, S., *The Works of Samuel Johnson LLD*, edited by A. Murphy, 12 vols (London, 1820)

Kalm, P., *Travels into North America*, translated into English by J. R. Forster FAS, 2 vols (London, 1771)

Kennedy, J., *Treatise upon Planting, Gardening and the Management of the Hot House*, 2nd edn, 2 vols (London, 1777)

Kirby, W., 'Observations upon certain fungi, which are parasites of the wheat', *Linnean Transactions*, 5 (1800), 112–25

Kirkland, T., *An Account of the Distemper Among the Horned cattle in Derbyshire in 1783* (Ashby-de-la-Zouch, 1783)

Knight, T. A., *A Treatise on the Culture of the Apple & Pear and on the Manufacture of Cider and Perry* (Ludlow, 1797)

Laird, F. C., *Topographical and Historical Description of the County of Nottingham* (London, 1820)

Laithwaite, P., *The History of the Lichfield Conduit Lands Trust, 1546–1946* (Lichfield, 1947)

Lambert, A. B., 'Description of the blight of wheat, *Uredo Frumenti*', *Transactions of the Linnean Society*, 4 (1798), 193–4

Lambert, A. B., *A Description of the Genus Cinchona* (London, 1797)

Lambert, A. B., *A Description of the Genus Pinus*, 2nd edn, 2 vols (London, 1828)

Lamport, W., *Cursory Remarks on the Importance of Agriculture* (London, 1784)

Lawrence, J., *A General Treatise on Cattle* (London, 1805)

Lawrence, J., *A Philosophical and Practical Treatise on Horses and on the Moral Duties of Man towards the Brute Creation*, 2 vols (London, 1796–8)

Leclerc, G. L., Comte de Buffon, *Natural History*, translated by W. Smellie, abridged, 2 vols (London, 1792)

Lichfield Botanical Society, *The Families of Plants ... Translated from the Last Edition of the 'Genera Plantarum'*, 2 vols (Lichfield, 1787)

Lichfield Botanical Society, *A System of Vegetables ... translated from the thirteenth edition of the Systema Vegetabilium of the late Professor Linneus*, 2 vols (Lichfield, 1783)

Lindley, J., *Flora Medica* (London, 1838)

Linnaeus, C., *Miscellaneous Tracts Relating to Natural History, Husbandry and Physick*, translated by B. Stillingfleet (London, 1759)

Loudon, J. C., *Arboretum et Fruticetum Britannicum*, 2nd edn, 8 vols (London, 1844)

Loudon, J. C., *An Encyclopaedia of Agriculture*, 5th edn (London, 1844)

Loudon, J. C., *The Encyclopaedia of Gardening*, 1st edn (London, 1822), 2nd edn (London, 1824)

Loudon, J. C., *Hints on the Formation of Gardens and Pleasure Grounds* (London, 1812)

Loudon, J. C., *Improvement in Hot-Houses* (Edinburgh, 1805)

Loudon, J. C., *Observations on the Formation and Management of Useful and Ornamental Plantations*, 2 vols (Edinburgh, 1804)

Lowe, R., *General View of the Agriculture of the County of Nottingham* (London, 1794)

Lysons, D. and S., *Magda Britannia*, vol. 5, *Derbyshire* (London, 1817)

Marshall, W., *On Planting and Rural Ornament: a Practical Treatise*, 2 vols (London, 1795)

Marshall, W., *The Rural Economy of Norfolk*, 2 vols (London, 1783)

Marshall, W., *The Rural Economy of the Midland Counties*, 2nd edn, 2 vols (London, 1796)

Marshall, W., *The Rural Economy of the West of England*, 2 vols (London, 1796)

Marshall, W., *The Rural Economy of Yorkshire*, 2 vols (London, 1788)

Martin, B. (attrib.), 'A description of a machine for a perpetual electrification', *The General Magazine of Arts and Sciences*, 1 (1755–6), 116–17

Meyrick, W., *The New Family Herbal; or Domestic Physician* (Birmingham, 1790)

Miller, P., *The Gardner's Dictionary*, edited by T. Martyn, 2 vols, in four parts (London, 1807)

Morgan, G. C., *Lectures on Electricity*, 2 vols (Norwich, 1794)

Mundy, F. N. C., *Needwood Forest* (Lichfield, 1776)

Newton, I., *Opticks: or a Treatise of the Reflections, Refractions, Inflections and Colours of Light*, 4th edn, 3 vols (London, 1730)

Nichols, J., *Literary Anecdotes of the Eighteenth Century* (London, 1812)

Nichols, J. and J. B., *Illustrations of the Literary History of the Eighteenth Century*, 8 vols (London, 1817–58)

Nixon, E., ed., *A Brief Memoir of the Life of John Gisborne, to which are added Extracts from His Diary* (London and Derby, 1852)

Paley, W., *Natural Theology: or Evidences of the Existence and Attributes of the Deity* (1802) in *The Works of William Paley* (London, 1849)

Paley, W., *The Works of William Paley* (London, 1849)

Parkinson, J., *Theatre of Plants or a Universal and Complete Herbal* (London, 1640)

Parkinson, R., *The Experienced Farmer*, 2 vols (London, 1798)

Pearson, G., *An Enquiry Concerning the History of the Cow-Pox* (London, 1798)

Pearson, G., 'Experiments and observations to investigate the composition of James's Powder', *Philosophical Transactions*, 81 (1791), 317–67

Pilkington, J., *A View Of the Present State of Derbyshire*, 2 vols (Derby, 1789)

Poey, A., 'Report on agricultural meteorology', *Report of the Commissioner of Agriculture for the Year 1869* (Washington DC, 1870), 152–7

Pontey, W., *The Forest Pruner; or Timber Owner's Assistant* (London, 1805)

Pontey, W., *The Profitable Planter: a Treatise on the Theory and Practice of Planting Forest Trees* (Huddersfield, 1808)

Power, J., 'Of the use of fermenting cataplasms in mortifications', *Medical Transactions of the College of Physicians*, 3 (1785), 47–53

Priestley, J., *History and Present State of Electricity*, 1st edn (London, 1767)

Priestley, J., *History and Present State of Electricity*, 2nd edn, 2 vols (London, 1775)

Rawson, J., *An Enquiry into the History and Influence of the Lichfield Waters* (Lichfield, 1840)

Rees, A., *Chambers' Cyclopaedia*, 4 vols (London, 1788)

Rees, A., *Cyclopaedia: or, Universal Dictionary of Arts and Sciences*, 39 vols (London, 1819)

Repton, H., *The Landscape Gardening and Landscape Architecture of the Late Humphry Repton Esquire*, edited by J. C. Loudon (London, 1840)

Repton, H., *Observations on the Theory and Practice of Landscape Gardening* (London, 1803)

Rhodes, E., *Peak Scenery* (London, 1824)

Rooke, H., *Descriptions and Sketches of some Remarkable Oaks in the Park at Welbeck* (London, 1790)

Roscoe, W., *An Address delivered before the Proprietors of the Botanic Garden in Liverpool* (Liverpool, 1802)

Roscoe, W., 'On artificial and natural arrangements of plants: and particularly on the systems of Linnaeus and Jussieu', *Transactions of the Linnaean Society*, 11 (1815), 50–78

Rousseau, J. J., *Rousseau Juge de Jean Jacques: Dialogues: Premier Dialogue; d'Après le Manuscrit de M. Rousseau, laissé entre les mains de M. Brooke Boothby* (Lichfield, 1780)

Rutt, J. T., ed., *The Theological and Miscellaneous Works of Joseph Priestley, LLD*, 25 vols (London, 1831)

Sachs, V. von, *History of Botany*, translated by H. E. F. Garnsey (London, 1890)

Sainbel, C. V. de., *Lectures on the Elements of Farriery; or the Art of Horse-Shoeing, and on the Diseases of the Foot* (London, 1793)

Sainbel, C. V. de., *The Posthumous Works of Charles Vial de Sainbel, Late Equery to the King* (London, 1795)

Seward, A., *Anna Seward's Life of Erasmus Darwin*, edited by P. K. Wilson, E. A. Dolan and M. Dick (Studley, 2010)

Seward, A., *The Complete Works of Anna Seward*, edited by W. Scott, 3 vols (Edinburgh, 1810)

Seward, A., *The Letters of Anna Seward Written between the years 1784 and 1807*, edited by A. Constable, 5 vols (Edinburgh, 1811)

Seward, A., *Memoirs of the Life of Dr. Darwin* (London, 1804)

Seward, A., 'On Mundy's Needwood Forest', in W. Scott ed., *The Poetical Works of Anna Seward*, 3 vols (Edinburgh, 1810), vol. 3, 394–7

Shaw, S., *The History and Antiquities of Staffordshire*, 2 vols (London, 1802)

Shenstone, W., *Works in Verse and Prose*, 2 vols (London, 1764)

Smith, J. E., 'Biographical memoirs of several Norwich botanists', *Transactions of the Linnean Society*, 7 (1804), 295–301

Smith, J. E., *An Introduction to Physiological and Systematical Botany*, 5th edn (London, 1825)

Smith, J. E., *Memoir and Correspondence of the late James Edward Smith*, 2 vols (London, 1832)

Smith, W. G., *Diseases of Field and Garden Crops* (London, 1884)

Snape, J., *A Plan of the City and Close of Lichfield* (Lichfield, 1781)

Sowerby, J., *Coloured Figures of English Fungi or Mushrooms*, 4 vols (London, 1791)

Stillingfleet, B., *Miscellaneous Tracts Relating to Natural History, Husbandry and Physick* (London, 1759)

Stokes, J., *A Botanical Materia Medica*, 4 vols (London, 1812)

Stringer, C. E., *A Short Account of the Ancient and Modern State of the City and Close of Lichfield* (Lichfield, 1819)

Stukeley, W., 'Concerning the causes of earthquakes', *Philosophical Transactions*, 46 (1749–50), 657–81

Tessier, H. A., *Trait des Maladies des Grains et des Farines* (Paris, 1783)

Thoreau, H. D., 'Night and moonlight', *Atlantic Monthly Magazine* (November 1863), 579–83

Thornton, R. J., *A Family Herbal: or Familiar Account of the Medical Properties of British and Foreign Plants* (London, 1814)

Tilley, J., *The Ballads and Songs of Derbyshire* (London, 1867)

Tissot, S. A. D., 'An account of the disease called ergot, in French, from its supposed cause, viz., vitiated rye', *Philosophical Transactions*, 55 (1765), 106–25

Tull, J., *Horse-Hoeing Husbandry; or an Essay on the Principles of Vegetation and Tillage*, 3rd edn (London, 1751), 4th edn (London, 1762)

Universal British Directory of Trade, Commerce and Manufacture (London, 1791–7)

Wagstaff, J., 'On field mice and on the transplantation of wheat', *Letters and Papers on Agriculture, Planting, etc. Selected from the Correspondence of the Bath and West of England Society*, 6 (1792), 127–31

Wakefield, P., *An Introduction to Botany, in a Series of Familiar Letters*, 5th edn (London, 1807)

Walker, J., 'Experiments on the motion of sap in trees', *Transactions of the Royal Society of Edinburgh*, 1 (1788), no. 2, 3–40

Wall, M., *Clinical Observations on the use of Opium in Low Fevers and in the Synochus* (Oxford, 1786)

Watson, W., 'An account of some experiments by Mr. [Charles] Miller of Cambridge, on the sowing of wheat', *Philosophical Transactions*, 58 (1768), 203–6

Wesley, J., *The Desideratum, or Electricity Made Plain and Useful* (London, 1760)

Wesley, J., *Primitive Physic: or an Easy and Natural Way of Curing Most Diseases* (London, 1747)

Weston, R., *Tracts on Practical Agriculture and Gardening*, 2nd edn (London, 1773)

Whatley, S., *England's Gazetteer; or an Accurate Description of all the Cities, Towns and Villages of the Kingdom*, 3 vols (London, 1751)

White, G., *The Natural History and Antiquities of Selbourne*, edited by E. T. Bennett and J. E. Harting (London, 1876 [1788])

Whitehurst, J., *Enquiry into the State and Formation of the Earth*, 1st edn (London, 1778)

Whitehurst, J., *Observations on the Ventilation of Rooms, on the Construction of Chimneys and on Garden Stoves* (London, 1794)

Whitehurst, J., *Tracts; Philosophical and Mechanical by John Whitehurst FRS* (London, 1792)

Wildman, T., *A Treatise on the Management of Bees* (London, 1768)

Wilmer, B., *Observations on the Poisonous Vegetables which are either Indigenous in Great Britain or Cultivated for Ornament* (London, 1781)

Withering, W., *An Account of the Foxglove and some of its Medical Uses* (Birmingham, 1785)

Withering, W., *A Botanical Arrangement of all the Vegetables naturally growing in Great Britain, with descriptions of all the Genera and Species according to the celebrated system of Linnaeus*, 1st edn, 2 vols (Birmingham, 1776), 2nd edn in 3 vols (Birmingham, 1792)

Withering, W., *The Miscellaneous Tracts of the late William Withering MD FRS to which is prefixed a Memoir*, edited by W. Withering junior, 2 vols (London, 1822)

Withering, W., *A Systematic Arrangement of British Plants*, 5th edn, 4 vols (London, 1812)

Woodward, J., *An Essay Toward a Natural History of the Earth* (London, 1695)

Young, A., *The Farmer's Calendar*, 6th edn (London, 1805)

Young, A., *The Farmer's Tour through the East of England*, 4 vols (London, 1771)

WORKS PRINTED AFTER c. 1900

Agar, J. and Smith C., eds, *Making Space for Science: Territorial Themes in the Shaping of Knowledge* (London, 1998)

Ainsworth, G. C., *Introduction to the History of Plant Pathology* (Cambridge, 1981)

Alberti, S. M. M., ed., *The Afterlives of Animals: A Museum Menagerie* (Charlottesville, 2011)

Alberti, S. M. M., 'Placing nature: natural history collections and their owners in nineteenth-century provincial England', *British Journal for the History of Science*, 35 (2002), 291–311

Allan, M., *Darwin and His Flowers: The Key to Natural Selection* (London, 1977)

Allen, D. E., *The Naturalist in Britain: A Social History* (London, 1976)

Anderson, V., *Creatures of Empire? How Domestic Animals Transformed Early America* (Oxford, 2004)

Ashmun, M., *The Singing Swan* (New Haven, 1931)

Auricchio, L., Cook, E. H. and Pacini, G., eds, *Invaluable Trees: Cultures of Nature, 1660–1830* (Oxford, 2012)

Averley, G., 'English Scientific Societies of the Eighteenth and Early Nineteenth Centuries', unpublished PhD thesis, University of Teeside Polytechnic and Durham University (Durham, 1989)

Ayres, P., *The Aliveness of Plants: The Darwins at the Dawn of Plant Science* (London, 2008)

Baker, K. H. and Reill R. M., eds, *What's Left of Enlightenment? A Postmodern Question* (Stanford, 2001)

Barfoot, M., 'Hume and the culture of science in the early eighteenth century', in M. A. Stewart ed., *Oxford Studies in the History of Philosophy* (Oxford, 1990), 151–90

Barker, H., 'Medical advertising and trust in late-Georgian England', *Urban History*, 36 (2009), 379–98

Barker-Benfield, G., *The Culture of Sensibility: Sex and Society in Eighteenth-century Britain* (Chicago, 1996)

Barlow J. and Elliott, P., 'A brush with the doctor: Samuel James Arnold's account of painting Erasmus Darwin's portrait', forthcoming.

Barnard, T., *Anna Seward: A Constructed Life: A Critical Biography* (Farnham, 2009)

Barnatt, J. and Bannister, N., *Chatsworth and Beyond: The Archaeology of a Great Estate* (Oxford, 2009)

Barnatt, J. and Smith, K., *The Peak District: Landscapes Through Time* (Macclesfield, 2004)

Barnatt, J. and Williamson, T., *Chatsworth: A Landscape History* (Macclesfield, 2005)

Barre, D., *Historic Gardens and Parks of Derbyshire* (Oxford, 2017)

Barrell, J., *The Dark Side of the Landscape: the Rural Poor in English Paintings, 1730–1840* (Cambridge, 1980)

Bartel, R., 'Shelley and Burke's swinish multitude', *Keats-Shelley Journal*, 18 (1969), 4–9

Bate, J., *Romantic Ecology: Wordsworth and the Environmental Tradition* (London, 1991)

Bate, J., *The Song of the Earth* (London, 2000)

Bazin, H., *The Eradication of Smallpox* (San Diego, 2000)

Beckett, J. V., *The Agricultural Revolution* (Oxford, 1990)

Beckett, J. V., *The Aristocracy in England, 1660–1914*, rev. edn (Oxford, 1989)

Beckett, J. V., *The East Midlands since AD 1000* (London, 1988)

Beckett, J. V. and Heath, J. E., 'When was the industrial revolution in the East Midlands?' *Midland History*, 13 (1988), 77–94

Bending, S., *Green Retreats: Women, Gardens and Eighteenth-century Culture* (Cambridge, 2013)

Bending, S., 'The improvement of Arthur Young: agricultural technology and the production of landscape in eighteenth-century England', in D. E. Nye ed., *Technologies of Landscape: From Reaping to Re-Cycling* (Boston MA, 1999), 241–53

Berg, M., *Luxury and Pleasure in Eighteenth-century Britain*, 2nd edn (Oxford, 2007)

Bermingham, A., *Landscape and Ideology: the English Rustic Tradition, 1740–1860* (Berkeley, 1986)

Bertucci, P., 'Sparks of Life: Medical Electricity and Natural Philosophy in England, c.1746–1792', DPhil thesis, University of Oxford (Oxford, 2001)

Bewell, A., '"Jacobin plants": botany as social theory in the 1790s', *Wordsworth Circle*, 20 (1989), 132–39

Black, J., *George III: America's Last King* (New Haven, 2006)

Blackwell, M., ed., *The Secret of Things: Animals, Objects and It-narratives in Eighteenth-century England* (Lewisburg, 2007)

Blatchley, J. M. and James, J., 'The Beeston-Coyte *Hortus Botanicus Gippovicensis* and its printed catalogue', *Proceedings of the Suffolk Institute for Archaeology and History*, 39 (1999), 339–52

Blunt, W., *The Art of Botanical Illustration* (London, 1950)

Bonehill, J. and Daniels, S., eds, *Paul Sandby: Picturing Britain* (London, 2009)

Borsay, P., *The English Urban Renaissance: Culture and Society in the Provinical Town, 1660–1770* (Oxford, 1991)

Borsay, P., 'The rise of the promenade: the social and cultural use of space in the English provincial town, c. 1660–1800', *British Journal for Eighteenth-Century Studies*, 9 (1986), 125–40

Bowerbank, S., *Speaking for Nature: Women and Ecologies of Early-Modern England* (Baltimore, 2004)

Bowler, P., *History of the Environmental Sciences* (London, 1992)

Brantz, D., *Beastly Natures: Animals, Humans and the Study of History* (London, 2010)

Brewer, J., *The Pleasures of the Imagination: English Culture in the Eighteenth Century* (London, 1997)

Brighton, T., *The Discovery of the Peak District* (Chichester, 2004)

Broad, J., 'Cattle plague in eighteenth-century England', *The Agricultural History Review*, 31 (1983), 104–15

Brown, D., 'Matthew Boulton, enclosure and landed society', in M. Dick ed., *Matthew Boulton: A Revolutionary Player* (Studley, 2009), 45–62

Brown, D. and Williamson, T., *Lancelot Brown and the Capability Men: Landscape Revolution in Eighteenth-century England* (London, 2016)

Brown, K. and Gilfoyle, D., eds, *Healing the Herd: Disease, Landscape Economies and the Globalisation of Veterinary Medicine* (Athens GA, 2010)

Browne, J., 'Botany for gentlemen: Erasmus Darwin and the *Loves of the Plants*', *Isis*, 80 (1989), 593–620

Budge, G., 'Erasmus Darwin and the poetics of William Wordsworth: "excitement without the application of gross and violent stimulants"', in G. Budge ed., 'Science and the Midlands Enlightenment', special issue, *Journal for Eighteenth-Century Studies*, 30 (2007), no. 2, 297–308

Budge, G., ed., 'Science and the Midlands Enlightenment', in special issue, *Journal for Eighteenth-Century Studies*, 30 (2007), no. 2, 157–308

Burnby, J. G. L., *A Study of the English Apothecary from 1660 to 1760*, Medical History, supplement no. 3 (1983)

Burroughs, Wellcome and Co, *The History of Inoculation and Vaccination for the Prevention and Treatment of Disease* (London, 1913)

Burton, E., *The Georgians at Home* (London, 1973)

Cantor, G. N., 'The theological significance of ethers', in G. N. Cantor and M. J. S. Hodge eds, *Conceptions of Ether: Studies in the History of Ether Theories, 1740–1900* (Cambridge, 1981), 135–56

Caradonna, J. L., 'Conservationism *avant la lettre*? Public essay competitions on forestry and deforestation in eighteenth-century France', in L. Auricchio, E. Heckendorn Cook and G. Paccini eds, *Invaluable Trees: Cultures of Nature, 1660–1830* (Oxford, 2012), 39–54

Carefoot, G. L. and Sprott, E. R., *Famine on the Wind: Plant Diseases and Human History* (Chicago, 1967)

Carney, J. and Rosomoff, R. N., *In the Shadow of Slavery: Africa's Botanical Legacy in the Atlantic World* (Berkeley, 2011)

Carter, H., *Sir Joseph Banks, 1743–1820* (London, 1988)

Cassidy, A., Dentinger, R. M., Scoefert, K. and Woods, A., 'Animal roles and traces in the history of medicine', in A. Rees ed., 'Animal agents: the non-human in the history of science', *British Journal for the History of Science*, themes 2 (2017), 11–33

Chakrabarti, P., *Materials and Medicine: Trade, Conquest and Therapeutics in the Eighteenth Century* (Manchester, 2010)

Chambers, J. D. and Mingay, G. E., *The Agricultural Revolution* (London, 1966)

Clark, P., *British Clubs and Societies 1580–1800: The Origins of an Associational World* (Oxford, 2000)

Clark, P., *The English Alehouse: A Social History, 1200–1830* (London, 1983)

Clark, P. and Houston R. A., 'Culture and leisure, 1700–1840', in P. Clark ed., *The Cambridge Urban History*, vol. 2 (Cambridge, 2000), 575–613

Clark, W., Golinski, J. and Schaffer, S., eds, *The Sciences in Enlightened Europe* (Chicago, 1999)

Clark-Kennedy, A. E., *Stephen Hales, DD, FRS: An Eighteenth-century Biography* (Cambridge, 1929)

Clarke, J. C. D., *English Society, 1660–1832: Religion, Ideology and Politics during the Ancien Regime*, 2nd edn (Cambridge, 2000)

Clow, A. and N., *The Chemical Revolution: A Contribution to Social Technology* (London, 1952)

Coffey, D., 'Protecting the botanic garden: Seward, Darwin, and Coalbrookdale', *Women's Studies*, 31 (2002), 141–64

Cole, L., 'Introduction: human–animal studies and the eighteenth century', *The Eighteenth Century*, 52 (2011), 1–10

Colvin, H., *Calke Abbey, Derbyshire: A Hidden House Revealed* (London, 1986)

Connolly, T., 'Flowery porn: form and desire in Erasmus Darwin's *The Loves of the Plants*', *Literature Compass*, 13 (2016), 604–16

Cook, G. C., 'Dr. Erasmus Darwin MD FRS (1731–1802): England's greatest physician?', in C. U. M. Smith and R. Arnott eds, *The Genius of Erasmus Darwin* (Aldershot, 2005), 47–62

Cooper, B., *Transformation of a Valley: The Derbyshire Derwent* (London, 1983)

Copeman, W. S. C., *The Worshipful Society of Apothecaries of London: A History* (London, 1967)

Corfield, P. J., 'From poison peddlers to civic worthies: the reputation of the apothecaries in Georgian England', *Social History of Medicine*, 22 (2009), 1–21

Corfield, P., *Power and the Professions in Britain, 1700–1850* (London, 1995)

Craven, M., *Distinguished Derbeians* (Derby, 1998)

Craven, M., *John Whitehurst: Innovator, Scientist, Geologist and Clockmaker*, 2nd edn (Croyden, 2015)

Craven, M. and Stanley, M., *Derbyshire Country House* (Ashbourne, 2001)

Creighton, C., *A History of Epidemics in Britain*, 2 vols (London, 1894)

Curth, L. H., *The Care of Brute Beasts: A Social and Cultural Study of Veterinary Medicine in Early Modern England* (Leiden, 2009)

Curth, L. H., ed., *From Physick to Pharmacology: Five Hundred Years of British Drug Retailing* (Aldershot, 2006)

Daniels, S., *Humphry Repton: Landscape Gardening and the Geography of Georgian England* (New Haven, 1999)

Daniels, S., *Joseph Wright* (London, 1999)

Daniels, S., 'Loutherbourg's chemical theatre: Coalbrookdale by night', in J. Barrell ed., *Painting and the Politics of Culture: New Essays on British Art, 1700–1850* (Oxford, 1992), 195–230

Daniels, S., 'The political iconography of woodland in later Georgian England', in D. Cosgrove and S. Daniels eds, *The Iconography of Landscape* (Cambridge, 1988), 43–82

Daniels, S., Seymour, S. and Watkins, C., 'Enlightenment, improvement and the geographies of horticulture in later Georgian England', in D. N. Livingstone and C. Withers eds, *Geography and Enlightenment* (Chicago, 1999), 345–71

Daston, L., ed., *Biographies of Scientific Objects* (Chicago, 2000)

Daston, L. and Park, K., *Wonders and the Order of Nature, 1150–1750* (New York, 1998)

Daunton, M. J., *Progress and Poverty: An Economic and Social History of Britain, 1700–1850* (Oxford, 1995)

DeLacy, M., *Contagionism Catches on: Medical Ideology in Britain, 1730–1800* (Cham, 2017)

Dickinson, H. T., *Liberty and Property: Political Ideology in Eighteenth-Century Britain* (London, 1979)

Dixon Hunt, J., *The Picturesque Garden in Europe* (London, 2002)

Donald, D., *Picturing Animals in Britain, 1750–1850* (New Haven, 2007)

Drayton, R., *Nature's Government: Science, Imperial Britain and the 'Improvement' of the World* (New Haven, 2000)

Drewitt, F. D., *The Romance of the Apothecaries' Garden at Chelsea*, 3rd edn (Cambridge, 1928)

Duckworth, A. M., *The Improvement of the Estate: A Study of Jane Austin's Novels* (Baltimore, 1971)

Dunlop, R. A. and Williams, D. J., *Veterinary Medicine: An Illustrated History* (Mosby, 1995)

Edwards, K. C., *The Peak District* (London, 1973)

Elliott, P. A., 'The Derby Arboretum (1840): The first specially designed municipal public park in Britain', *Midland History*, 26 (2001), 144–76

Elliott, P. A., *The Derby Philosophers: Science and Culture in British Urban Society, 1700–1850* (Manchester, 2009)

Elliott, P. A., *Enlightenment, Modernity and Science: Geographies of Scientific Culture and Improvement in Georgian England* (London, 2010)

Elliott, P. A., 'Enlightenment science, technology and the Industrial Revolution: a case study of the Derby philosophers, c1750–1820', in C. Wrigley ed., *Industry and the British People: Cromford, the Derwent Valley and Beyond* (Cromford, 2020), 79–140

Elliott, P. A., 'Medicine, scientific culture and urban improvement in early nineteenth-century England: the politics of the Derbyshire General Infirmary', in J. Reinarz ed., 'Medicine and the Midlands, 1750–1950', special issue of *Midland History* (2007), 27–46

Elliott, P. A., '"More subtle than the electric aura": Georgian medical electricity, the spirit and animation and the development of Erasmus Darwin's psychophysiology', *Medical History*, 52 (2008), 195–220

Elliott, P. A., 'The politics of urban improvement in Georgian Nottingham: the enclosure dispute of the 1780s', *Transactions of the Thoroton Society*, 110 (2006), 87–102

Elliott, P. A. (with C. Watkins and S. Daniels), eds, *The British Arboretum: Trees, Science and Culture in the Nineteenth Century* (London, 2011)

Emerson, R. L., 'The Enlightenment and social structures', in P. Fritz and D. Williams eds, *City and Society in the Eighteenth Century* (Toronto, 1973), 99–129

Estabrook, C., *Urbane and Rustic England* (Manchester, 1998)

Everett, N., *The Tory View of Landscape* (New Haven, 1994)

Everitt, A., 'Country, county and town: patterns of regional evolution in England', in P. Borsay ed., *The Eighteenth-century Town* (London, 1990), 83–115

Everitt, A., *Landscape and Community in England* (Oxford, 1985)

Fara, P., *Erasmus Darwin: Sex, Science and Serendipity* (Oxford, 2012)

Figlio, K. M., 'Theories of perception and the physiology of mind in the late eighteenth century', *History of Science*, 13 (1975), 177–212

Finger, S., *Doctor Franklin's Medicine* (Philadelphia, 2006)

Fitton, R. S., *The Arkwrights: Spinners of Fortune* (Manchester, 1989)

Fitton, R. S. and Wadsworth, A. P., *The Strutts and the Arkwrights, 1758–1830: A Study of the Factory System* (Manchester, 1958)

Fitzpatrick, M., 'Rational Dissent and the Enlightenment', *Faith and Freedom*, 38 (1985), 83–101

Ford, T. D., 'White Watson (1760–1835) and his geological tablets', *Mercian Geologist*, 13 (1995), 157–64

Ford, T. D. and Rieuwerts, J. H., eds, *Lead Mining in the Peak District*, 4th edn (Ashbourne, 2000)

Foust, C. M., *Rhubarb, the Wondrous Drug* (Princeton, 1992)

Fudge, E., *Perceiving Animals: Humans and Beasts in Early-Modern English Culture* (Champaign, 2002)

Fulford, T., 'Coleridge, Darwin, Linnaeus: the sexual politics of botany', *The Wordsworth Circle*, 28 (1997), 124–30

Fulton, J. F., 'Charles Darwin (1758–1778) and the history of the early use of digitalis', *Journal of Urban Health*, 76 (1934), 533–41

Fussell, G. E., 'Four centuries of farming systems in Derbyshire: 1500–1900', *Derbyshire Archaeological Journal*, 71 (1951), 1–37

Gage, A. T. and Stearn, W. T., *A Bicentenary History of the Linnean Society of London* (London, 2001)

Garrard, G., *Ecocriticism*, 2nd edn (Abingdon, 2012)

Gascoigne, J., *Joseph Banks and the English Enlightenment: Useful Knowledge and Polite Culture* (Cambridge, 1994)

Gately, I., *Drink: A Cultural History of Alcohol* (New York, 2008)

Gibbs, D., 'Physicians and physic in seventeenth and eighteenth-century Lichfield', in C. U. M. Smith and R. Arnott eds, *The Genius of Erasmus Darwin* (Aldershot, 2005), 35–46

Gibbs, D. D., 'Sir John Floyer', *British Medical Journal*, 1 (1969), 242–5

Goldstein, J., '*Terra economica*: waste and the production of enclosed nature', *Antipode*, 45 (2013), 357–75

Golinksi, J., *British Weather and the Climate of Enlightenment* (Chicago, 2007)

Golinski, J., *The Experimental Self: Humphry Davy and the Making of Man of Science* (Chicago, 2016)

Golinski, J., *Making Natural Knowledge: Constructivism and the History of Science* (Cambridge, 1998)

Golinski, J., *Science as Public Culture: Chemistry and Enlightenment in Britain, 1760–1820* (Cambridge, 1992)

Gould, S. J., *Dinosaur in a Haystack: Reflections on Natural History* (New York, 1995)

Greaves, R. W., *The Corporation of Leicester, 1689–1836*, 2nd edn (Leicester, 1970)

Green, J. Reynolds, *A History of Botany in the United Kingdom* (London, 1914)

Green, M., 'Blake, Darwin and the promiscuity of knowing: rethinking Blake's relationship to the Midlands Enlightenment', in G. Budge ed., 'Science and the Midlands Enlightenment', special issue, *Journal for Eighteenth-Century Studies*, 30 (2007), no. 2, 193–208

Greenslade, M. W., ed., *A History of the County of Stafford, Vol. 14, Lichfield* (London, 1990)

Greenslade, M. W. and Kettle, A. J., 'Forests', in *A History of the County of Stafford, Vol. 2* (London, 1967), 335–9

Grier, J., *A History of Pharmacy* (London, 1937)

Griggs, B., *New Green Herbal: the Story of Western Herbal Medicine*, 2nd edn (London, 1997)

Grigson, C., *Menagerie: The History of Exotic Animals in England, 100–1837* (Oxford, 2016)

Grove, R., *Green Imperialism: Colonial Expansion, Tropical Island Edens and the Origins of Environmentalism, 1600–1860* (Cambridge, 1995)

Gruber, H. E., *Darwin on Man: A Psychological Study of Scientific Creativity* (London, 1974)

Guenther, M., 'Tapping nature's bounty: science and sugar maples in the age of improvement', in L. Auricchio, E. Heckendorn Cook and G. Paccini eds, *Invaluable Trees: Cultures of Nature, 1660–1830* (Oxford, 2012), 135–49

Guerrini, A., *Experimenting with Humans and Animals: From Galen to Animal Rights* (London, 2003)

Habermas, J., *The Structural Transformation of the Public Sphere*, translated by T. Burger and F. Lawrence (Cambridge MA, 1989)

Hamblyn, R., *The Invention of Clouds* (London, 2001)

Handley, J. E., *The Agricultural Revolution in Scotland* (Glasgow, 1963)

Hankins, T. L., *Science and the Enlightenment* (Cambridge, 1985)

Harding Rains, A. J., *Edward Jenner and Vaccination* (London, 1974)

Harris, S., 'Myth and medicine in Erasmus Darwin's epic poetry', in R. U. M. Smith and R. Arnott eds, *The Genius of Erasmus Darwin* (Aldershot, 2005), 321–35

Hartley, H., *Humphry Davy* (Wakefield, 1972)

Hartwell, C., Pevsner, N. and Williamson, E., *The Buildings of England: Derbyshire* (New Haven, 2016)

Hatfield, G., *Memory, Wisdom and Healing: the History of Domestic Plant Medicine* (Stroud, 2005)

Heilbron, J. L., *Electricity in the 17th and 18th Centuries: A Study in Early Modern Physics*, 2nd edn (New York, 1999)

Henninger-Voss, M., ed., *Animals in Human Histories: The Mirror of Nature and Culture* (Rochester NY, 2002)

Hey, D., *Derbyshire: A History* (Lancaster, 2008)

Hoff, H. E., 'Galvani and the pre-Galvanian electrophysiologists', *Annals of Science*, 1 (1936), 157–72

Home, R. W., 'Electricity and the nervous fluid', *Journal of the History Biology*, 3 (1970), 235–5

Hopkins, M. A., *Dr. Johnson's Lichfield* (London, 1957)

Hoskins, W. G., *The Making of the English Landscape* (London, 1970)

Hudson, D. and Luckhurst, K. W., *The Royal Society of Arts, 1754–1954* (London, 1954)

Hudson, K., *Patriotism with Profit: British Agricultural Societies in the Eighteenth and Nineteenth Centuries* (London, 1972)

Inkster, I., *Scientific Culture and Urbanisation in Industrialising Britain* (Aldershot, 1997)

Inkster, I. and Morrell, J. B., eds, *Metropolis and Province: Science in British Culture* (London, 1983)

Israel, J. I., *Enlightenment Contested: Philosophy, Modernity and the Emancipation of Man* (Oxford, 2008)

Jacob, M. C., *The Cultural Meaning of the Scientific Revolution* (New York, 1988)

Jacyna, S. L., 'Galvanic influences: themes in the early history of British animal electricity', *Bologna Studies in History of Science*, 7 (1999), 167–85

Jankovic, V., 'The place of nature and the nature of place: the chorographic challenge to the history of British provincial science', *History of Science*, 38 (2000), 80–113

Jankovic, V., *Reading the Skies: A Cultural History of English Weather, 1650–1820* (Chicago, 2000)

Jardine, N., Secord, J. A. and Spary, C., eds, *Cultures of Natural History* (Cambridge, 1996)

Jarvis, C., *Order out of Chaos: Linnaean Plant Names and their Types* (London, 2007)

Johnson, N., *Nature Displaced, Nature Displayed: Order and Beauty in Botanical Gardens* (London, 2011)

Jones, P. M., *Agricultural Enlightenment: Knowledge, Technology and Nature, 1750–1840* (Oxford, 2016).

Jones, P. M., *Industrial Enlightenment: Science, Technology and Culture in Birmingham and the West Midlands, 1760–1820* (Manchester, 2008)

Kalof, L. and Resl, B., eds, *A Cultural History of Animals*, 6 vols (Oxford, 2007)

Kean, H. and Howell, P., eds, *The Routledge Companion to Animal–Human History* (London, 2019)

Kelley, T. M., *Clandestine Marriage: Botany and Romantic Culture* (Baltimore, 2012)

King, S., 'Accessing drugs in the eighteenth-century regions', in L. H. Curth ed., *From Physick to Pharmacology: Five Hundred Years of British Drug Retailing* (Aldershot, 2006), 49–78

King, S., 'The healing tree', in L. Auricchio, E. Heckendorn Cook and G. Pacini eds, *Invaluable Trees: Cultures of Nature, 1660–1830* (Oxford, 2012), 237–50

King-Hele, D., ed., *The Collected Letters of Erasmus Darwin* (Cambridge, 2007)

King-Hele, D., *Erasmus Darwin and Evolution* (Sheffield, 2014)

King-Hele, D., *Erasmus Darwin: A Life of Unequalled Achievement* (London, 1999)

King-Hele, D., *Erasmus Darwin and the Romantic Poets* (London, 1986)

Kirk, R. and Worboys, M., 'Medicine and species: one medicine, one history?' in M. Jackson ed., *The Oxford Handbook of the History of Medicine* (Oxford, 2011), 561–77

Knight, D., *Ordering the World: A History of Classifying Man* (London, 1981)

Laird, M., *A Natural History of English Gardening* (New Haven, 2015)

Laird, M. and Weisberg-Roberts, A., eds, *Mrs. Delany & Her Circle* (New Haven, 2009)

Landes, J. B., Lee, P. Y. and Youngquist, P., eds, *Gorgeous Beasts: Animal Bodies in Historical Perspective* (University Park, 2012)

Lane, J., 'Eighteenth-century medical practice: a case study of Bradford Wilmer, surgeon of Coventry, 1737–1818', *Social History of Medicine*, 3 (1990), 369–86

Langford, P., *A Polite and Commercial People: England, 1727–1783*, 2nd edn (Oxford, 1998)

Langford, P., *Public Life and the Propertied Englishman, 1689–1798* (Oxford, 1991)

Large, E. C., *The Advance of the Fungi* (London, 1940)

Latour, B. and Woolgar, S., *Laboratory Life: The Construction of Scientific Facts* (Princeton, 1981)

Le Rougetel, H., *The Chelsea Gardener: Philip Miller, 1691–1771* (London, 1990)

Leonard, D. C., 'Erasmus Darwin and William Blake', *Eighteenth-Century Life*, 4 (1978), 79–81

Levere, T. H., 'Dr Thomas Beddoes (1760–18108) and the Lunar Society of Birmingham: collaborations in medicine and science', in G. Budge, ed. 'Science and the Midlands Enlightenment', special issue, *Journal for Eighteenth-Century Studies*, 30 (2007), no. 2, 209–66

Livingstone, D., *Putting Science in its Place: Geographies of Scientific Knowledge* (Chicago, 2003)

Livingstone, D. N. and Withers, C. W. J., eds, *Geography and Enlightenment* (Chicago, 1999)

Logan, J. V., 'The Poetry and Aesthetics of Erasmus Darwin', *Princeton Studies in English*, 15 (1936), 46–92

Lovejoy, A., *The Great Chain of Being: A Study of the History of an Idea* (Boston MA, 1957)

Ludington, C., *The Politics of Wine in Britain: A New Cultural History* (Basingstoke, 2013)

Mabey, R., *Gilbert White: A Biography of the Author of the Natural History of Selbourne* (London, 1987)

McClellan III, J. E., *Science Reorganised: Scientific Societies in the Eighteenth Century* (New York, 1985)

McGann, J., *The Poetics of Sensibility: A Revolution in Literary Style* (Oxford, 1996)

MacGregor, A., *Curiosity and Enlightenment: Collectors and Collections from the Sixteenth to the Nineteenth Century* (New Haven, 2007)

McNeil, M., 'The scientific muse: the poetry of Erasmus Darwin', in L. J. Jordanova ed., *Languages of Nature: Critical Essays on Science and Literature* (London, 1986), 159–203

McNeil, M., *Under the Banner of Science: Erasmus Darwin and His Age* (Manchester, 1987)

Malcolmson, R. and Mastoris, S., *The English Pig: A History* (London, 2001)

Manuso, S. and Viola, A., *Brilliant Green: The Surprising History and Science of Plant Intelligence* (Washington DC, 2015)

Marshall, A., 'Erasmus Darwin contra David Hume', *British Journal for Eighteenth-Century Studies*, 20 (2007), 89–111

Marshall, P. J. and Williams, G., *The Great Map of Mankind: British Perceptions of the World in the Age of Enlightenment* (London, 1982)

Martin, M., 'Bourbon renewal at Rambouillet', in L. Auricchio, E. Heckendorn Cook and G. Pacini eds, *Invaluable Trees: Cultures of Nature, 1660–1830* (Oxford, 2012), 151–70

Mason, S., *The Hardwareman's Daughter: Matthew Boulton and his 'Dear Girl'* (Chichester, 2005)

Mathias, P., 'Agriculture and the brewing and distilling industries', in E. L. Jones ed., *Agriculture and Economic Growth in England, 1650–1815* (London, 1967), 80–93

Mayr, E., *The Growth of Biological Thought* (Cambridge MA, 1982)

Melton J. Van Horn, *The Rise of the Public in Enlightenment Europe* (Cambridge, 2001)

Menuge, A., 'The cotton mills of the Derbyshire Derwent and its tributaries', *Industrial Archaeology Review*, 16 (1993), 38–61

Merchant, C., *Reinventing Eden: The Fate of Nature in Western Culture* (New York, 2004)

Millburn, J. R., *Benjamin Martin: Author, Instrument Maker and 'Country Showman'* (Leiden, 1976)

Millburn, J. R., '*Martin's Magazine*: the *General Magazine of Arts and Sciences, 1755–65*', *The Library*, 28 (1973), 221–39

Miller, G., ed., *The Letters of Edward Jenner and other Documents Concerning the History of Vaccination* (Baltimore, 1983)

Millhauser, M., 'The scriptural geologists: an episode in the history of opinion', *Osiris*, 11 (1954), 65–86

Minter, S., *The Apothecaries' Garden: A History of the Chelsea Physic Garden* (Stroud, 2000)

Mitchison, R., *Agricultural Sir John: The Life of Sir John Sinclair of Ulbster, 1754–1835* (London, 1962)

Money, J., *Experience and Identity: Birmingham and the West Midlands, 1760–1800* (Montreal, 1977)

Morris, R. J., 'Clubs, Societies and Associations', in F. M. L. Thompson ed., *The Cambridge Social History of Britain, 1750–1950*, 3 vols (Cambridge, 1990), vol. 3, 395–443

Mortenson, T., 'British scriptural geologists in the first half of the nineteenth century: part 6: Thomas Gisborne (1758–1846)', *Technical Journal*, 14 (2000), 75–80

Mottelay, P. F., *Bibliographical History of Electricity and Magnetism* (London, 1922)

Munck, T., *The Enlightenment: A Comparative Social History, 1721–1794* (London, 2000)

Musson, A. E. and Robinson, A. E., *Science and Technology in the Industrial Revolution* (Manchester, 1969)

Nance, S., *The Historical Animal* (New York, 2015)

Naylor, P. J., *A History of the Matlocks* (Ashbourne, 2003)

Naylor, S., 'Historical geographies of science: places, contexts, cartographies', *British Journal for the History of Science*, 38 (2005), 1–12

Neeson, J. M., *Commons: Common Right, Enclosure and Social Change in England, 1700–1820* (Cambridge, 1993)

Nicholson, B., *Joseph Wright of Derby, Painter of Light*, 2 vols (New York, 1967)

Nixon, F., *The Industrial Archaeology of Derbyshire* (Newton Abbot, 1969)

Ottum, L. and Reno, S. T., ed., *Wordsworth and the Green Romantics* (Lebanon NH, 2016)

Packham, C., 'The science and poetry of animation: personification, analogy, and Erasmus Darwin's *Loves of the Plants*', *Romanticism*, 10 (2004), 191–208

Page, M., 'The Darwin before Darwin: Erasmus Darwin, visionary science, and romantic poetry', *Papers on Language and Literature*, 41 (2005), 146–69

Page, W. and Tringham, N. J., *A History of the County of Stafford, Vol. 10, Tutbury and Needwood Forest* (London, 2007)

Pancaldi, G., *Volta: Science and Culture in the Age of Enlightenment* (Princeton, 2003)

Pavord, A., *The Naming of Names: The Search for Order in the World of Plants* (New York, 2005)

Pawson, H. C., *Robert Bakewell: Pioneer Livestock Breeder* (London, 1957)

Pearson, H., *Doctor Darwin* (London, 1933)

Pearson, H., *The Swan of Lichfield* (London, 1936)

Pearson, J., *Stags and Serpents: The Story of the House of Cavendish and the Dukes of Devonshire* (London, 1983)

Pearson, K., *The Life, Letters and Labours of Francis Galton*, 3 vols (London, 1914–30)

Penoyre, J. and J., *Houses in the Landscape: A Regional Study of Vernacular Building Styles in England and Wales* (London, 1978)

Pera, M., *The Ambiguous Frog: The Galvani-Volta Controversy*, translated by J. Mandelbaum (Princeton, 1992)

Pitt, W., *General View of the Agriculture of the County of Stafford*, 2 vols (London, 1796)

Plumb, C., *The Georgian Menagerie: Exotic Animals in Eighteenth-century London* (London, 2015)

Porter, R., ed., *The Cambridge History of Science, vol. 4: Eighteenth-century Science* (Cambridge, 2003)

Porter, R., *England in the Eighteenth Century* (London, 1998)

Porter, R., *Enlightenment: Britain and the Creation of the Modern World* (London, 2000)

Porter, R., 'Erasmus Darwin: doctor of evolution?' in J. R. Moore ed., *History, Humanity, and Evolution: Essays for John C. Greene* (Cambridge, 1989), 39–69

Porter, R., *Flesh in the Age of Reason* (London, 2003)

Porter, R., *Health for Sale: Quackery in England, 1660–1850* (Manchester, 1989)

Porter, R., *The Making of the Science of Geology* (Cambridge, 1977)

Porter, R., 'Science, provincial culture and public opinion in Enlightenment England', *British Journal for Eighteenth Century Studies*, 3 (1980), 20–46

Porter, R. and Rousseau, G. S., *Gout: The Patrician Malady* (New Haven, 1998)

Porter, R. and Teich, M., eds, *The Enlightenment in National Context* (Cambridge, 1991)

Priestman, M., *The Poetry of Erasmus Darwin: Enlightened Spaces, Romantic Times* (Farnham, 2013)

Prothero, R. E. (Lord Ernle), *English Farming, Past and Present*, 6th edn with introductions by G. E. Fussell and O. R. McGregor (London, 1961)

Pyenson, L. and Sheets-Pyenson, S., *Servants of Nature: A History of Scientific Institutions, Enterprises and Sensibilities* (London, 2000)

Rackham, O., *Trees and Woodland in the British Landscape*, rev. edn (London, 2001)

Rackham, O., *Woodlands* (London, 2012)

Raven, C. E., *John Ray, Naturalist: His Life and Works* (Cambridge, 1950)

Razzell, P., *The Conquest of Smallpox: The Impact of Inoculation on Smallpox Mortality in Eighteenth-century Britain* (Lewes, 1977)

Redman, N., *Illustrated History of Breadsall Priory* (Derby, 1998)

Reed, M., 'The transformation of urban space, 1700–1840', in P. Clark ed., *The Cambridge Urban History*, vol. 2 (Cambridge, 2000), 615–40

Richardson, T., *The Arcadian Friends: Inventing the English Landscape Garden* (London, 2008)

Riden, P., *A History of Chesterfield Grammar School* (Cardiff, 2017)

Ritvo, H., *The Animal Estate: The English and other Creatures in the Victorian Age* (Harmondsworth, 1990)

Ritvo, H., *The Platypus and the Mermaid and other Figments of the Classifying Imagination* (Cambridge MA, 1997)

Robertson, B. P., ed., *Romantic Sustainability: Endurance and the Natural World, 1780–1830* (Lanham, 2016)

Robertson, F. W., *Early Scottish Gardeners and their Plants, 1650–1750* (East Linton, 2000)

Rohde, E. S., *The Old English Herbals* (London, 1922)

Rousseau, G., '"Brainomania": brain, mind and soul in the long eighteenth century', in G. Budge ed., 'Science and the Midlands Enlightenment', special issue, *Journal for Eighteenth-Century Studies*, 30 (2007), no. 2, 161–91

Rousseau, G. S. and Porter, R., eds, *The Ferment of Knowledge* (Cambridge, 1980)

Rude, G., *The Crowd in History, 1730–1848* (London, 1981)

Ruse, M., *Mystery of Mysteries: Is Evolution a Social Construction?* (Cambridge MA, 1999)

Russell, C., *Science and Social Change, 1700–1900* (Basingstoke, 1983)

Russell, E. J., *A History of Agricultural Science in Great Britain, 1620–1954* (London, 1966)

Russell, N., *Like engendr'ing like: Heredity and Animal Breeding in Early-Modern England* (Cambridge, 1986)

Ruston, S., 'Shelley's links to the Midlands enlightenment: James Lind and Adam Walker', in G. Budge ed., 'Science and the Midlands Enlightenment', special issue, *Journal for Eighteenth-Century Studies*, 30 (2007), no. 2, 227–41

Sachs, J. von, *History of Botany (1530–1860)*, translated by E. F. Garnsey and revised by I. B. Balfour (Cambridge, 1890)

Sargent II, F., *Hippocratic Heritage: History of Ideas about Weather and Human Health* (New York, 1982)

Schaffer, S., 'The Earth's fertility as a social fact in early modern Britain', in M. Teich, R. Porter and B. Gustafsson eds, *Nature and Society in Historical Context* (Cambridge, 1997), 124–47

Schaffer, S., 'Natural philosophy and public spectacle in the eighteenth century', *History of Science*, 21 (1983), 1–43

Schaffer, S. and Shapin, S., *Leviathan and the Air-Pump: Hobbes, Boyle, and the Experimental Life* (Princeton, 1985)

Schiebinger, L., *The Mind Has no Sex? Women and the Origins of Modern Science* (Cambridge MA, 1989)

Schiebinger, L., *Plants and Empire: Colonial Bioprospecting in the Atlantic World* (Cambridge MA, 2004)

Schiebinger, L., 'The private life of plants: sexual politics in Carl Linnaeus and Erasmus Darwin', in M. Benjamin ed., *Science and Sensibility: Gender and Scientific Enquiry, 1780–1945* (Oxford, 1991), 121–43

Schiebinger, L., *Secret Cures of Slaves: People, Plants and Medicine in the Eighteenth-century Atlantic World* (Chicago, 2017)

Schiffer, B., *Draw the Lightning Down: Benjamin Franklin and Electrical Technology in the Age of Enlightenment* (Berkeley, 2003)

Schofield, R. E., *The Lunar Society of Birmingham: A Social History of Provincial Science and Industry in Eighteenth-century England* (Oxford, 1963)

Seth, S., *Difference and Disease: Medicine, Race and the Eighteenth-century British Empire* (Cambridge, 2018)

Seymour, S. and Calvocoressi, R., 'Landscape parks and the memorialisation of empire: the Pierreponts' "naval seascape" in Thoresby Park, Nottinghamshire during the French Wars, 1793–1815', *Rural History*, 18 (2007), 95–118

Shapin, S., *The Scientific Revolution* (Chicago, 1998)

Shapin, S., *A Social History of Truth: Science and Civility in Seventeenth-century England* (Chicago, 1995)

Shapin, S. and Schaffer, S., *Leviathan and the Air Pump: Hobbes, Boyle and the Experimental Life* (Princeton, 1985)

Sheets-Pyenson, L. and Sheets-Pyenson, S., *Servants of Nature: A History of Scientific Institutions, Enterprises and Sensibilities* (London, 1999)

Sheldon, P., *The Life and Times of William Withering: His Work, His Legacy* (Studley, 2004)

Shteir, A. B., *Cultivating Women, Cultivating Science: Flora's Daughters and Botany in England, 1760–1860* (Baltimore, 1996)

Shuttleworth, S., *The Mind of the Child: Child Development in Literature, Science and Medicine, 1840–1900* (Oxford, 2010)

Shuttleworth, S., 'The psychology of childhood in Victorian literature and medicine', in H. Small and T. Tate eds, *Literature, Science, Psychoanalysis, 1830–1970: Essays in Honour of Gillian Beer* (Oxford, 2003), 86–101

Simo, M. L., *Loudon and the Landscape: From Country Seat to Metropolis* (New Haven, 1988)

Sitter, J., 'Sustainability Johnson', in A. W. Lee ed., *New Essays on Samuel Johnson: Revaluation* (Lanham, 2018), 111–30

Sloan, A. W., *English Medicine in the Seventeenth Century* (Durham, 1996)

Sloan, E., *The Landscape Studies of Hayman Rooke (1723–1806): Antiquarianism, Archaeology and Natural History in the Eighteenth Century* (Woodbridge, 2019)

Smith, C. U. M. and Arnott, R., eds, *The Genius of Erasmus Darwin* (Aldershot, 2005)

Smith, R., 'The background of physiological psychology in natural philosophy', *History of Science*, 11 (1973), 75–123

Sorrenson, R., 'Towards a history of the Royal Society in the eighteenth century', *Notes and Records of the Royal Society of London*, 50 (1996), 29–46

Spadafora, D., *The Idea of Progress in Eighteenth-century Britain* (New Haven, 1990)

Spary, E., *Utopia's Garden: French Natural History from Old Regime to Revolution* (Chicago, 2000)

Spinage, C. A., *Cattle Plague: A History* (New York, 2003)

Stevens, P. F., *The Development of Biological Systematics: Antoine-Laurent de Jussieu, Nature and the Natural System* (New York, 1994)

Stockwell, C., *Nature's Pharmacy: a History of Plants and Healing* (London, 1989)

Swabe, J., *Animals, Disease and Human Society: Human–Animal Relations and the Rise of Veterinary Medicine* (London, 1999)

Sweet, R., *Antiquaries: The Discovery of the Past in Eighteenth-century Britain* (London, 2004)

Sweet, R., *The English Town, 1680–1840* (London, 1999)

Taylor, K. and Peel, R., *Passion, Plants and Patronage: 300 Years of the Bute Family Landscapes* (London, 2012)

Teeter Dobbs, B. J. and Jacob, M. C., *Newton and the Culture of Newtonianism* (Atlantic Highlands, 1995)

Teute, F. J., 'The Loves of the Plants; or, the cross-fertilization of science and desire at the end of the eighteenth century', *The Huntington Library Quarterly*, 63 (2000), 319–45

Thacker, C., *The Genius of Gardening* (London, 1994)

Thomas, K., *Man and the Natural World: Changing Attitudes in England, 1500–1800* (London, 1983)

Thomas, M., 'The rioting crowd in eighteenth-century Derbyshire', *Derbyshire Archaeological Journal*, 95 (1975), 40–5

Thompson, E. P., *Customs in Common* (Harmondsworth, 1991)

Thompson, K., *Darwin's Most Wonderful Plants: Darwin's Botany Today* (London, 2019)

Trinder, B., *The Making of the Industrial Landscape* (Gloucester, 1982)

Tringham, N., 'The 1655 petition against the enclosure of Needwood Forest', *Transactions of the Staffordshire Archaeological and Historical Society*, 96 (2013), 72–96

Tringham, N., 'Needwood Forest Surveys of 1649–50 and attempted enclosure', *Transactions of the Staffordshire Archaeological and Historical Society*, 94 (2010), 28–70

Trott, N., 'Wordsworth's Loves of the Plants', in N. Trott and S. Perry eds, *1800: The New Lyrical Ballads* (Basingstoke, 2001), 141–68

Tsoupoulos, N., 'The influence of John Brown's ideas in Germany', in W. F. Bynum and R. Porter eds, 'Brunonianism in Britain and Europe', *Medical History* supplement no. 8 (1988), 63–74

Turbutt, G., *A History of Derbyshire*, 4 vols (Cardiff, 1999)

Turner, M., *English Parliamentary Enclosure: Its Historical Geography and Economic History* (Folkestone, 1980)

Uglow, J., 'But what about the women? The Lunar Society's attitude to women and science, and to the education of girls', in C. U. M. Smith and R. Arnott eds, *The Genius of Erasmus Darwin* (Aldershot, 2005), 179–94

Uglow, J., *The Lunar Men: The Friends who made the Future, 1730–1810* (London, 2002)

Vickery, A., *The Gentlemen's Daughter: Women's Lives in Georgian England* (New Haven, 1998)

Wade, E., *A Proposal for Improving and Adorning the Island of Great Britain for the Maintenance of our Navy and Shipping* (London, 1755)

Wade Martins, S., *Coke of Norfolk (1754–1842): A Biography* (Woodbridge, 2009)

Wade Martins, S., *Farmers, Landlords and Landscapes: Rural Britain, 1720 to 1870* (Macclesfield, 2004)

Walker, M., *Sir James Edward Smith (1759–1828): First President of the Linnean Society of London* (London, 1988)

Walker, W. C., 'Animal electricity before Galvani', *Annals of Science*, 2 (1937), 84–113

Wallis, P., 'Consumption and cooperation in the early modern medical economy', in M. S. R. Jenner and P. Wallis eds, *Medicine and the Market in England and its Colonies, c1450–1850* (Basingstoke, 2007), 47–68

Wallis, P., 'Consumption, retailing and medicine in early-modern London', *Economic History Review*, 61 (2008), 26–53

Wallis, P., 'Exotic drugs and English medicine: England's drug trade, c1550–c1800', *Social History of Medicine*, 25 (2012), 20–46

Walters, A. N., 'Science and politeness in eighteenth-century England', *History of Science*, 35 (1997), 121–54

Ward, H. M., *Disease in Plants* (London, 1901)

Watson, J. S., *The Reign of George III* (London, 1960)

Watkins, C., *Trees in Art* (London, 2018)

Watkins, C., *Trees, Woods and Forests: A Social and Cultural History* (London, 2014)

Watkins, C. and Cowell, B., *Uvedale Price (1747–1829): Decoding the Picturesque* (Woodbridge, 2012)

Watts, R., *Women in Science: A Social and Cultural History* (Abingdon, 2007)

Weatherill, L., *Consumer Behaviour and Material Culture in Britain, 1660–1760*, 2nd edn (London, 1996)

Whetzel, H. H., *An Outline of the History of Phytopathology* (Philadelphia, 1918)

Whittle, E., *The Historic Gardens of Wales* (London, 1992)

Wilkinson, L., *Animals and Disease: An Introduction to the History of Comparative Medicine* (Cambridge, 2005)

Williams, R., *The Country and the City* (Oxford, 1975)

Williamson, T., 'The management of trees and woods in eighteenth-century England', in L. Auricchio, E. H. Cook and G. Pacini eds, *Invaluable Trees: Cultures of Nature, 1660–1830* (Oxford, 2012), 221–35

Williamson, T., *Polite Landscapes: Gardens and Scenery in Eighteenth-century England* (Stroud, 1995)

Williamson, T., *The Transformation of Rural England: Farming and the Landscape, 1700–1870* (Exeter, 2002)

Wilmot, S., *'The Business of Improvement': Agriculture and Scientific Culture in Britain, c. 1700–1870*, Historical Geography Research Series, 24 (1990)

Withers, C., 'On georgics and geology: James Hutton's "Elements of Agriculture" and agricultural science in eighteenth-century Scotland', *Agricultural History Review*, 42 (1994), 38–48

Withers, C., *Placing the Enlightenment: Thinking Geographically about the Age of Reason* (Chicago, 2007)

Withers, C., 'William Cullen's agricultural lectures and writings and the development of agricultural science in eighteenth-century Scotland', *Agricultural History Review*, 37 (1989), 145–7

Withers, C. W. J. and Ogborne, M., eds, *Georgian Geographies: Essays on Space, Place and Landscape in the Eighteenth Century* (Manchester, 2004)

Wokler, R., 'Apes and races in the Scottish Enlightenment: Monboddo and Kames on the nature of man', in P. Jones ed., *Philosophy and Science in the Scottish Enlightenment* (Edinburgh, 1988), 145–68

Wolf, A., *A History of Science, Technology and Philosophy in the 18th Century* (London, 1938)

Wood, A., *The Politics of Social Conflict: The Peak Country, 1520–1770* (Cambridge, 1999)

Woodham-Smith, C., *The Great Hunger: Ireland, 1845–9* (London, 1962)

Woods, A., 'From one medicine to two: the evolving relationship between human and veterinary medicine in England, 1791–1835', *Bulletin of the History of Medicine*, 91 (2017), 494–523

Woods, A., Bresalier, M., Cassidy, A. and Dentinger, R. M., *Animals and the Shaping of Modern Medicine: One Health and its Histories* (Cham, 2017)

Worrall, D., 'William Blake and Erasmus Darwin's *Botanic Garden*', *Bulletin of the New York Public Library*, 79 (1975), 397–417

Worsley, P., *The Darwin Farms: The Lincolnshire Estates of Charles and Erasmus Darwin and their Family* (Lichfield, 2017)

Yelling, J. A., *Common Field and Enclosure in England, 1450–1850* (London, 1977)

Zonneveld, J., *Sir Brooke Boothby* (Voorburg, 2003)

INDEX

Garden and Landscape History

Previously Published

Designs Upon the Land: Elite Landscapes of the Middle Ages
Oliver H. Creighton

Richard Woods (1715–1793): Master of the Pleasure Garden
Fiona Cowell

Uvedale Price (1747–1829): Decoding the Picturesque
Charles Watkins and Ben Cowell

Common Land in English Painting, 1700–1850
Ian Waites

Observations on Modern Gardening, *by Thomas Whately:*
An Eighteenth-Century Study of the English Landscape Garden
Michael Symes

The Landscape Studies of Hayman Rooke (1723–1806): Antiquarianism, Archaeology and
Natural History in the Eighteenth Century
Emily Sloan

Transhumance and the Making of Ireland's Uplands, 1550–1900
Eugene Costello

The Foldcourse and East Anglian Agriculture and Landscape, 1100–1900
John Belcher

www.ingramcontent.com/pod-product-compliance
Ingram Content Group UK Ltd.
Pitfield, Milton Keynes, MK11 3LW, UK
UKHW050050280426
470336UK00021B/87